Polypropylene Cable Insulation

IEEE Press
445 Hoes Lane
Piscataway, NJ 08854

Polypropylene Cable Insulation

Boxue Du
Tianjin University
China

Zhonglei Li
Tianjin University
China

IEEE Press

WILEY

Published by John Wiley & Sons, Inc., Hoboken, New Jersey.
Published simultaneously in Canada.

Library of Congress Cataloging-in-Publication Data Applied for
Hardback ISBN: 9781394234431

Cover Design: Wiley
Cover Images: © Storman/Getty Images

Set in 9.5/12.5pt STIXTwoText by Straive, Pondicherry, India

Contents

About the Author

Boxue Du

He received the ME degree in Electrical Engineering from Ibaraki University, Hitachi, Japan, in 1993, and his PhD degree from Tokyo University of Agriculture and Technology, Tokyo, Japan, in 1996.

From 1996 to 2002, he was an Associate Professor at the Niigata Institute of Science and Technology, Niigata, Japan. Concurrently, from 2000 to 2002, he served as a Visiting Scientist at Niigata University, Niigata, Japan. Since 2002, he has been a Professor and the Director-Founder of the Institute of High Voltage, within the School of Electrical and Information Engineering at Tianjin University, Tianjin, China. He is currently a Chair Professor and has been ranked the #1 Surface Charge by ScholarGPS.

His research interests primarily encompass advanced insulation materials for electrical equipment, polymer dielectrics for energy storage, and high-voltage insulation technology. Prof. Du is deeply engaged in both teaching and research in high-voltage technology. He has spearheaded Key Projects, Joint Funds, and Funds for International Cooperation by the National Natural Science Foundation of China.

Prof. Du has published 8 monographs and over 900 academic papers, including more than 275 articles in *IEEE Transactions*, accumulating over 8800 citations. He holds over 30 authorized invention patents and has been recognized as one of the highly cited scholars by Elsevier and one of the top 100,000 scientists globally. His editorial contributions include roles as Contributing Editor-in-Chief and Associate Editor for IEEE and IET journals, as well as membership on the editorial boards of journals such as *High Voltage, CJEE, Scientific Reports, and IET*

Nanodielectrics. He has been honored by the IEEE International Standards Committee with the Special Contribution Award and the Special Advancement Award for his exceptional contributions. Prof. Du is an active member of IEEE DEIS TC and the CIGRE D1.64 Committee.

Zhonglei Li

He received the PhD degree in electrical engineering from Tianjin University, Tianjin, China, in 2016. He is currently an Associate Professor with the Key Laboratory of Smart Grid of Education Ministry, School of Electrical and Information Engineering, Tianjin University. His research focuses on addressing the strategic demands of HVDC power cable transmission for large-scale offshore wind power in China. Specifically, he investigates key issues related to improving the electrical properties of DC cable insulation and regulating electric fields in cable accessories.

He has led several research projects funded by the National Natural Science Foundation of China, including a Young Scientists Fund, a General Program, a subproject of a Key Program, a subproject of the Smart Grid Joint Funds of the NSFC, and a subproject of the China National Key R&D Program. He has authored over 80 SCI-indexed papers, including publications in *IEEE Transactions on Dielectrics and Electrical Insulation* (IEEE TDEI), and has an H-index of 26. Additionally, he has published two monographs: "DC Cable Accessory Insulation" by Science Press.

He has been recognized by several prestigious programs, including the Young Elite Scientists Sponsorship Program by the China Association for Science and Technology (CAST), and the World's Top 2% Scientists list published by Elsevier and Stanford University. He has also been selected for the Technology Leading Talent Plan of Tianjin University. His future research aims to further explore methods for suppressing space charges and enhancing dielectric properties.

Preface

Power cables are essential components for electrical energy transmission within power systems. With the advancement of new energy sources, such as offshore wind power, submarine cables have become critical for energy transmission. Over the past few decades, extruded cables with polymer insulation have emerged as the predominant choice for both submarine and terrestrial applications. The continuous pursuit of stronger insulation materials to achieve higher voltage levels and increased transmission capacity has been relentless. This pursuit serves as the impetus for the publication of this book.

Polypropylene, renowned for its excellent electrical performance, is considered an environmentally friendly cable insulation material for the next generation of power systems. The use of polypropylene insulation, which does not require a cross-linking process during production, represents a significant advancement in the global drive for green energy development and the pursuit of carbon neutrality.

This book focuses on polypropylene cable insulation, offering an in-depth examination of polypropylene insulation for high-voltage DC cables and high-voltage AC cables in Part I and Part II, respectively. It discusses many thought-provoking and critical topics, including space charge, electrical tree, lifetime modeling, and insulation breakdown. The methodologies and innovations discussed within these pages will play a pivotal role in shaping the future of energy transmission, facilitating the transition to renewable energy sources and enhancing the reliability of power systems globally. It is hoped that this book will provide valuable guidance for the development of future high-voltage cables for power systems.

This specialized book in the field of electrical engineering is particularly geared towards professionals involved in high-voltage power transmission. It is also suitable for researchers, graduate students, and advanced undergraduate students working in the area of high-voltage cable transmission.

In conclusion, this book not only addresses the technical aspects of polypropylene cable insulation but also underscores its critical role in achieving sustainable and efficient power systems. It bridges the gap between theoretical research and practical application, offering valuable guidance for the next generation of electrical engineers. As we move towards a future increasingly reliant on renewable energy, the knowledge contained within these pages will prove invaluable. It is with great anticipation that I invite you to delve into this work, confident that it will inspire innovation and progress in the field of high-voltage power transmission.

Boxue Du
Zhonglei Li
Tianjin University

Acknowledgements

The successful publication of this book has been made possible through the invaluable assistance and support of numerous individuals and organizations. The academic scholars, including Prof. Weijiang Chen, Prof. Yoshimichi Ohki, Prof. Jinliang He, Prof. Shengtao Li, Prof. Hong Zhao, and Prof. Pingkai Jiang, contributed significantly to the project and contents of this book.

This book has also received significant supports from the industry. We extend our profound gratitude to the State Grid Electric Power Research Institute, State Grid Smart Grid Research Institute, China Electric Power Research Institute, State Grid Tianjin Electric Power Research Institute, State Grid Shandong Electric Power Research Institute, State Grid Liaoning Electric Power Research Institute, and State Grid Jiangsu Electric Power Research Institute for their essential contributions to collaborative researches and applications of the polypropylene cable insulation.

Our heartfelt thanks also go to Jiangsu Shangshang Cable Group, Kunming Cable Group, Ningbo Orient Wires & Cables Group, Zhongtian Technology Group, and Far East Cable Co., LTD., whose unwavering support enriched the technical content of this book.

Finally, we express our deep appreciation to Wiley Press for their crucial support in bringing this publication to fruition.

1

Introduction

1.1 Background

Power cables are key equipment for achieving high-capacity power transmission. The earliest cables can be traced back to the 19th century. In its development history of over 150 years, the voltage level and transmission capacity of cables have continuously increased, making important contributions to the development of society and economy. Especially after entering the 21st century, with the advancement of internationalization and the development of new energy sources such as offshore wind power, submarine cables are considered the most important means of transmission technology. Figure 1.1 is the schematic diagram of the grid connection and transmission of large-scale offshore wind power based on the high-voltage and large-capacity power cable systems.

Power cable insulation is a protective medium that isolates high potential and is crucial for ensuring the safe and reliable operation of cable transmission systems. In other words, cable insulation needs to have high breakdown field strength, low dielectric loss, resistance to electrical tree discharge, and long service life, as well as good heat resistance, flexibility, and mechanical strength. Cable insulation materials are constantly advancing and developing in pursuit of better performance. The initial cable insulation used oil-filled (OF)-insulated cables and later developed into mass-impregnated (MI) traditional or polypropylene-laminated paper (PPLP)-insulated cables. OF-insulated cables have the advantages of safety, stability, reliability, and long service life. However, the length of OF-insulated cables is greatly limited due to the need for oil supply equipment, and there are also environmental pollution hazards. Compared with OF-insulated cables, MI-insulated cables do not require oil supply equipment, break the length limit of OF-insulated cables, and have no environmental hazards. They have higher insulation performance and heat resistance and are recognized as the most reliable type of high-voltage cable insulation at that time.

Polypropylene Cable Insulation, First Edition. Boxue Du and Zhonglei Li.

Published 2025 by John Wiley & Sons, Inc.

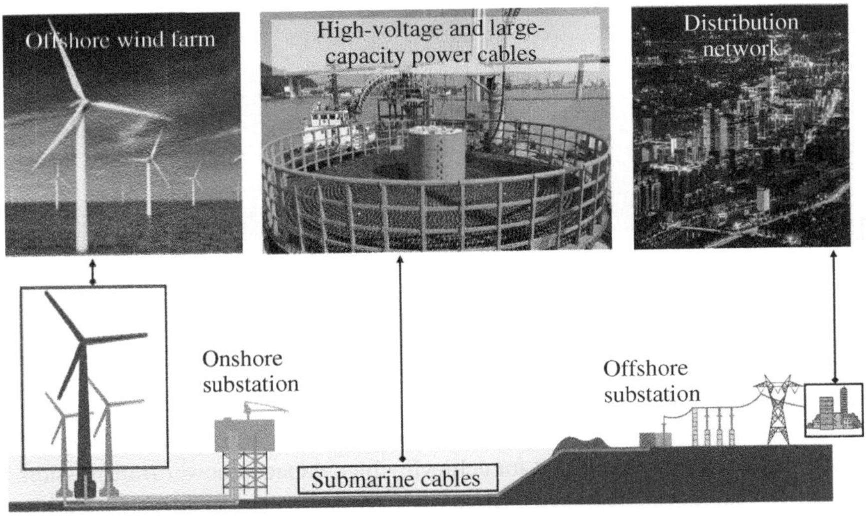

Figure 1.1 Schematic diagram of the grid connection and transmission of large-scale offshore wind power based on the high-voltage and large-capacity power cable systems.

Until the 1950s, with the emergence of plastic extruded cable insulation represented by cross-linked polyethylene (XLPE), cable insulation had a new development direction. Compared to the first two, extruded insulation has advantages such as high-temperature resistance, high transmission power density, high strength, lightweight, environmental protection, and easy installation. Therefore, it has been widely applied and developed after its release. In 1957, General Electric (GE) Company produced the earliest XLPE AC cable by using the advanced production processes in chemistry. In 1999, the world's first commercialized XLPE-extruded DC cable was applied in the Gotland Island II Project in Sweden. The voltage level of this DC cable was ± 80 kV, the transmission power was 50 MW, and the cable length was 140 km. The advantages and disadvantages of the three types of cable insulation are compared in Table 1.1, and Figure 1.2 indicates the cable typologies that can be conveniently used in relation to the length, power, and voltage of the link and the type of transmission system.

At present, the most commonly used extruded insulation material for power cables is cross-linked polyethylene insulation. Its main producers include Dow Corporation in the United States and Nordic Chemical Company, which basically occupy the world market for DC cable insulation materials. The "superclean" XLPE insulation material launched by Nordic Chemical can meet the insulation material performance requirements of 500 kV high-voltage AC and ±640 kV high-voltage DC cables [1].

Table 1.1 Advantages and disadvantages of OF-, MI-, and XLPE-insulated cables.

Cable insulation	Advantage	Disadvantage
OF	① Safe and reliable operation ② Long service life ③ High conveying power	① Complex installation ② Oil supply equipment needs to be installed ③ Potential environmental pollution
MI	① Safe and reliable operation ② No oil supply equipment required ③ Cable length unlimited ④ No potential environmental pollution	① Complex installation ② Restricted working temperature
Extruded XLPE	① High operating temperature ② High-rated capacity ③ Lightweight and low loss ④ Easy installation	① Space charge injection and accumulation ② Thermosetting materials cannot be melted and reused

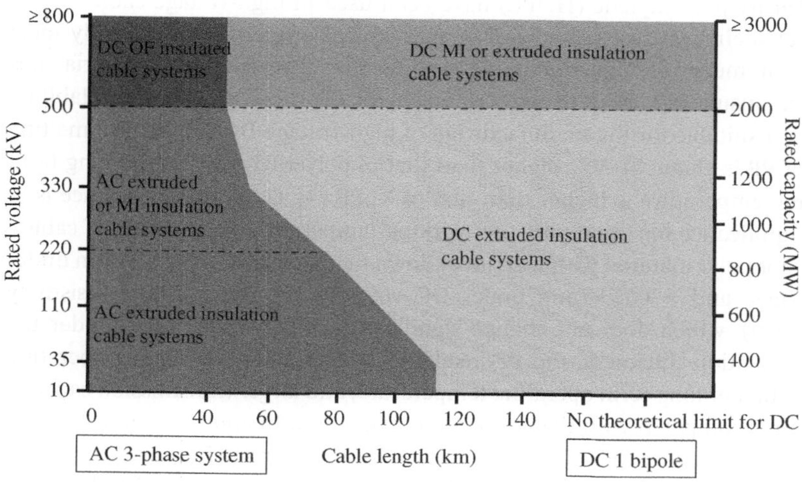

Figure 1.2 Cable typologies that can be conveniently used in relation to the length, power, and voltage of the link and the type of transmission system.

However, with the widespread application of XLPE, its shortcomings are constantly exposed. (i) The production process of XLPE high-voltage DC cables requires controlling the environment and temperature of the cross-linking process, resulting in complex production processes, high-energy consumption, and low-production efficiency. (ii) After the XLPE cable is extruded, a long-term degassing treatment is required to reduce the impact of crosslinking by-products on DC conductivity and space charge accumulation. (iii) The cross-linking by-products, mainly including acetophenone, cumyl alcohol, and α-methylstyrene, accelerates the aging of cable insulation, which makes it difficult to achieve a lifetime of 40 years [2]. (iv) XLPE is a thermosetting polymer, and after cable retirement, the insulation cannot be melted again for processing and utilization [3], making it difficult to recover and treat, which has a significant adverse impact on the environment. Given the many issues mentioned above, the application of XLPE insulation material in high-voltage DC cables has significant limitations, and the development of thermoplastic insulation materials is a new research direction.

Thermoplastic-insulated cables have many advantages, such as no cross-linking process is required in the production process, simple process, and low-energy consumption, no need for degassing, shortened production cycle, no cross-linking by-products (which does not affect the electrical performance of insulation), and recyclable by melting. Since the 1970s, thermoplastic insulation materials such as low-density polyethylene (LDPE), linear low-density polyethylene (LLDPE), and high-density polyethylene (HDPE) have been used in high-voltage cable insulation, but their heat resistance level or mechanical properties do not fully meet the requirements of high-voltage DC cables. Polypropylene (PP) material has excellent insulation, heat resistance, mechanical properties, and recyclability, making it suitable for the manufacturing of high-voltage DC cables. The melting point of PP is about 40–50% higher than that of polyethylene, and the long-term working temperature is higher than that of XLPE [4]. Good heat resistance is of great significance for improving the working temperature and voltage of cables. PP is a nonpolar material with high breakdown field strength ($\geq$100 kV/mm under AC voltage and $\geq$300 kV/mm under DC voltage) and high-volume resistivity ($\sim 10^{16}\ \Omega$ m), which does not change significantly with temperature. Under the same insulation thickness, the PP-insulated cable system has higher operating voltage, higher long-term operating temperature, and larger transmission capacity of the line, compared with that with XLPE insulation. In another word, PP cable insulation can reduce the thickness of the insulation layer and increase the current carrying capacity of the cable system. PP has a high charge injection threshold and less space charge accumulation [5]. PP-insulated cables can reduce the energy consumption of the cable cross-linking process by 1000 kWh/km and reduce carbon emissions by 80% throughout their entire life cycle, according to data from Prysmian. In summary, studying PP as an insulation material for high-voltage DC cables has broad application prospects.

As early as the 1990s, scholars began to study the electrical properties of PP as an insulation material for high-voltage DC cables. Scholars from Osaka University collaborated with Mitsubishi Cable to prepare 600 V and 22 kV cables using PP as insulation. The study found that the AC breakdown field strength and dielectric loss of PP cables at different temperatures met the requirements of the cables. Further research on 22 kV cables found that the lifespan of cables with PP as the main insulation is higher than that of XLPE, and the suppression effect on water trees is better than that of XLPE [6]. In 2010, Prysmian released a high-performance thermoplastic elastomer (HPTE) insulation material developed based on PP materials. P-Laser cables prepared using HPTE as the matrix have better electrical and mechanical properties than traditional XLPE cables [7]. In 2015, Prysmian developed a 320 kV DC high-voltage cable system, and, in 2016, it announced the successful development of 525 kV/2.6 GW and 600 kV P-laser-insulated cables, but none of them have been put into engineering application. In 2015, A. Vaughan et al. from the University of Southampton [8] prepared a model cable by blending PP with propylene–ethylene copolymer. The study found that the mechanical and electrical properties of the cable were better than those of XLPE cable. In 2019, P. K. Jiang from Shanghai Jiao Tong University and Shanghai Huapu Cable Company jointly developed 35 kV and below PP power cables and announced their successful development [9]. In 2023, the authors of this book collaborated with Nanrui Group Corporation to develop functionalized-graft-modified PP cable insulation. In 2023, Prysmian will supply the cable system for the BalWin offshore wind farms in the North Sea of Germany, where P-laser cable based on PP-based insulation will be used for land cables.

PP is polymerized from propylene monomers, and there are a large number of asymmetric carbon atoms on the main chain of PP. According to the different arrangement of methyl groups on the main chain in space, PP can be divided into three molecular configurations, as shown in Figures 1.3. Isotactic polypropylene (iPP) has all methyl groups located on one side of the molecular chain, while syndiotactic polypropylene (sPP) has alternating methyl groups located on both sides

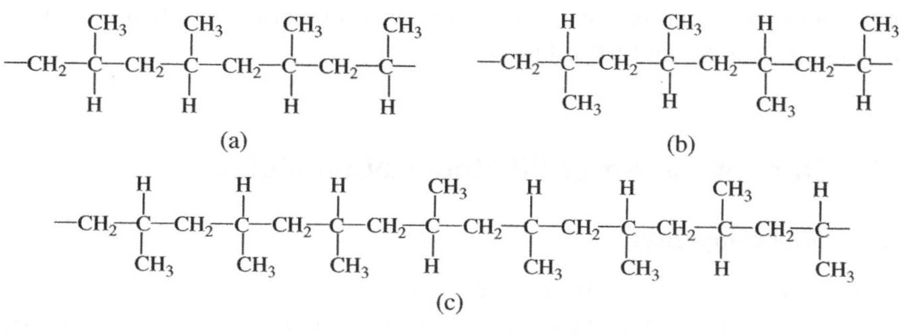

Figure 1.3 Molecular configurations of PP. (a) iPP, (b) sPP, (c) aPP.

of the molecular chain. The irregularly located methyl groups on both sides of the molecular chain are called atactic polypropylene (aPP). Among them, iPP and sPP are crystalline materials, while aPP is an amorphous material. The density of iPP is 0.92–0.94 g/cm^3, and the melting point is approximately 165 °C. The density of sPP is 0.89–0.91 g/cm^3, and the melting point is approximately 135 °C. The density of aPP is 0.90–0.91 g/cm^3, and the softening temperature is 90–130 °C. Among the three molecular configurations of PP, aPP is an amorphous polymer and is not suitable for use in insulation materials. The crystallinity of sPP is relatively low, and, in the presence of nucleating agents, the crystallinity is only 30–42%. Moreover, the crystallization rate is very slow, and it depends heavily on the regularity of the opposite structure, resulting in poor processability [10]. At the same time, the thermal deformation temperature of sPP is about 30 °C lower than that of iPP. The most common type of iPP, with high melting point, high thermal deformation temperature, excellent processability, and a wide range of adjustable spherulite size, is suitable as an insulation material for high-voltage DC cables.

However, iPP also has drawbacks that limit its ability to become cable insulation, including strong rigidity, high brittleness, insufficient flexibility at room temperature, poor low-temperature impact performance, and the tendency of iPP to form large spherulites, leading to a significant reduction in the strength of the breakdown field. These drawbacks result in the iPP substrate cable being prone to cracking under stress, making it difficult to directly produce and process, and making it difficult to bend after cable formation, which is not conducive to cable laying and use. So it is necessary to modify iPP to improve its mechanical toughness, low-temperature impact, and breakdown field strength before it can be applied in power cable insulation. However, most toughening and modification methods will inevitably lead to varying degrees of degradation in electrical and heat resistance properties. In recent years, scholars from various countries have conducted research on the modification of polypropylene insulation, with the goal of synergistically regulating the electrical, mechanical, and heat resistance properties of polypropylene insulation. This book will review the state-of-the-art of PP-based cable insulation and discuss its dielectric properties and their modification methods for AC and DC cables, respectively.

1.2 State of the Art of PP Modification Method

1.2.1 Nanocomposites

Extensive research efforts have been dedicated to the enhancement of the dielectric properties of polypropylene (PP) insulations. With the laboratory research and industrial application of nanomaterials, nano addition has become a commonly used method of PP modification [11, 12]. Nanoparticles refer to groups of atoms or molecules with a size of nanometers, which have small particle size and large specific surface area, and show significant nano-effects, such as quantum size effect and volume effect. In addition to their own special properties, nanoparticles can also endow materials with different functions, such as optoelectronic, magnetic, and dielectric and anti-aging properties [13]. Nonetheless, it is imperative to acknowledge that nanoparticles are burdened by exceedingly high surface energy, rendering them prone to substantial agglomeration tendencies.

The addition of nano MgO particles could significantly inhibit the injection and accumulation of space charges in PP material. The composite material with 3 phr MgO had improved DC breakdown strength, with an increase of 29.3% compared to pure PP [14]. The study found that the addition of nano MgO reduced the space charge density of PP/POE blends and increased the strength of the DC lightning breakdown voltage [15]. In the European project GRIDABLE, experimental data showed that compared with pure PP (6.9×10^{-16} S/m), 4.5 wt% PP/SiO_2 composite insulation had lower conductivity (3.8×10^{-16} S/m). This could be attributed to the ability of nano SiO_2 particles to inhibit charge accumulation and improve trap site distribution [16]. There were attempts on small-scale production of inorganic nano/PP composite insulations. The doping of SiO_2 particles could only improve the performance of low-probability breakdown areas at room temperature, making the breakdown data more concentrated [17]. There were also some studies focusing on the effect of nano particles on the polymer microstructures. Inorganic nanoparticles would affect the crystalline morphology, crystallization rate, and crystallinity of the PP, resulting in changes in the properties of PP composites. The changes of the crystallization behavior of PP/ZnO nanocomposites were studied in [18]. The results showed that nano-ZnO could promote the nucleation of PP, and the crystallization temperature increased by 7 °C with the content of 0.06 wt%. The crystal morphology results proved that nano-ZnO played a positive role in refining spherulites. Moreover, it was found that nano-$CaCO_3$ can help to form a uniform spherulite structure and eliminate the boundary between spherulites, confirmed by a scanning electron microscope. At the same time, the crystallization time of the composite material was significantly shortened, indicating that the crystallization rate was improved [19].

However, due to the large difference in polarity and high surface energy between nanoparticles and PP, nanoparticles may exhibit a strong tendency to agglomerate in the PP matrix, resulting in an increase in interfacial defects and ultimately the deterioration of the composite properties [20]. In [21], the PP composite insulation with 2 wt% nano-clay had a shorter service life, mainly due to the different thermal properties of the nanoparticles and the polymer matrix, and the weak coupling effect between the nanofillers and the matrix. Under long-term electric field and higher temperature, the dissociation of the nanoparticles may be caused. Inorganic nanoparticles are prone to cause agglomeration and separation phenomenon due to dispersibility problems, which may form micropores inside the insulation. Such electrical weaknesses can lead to a decrease in breakdown field strength.

To improve the dispersion of nanoparticles in PP matrix, some surface treatment or modification methods were used. The nano SiO_2 was modified with trimethylsilane, which could fill the gaps of spherulites inside PP and block the charge transport. The AC and DC breakdown strengths were, respectively, increased by 19.9% and 52.3%, respectively, compared to those of the pure PP [22]. In [23], two-dimensional hexagonal boron nitride (h-BN) nanosheets were coated by non-covalent polymerization of dopamine hydrochloride (PDA) to obtain BNNSs@PDA. The results showed that the expansion of electrical trees was effectively hindered under the electric field. The dielectric loss was only 0.006, and the theoretical energy storage density reached as high as 7.42 J/cm^3. In addition, organic–inorganic hybrid nanofillers have attracted the attention of many researchers because of its better compatibility compared with inorganic particles. Polyhedral oligomeric silsesquioxanes (POSSs) had a cage structure consisting of a skeleton alternately connected by Si–O. The groups connected to Si atoms on the top corners could be reactive or inert groups, which mainly determined the chemical or physical interaction between POSS and the polymer matrix. Compared with traditional fillers such as SiO_2 and Al_2O_3, nano POSS was a type of organic–inorganic hybrid particles with a strong comprehensive performance due to the remarkable small-size effect, close bonding interface, and good compatibility, which had capacity in the regulation of polymer properties [24, 25]. Some studies showed that POSS helped to improve the corona resistance and suppress partial discharges (PDs) of the insulation material [26]. After adding 3 wt% of octamethyl-POSS, the average AC breakdown strength of PP increased by 22% [27].

However, in the field of electrical properties improvement of polymers, the application of other kinds of organic–inorganic hybrid nano particles has not been found yet. The feasibility of doping such nanofillers to thin insulations has not been verified. The effects of nanofiller type on the crystallization, melting, and other behaviors of PP insulations are not clear enough. The relationship among the molecular structures of organic–inorganic hybrid nanofillers, the micro morphologies, and dielectric properties of PP insulations needs further research. At the same time, the complex voltage stresses under operating conditions have not been considered when evaluating the breakdown performances of modified PP insulations.

1.2.2 Polymer Blending

The properties of different polymers can be synthesized through proper physical blending, which is a widely used material modification method. After PP and ethylene–propylene copolymer were blended, when the mass fraction of the ethylene–propylene copolymer is 50% and the ethylene content is 9 mol%, the breakdown field strength of the blend increased [28]. Studies showed that the trap density of PP/ethylene–octene (EO) blends was twice that of PP/ethylene–propylene (EP) blends, which effectively reduced electric field distortion inside the blends [29]. When PP/ethylene–octene copolymer (POE)/styrene–ethylene–butylene–styrene block copolymer (SEBS) was (100/10/10), the low-temperature impact performance of the blend increased by 1.82 times and the elongation at break raised to ~614.1% [30]. In [31], PP/ULDPE blends with 15 wt% ULDPE had excellent mechanical toughness. In order to further improve the electrical characteristics, the doping of 0.01 wt% nano-graphene introduced a large number of deep traps in the composite. The space charge accumulation of the PP/ULDPE/graphene composite material was reduced, and the breakdown strength increased by ~14.3% compared with that of the pure PP at 30 °C. Polyvinylidene fluoride (PVDF) with low molecular weight had good compatibility with PP. Although the dielectric constant was elevated to 3.6, but the breakdown strength of the composite insulation was only about 439 kV/mm. Furthermore, boron nitride nanosheets (BNNSs) helped to improve the compatibility, resulting in higher Young's modulus and tensile strength [32]. It was worth noting that the dielectric loss of PP/PFDF composite insulation reached 0.54%, which was usually 1–2 orders of magnitude higher than that of the pure PP, too high to be suitable for applications in power systems [33, 34]. At the same time, the two-phase interface produced by different polymers would increase the dielectric loss and affect the space charge accumulation. Comparing the effects of EPDM, POE, and EVA on the electrical and mechanical properties of PP composites, it was found that the parameters such as melting temperature and crystal form of the blends did not change much, and their toughness was significantly improved. However, there were a large number of interfaces between different materials, which led to serious accumulation of space charges in the blends and a decrease of 5–15% in breakdown performance [35]. What's more, the crystallization behavior of PP was significantly affected in the blend, which may reduce the crystallinity of the polymer insulation.

It can be seen that in order to improve polymer properties through blending, not only it is necessary to further explore the optimal mixing ratio, but also the coordination relationship between various properties is also crucial. How to realize the regulation of the long-term electrical resistance and loss characteristics of composite materials still needs further research.

1.2.3 Chemical Copolymerization and Grafting

Copolymerization refers to the polymerization of two or more compounds into one substance under certain conditions. Based on the study of the thermal and dielectric properties of ethylene–propylene block copolymer, ethylene–propylene random copolymer, and ethylene–propylene–butene random copolymer, it was found that random copolymer had a higher melting point. And, the limited space charge injection resulted in relatively high breakdown field strength [36]. Some scholars compared the dielectric properties of polypropylene and ethylene–propylene block copolymer polypropylene (EPC). Results showed that the DC breakdown field strength of EPC at high temperatures was more stable than that of the PP, and the amount of space charge accumulation was less [37]. The melting and crystallization peak temperatures of copolymerized PP were lower than those of homopolymerized PP. This was because the chemical structures of the molecular chain of the former had changed greatly compared with the latter, which directly affected the arrangement of the molecular chain and the crystallization behavior, further affecting its macroscopic thermal parameters [38]. However, the copolymerization process is complicated and the by-products are difficult to control. Besides, PP molecules are easily entangled with other polymer molecules, which may form smashed crystals. Or, the copolymerization chemistry will reduce the molecular chain regularity, adversely affecting the crystallinity of the material and even destroying the mechanical or electrical properties.

Scholars at home and abroad used chemical methods to graft maleic anhydride onto PP molecules, aiming to introduce deep traps to inhibit space charge injection and accumulation. It was proved that maleic anhydride-grafted polypropylene (PP-g-MAH) could significantly promote the dispersibility of ZrO_2 nanoparticles. The DC breakdown strength of the ternary system composed of PP, PP-g-MAH, and ZrO_2 was elevated by 43.3% compared with that of unmodified PP [39]. After the antioxidant-hindered phenol group (AO) was grafted to the main chain of the PP molecules using esterification reaction, the thermal oxidation stability of the material was significantly enhanced [40]. Studies showed that by grafting styrene copolymer to PP, the chain segment relaxation of the dielectric could be improved by heterogeneous nucleation. The reduction of dielectric loss reached approximately 37.6% [41]. The benzoyl peroxide (BPO) was used to initiate the graft reaction of PP molecular chain and styrene monomer. The graft modification had no obvious effect on the thermal parameters of the material, and the breakdown strength was ~ 498 kV/mm at 30 °C. The benzene ring group grafted on the PP molecular chain had a large volume, and the rigid group could not be rotated. The steric hindrance effect hindered the movement and orderly folding of the PP molecular chains, and, finally, reduced the crystallinity [38].

There are also some studies on the chemical graft modification of the surface layer of polymers. After fluorine was grafted onto the surface, the trap distribution of the PP insulation could be controlled. Charges were mainly bound in the fluorinated layer, which was difficult to inject into the insulation. At the same time, the electric field formed by the charge accumulation inside the fluorinated layer would weaken the external electric field and further reduce the charge injection [42]. Under the catalysis of ultraviolet light, the C–H of the surface layer in the PP insulation was replaced by C–OH [43]. The grafted PP–OH insulation always had better insulation resistance than that of the unmodified insulation. The chemical deep trap introduced by the surface hydroxyl (–OH) layer helped to capture the injected charge, reduce the number of carriers, and prevent the internal degradation of the insulation by preventing the carriers from hitting the molecular chain [44].

It can be seen that chemical treatment methods have the potential to improve the dielectric properties of insulations, but such research is still in the laboratory-level preparation stage. At present, due to the low success rate and by-products of chemical graft technology, how to achieve precise and efficient control is the main problem to be solved.

1.2.4 Crystallization Regulation

Research has shown that there is a strong relationship between the microscopic characteristics and the macroscopic properties of the PP insulation. In addition to the influence of molecular weight and molecular chain regularity, the crystallization behavior is also controlled by heterogeneous nucleus and crystallization conditions. The control of the crystallization process was evaluated and showed the potential for the application to the structure optimization of PP insulations [45]. The low crystallinity of PP induced an increase in electron transport in the amorphous region, which resulted in an increase in impact ionization. Furthermore, the low density of the spherulite polymer caused a decrease in the dielectric breakdown strength of the PP material [46, 47].

Under different crystallization conditions, PP molecular chains showed different spatial arrangements in the crystal lattice and could form five crystalline forms including α, β, γ, δ, and pseudo-hexagonal state [48–50]. Among them, the α-crystal form belonged to the monoclinic crystal system, which is the most thermodynamically stable crystal form. PP generally tended to crystallize to form the α-crystal form with normal preparation conditions. α-Nucleating agents could induce α-crystals in PP, which were divided into three categories: inorganic, organic, and polymer types. The inorganic nucleating agents mainly included mica, titanium dioxide, and calcium carbonate, which had cheap price but poor compatibility with polymers, so could not be used in mass production [51].

Organic nucleating agents, such as dibenzylidene sorbitol, aromatic carboxylates, and substituted aryl heterocyclic phosphates, had good dispersibility, showed high nucleation efficiency with low contents and developed most rapidly among all α-nucleating agents. The main representatives of polymer nucleating agents were polyvinyl naphthenic compounds, but the related technology was not yet mature, and there was almost no report on the advent of industrialized products.

The results showed an increase in the crystallization peak temperature of the composites and a preferred orientation of the polymer crystallites as the concentration of the nucleating agent content increased [52]. When the cooling rate was 10 °C/min and the concentration of 1,3:24-bis (3,4-dimethylobenzylideno) sorbitol (DMDBS) was 0.2 wt%, the crystallization peak temperature of the isostatic PP could increase from 118.7 °C for the original PP to 130.1 °C [53]. In addition, PP with 0.15% 1,2,3-tridesoxy-4,6:5,7-bis-O-[(4-propylphenyl) methylene] nonitol (TBPMN), in a certain cooling rate range, crystallized at a temperature lower than the crystallization temperature of nonnuclear PP [54]. Quenched PP samples had electrical properties that were superior to isothermally crystallized or naturally cooled samples [55]. Studies showed that with the doping of dibenzylidene sorbitol (DBS), the crystallization rate of PP composites increased compared with that of the pure sample [56]. Aryl heterocyclic organic phosphates were mainly used for rigidity modification for polyolefin. Studies have shown that aryl heterocyclic phosphoric acid monovalent metal salt nucleating agents could significantly increase the crystallization rate of and the crystallization peak temperature of PP. It could reduce the folding surface free energy and the energy required to reach the critical size of the crystal nucleus, promoting the nucleation and crystallization of PP. The usual addition content was 0.1–0.3% for improvement of the crystallization and optical and mechanical properties. In [57], a dispersed system of sodium 2,2′-methylene-bis (4,6-di-tertbutylphenyl) phosphate (NA11) was constructed to avoid severe agglomerations, which had an excellent nucleating performance in the PP matrix. The doping of the organic phosphorus nucleating agent content ranged from 0% to 0.8% in [58]. The crystallization peak temperature increased by 10 °C, and the nucleation density was highly elevated. The tensile and flexural strengths of the nucleated composite increased by 15%, indicating improved mechanical properties. In summary, there is a potential to regulate the performances of polymer materials by crystallization control. However, the effects of α-nucleating agents on the electrical performances are almost neglected. The polymer modification used for electrical equipment by changing a-type crystal morphology has not been attempted.

The β-crystal form was metastable state, which was only obtained under low subcooling, temperature gradient, shear force, or the effect of β-nucleating agent [59–63]. The doping of β-nucleating agent was the most effective way to obtain PP with high β-crystal content [64]. The β-nucleating agent had the effect of inducing the formation of β-crystals, which could effectively improve the toughness and heat resistance of the material. Compared with a-nucleating agent, the research and development technology of β-nucleating agent was still immature, and the discovery of its composition and structure basically depends on experience and practice. TMB-5 was aromatic amine compound which can induce the formation of β-crystals. For pure PP, as the temperature decreases, the spherulites grow radioactively with the crystal nucleus as the center, and unique Maltese cross image appeared between the two orthogonal polarizers of the polarizing microscope. The crystallization process of the samples after adding β-nucleating agent was significantly different. The size of spherulites was smaller than α-spherulites and the gap between spherulites decreased. In [65], to introduce β-crystals, the nucleating agent *N*,*N*,0-dicyclohexylterephthalamide (DCTH) was used, showing a notable ability to suppress space charges. By doping with 0.1% rare earth β-nucleating agent (WBG), the DC breakdown strength of PP was improved by 16%, which was attributed to the chain coupling between β-crystals and the amorphous phase [66]. In addition, the introduction of β-crystals enhanced the charge-blocking effect in the PP material, which was beneficial to reduce the average free path of charges and limit impact ionization, thereby inhibiting the growth and development of electrical trees [67]. The existing problem is that β-crystal is unstable. In practical applications, the parameters of the electrothermal field keep changing all the time, and it is uncertain whether β-PP can still play the same role as that obtained in the existing research. Therefore, the modification ideas based on crystal morphology should focus on the α-crystal form.

Previous studies on the crystalline morphology of PP mainly focus on how to make it suitable for HVDC cable materials [66, 67]. The emphasis lies in the aspects of mechanical properties and space charge accumulation. For PP insulations with a micron-level thickness, the effects of molecular chains and aggregate structures on insulation properties lack research. It is not yet known whether these research results can be directly applied to the study of PP insulations. In addition, under such severe operating conditions, whether the change of the micromorphology will affect the breakdown characteristics and loss of PP insulations is also urgently needed to be explored.

1.3 Effect of Microstructures on Dielectric Properties

1.3.1 Effect of Molecular Chain Structures

PP material is a kind of semicrystalline polymer, whose molecular chain and aggregate structures are important internal factors influencing the insulation failure process of PP insulations. When the orientation of the molecular chain was too small, the chain entanglement and bending were caused, which directly decreased the breakdown strength. When the distance between molecular chains in the PP/POE composite sample increased slightly, the free volume became larger, which was conducive to charge transfer. But, when the free volume further increased, the change of the structure would increase the trap depth inside the sample [68]. The carbon atoms on the backbone of the main chain linked to the pendant methyl groups were tertiary carbon atoms, a large number of which in the amorphous region were easily dehydrogenated into free radicals under strong electric field. The highly active free radicals were combined with the PP molecular chain, leading to chain break and reduction in the breakdown strength [69].

In the PP molecular chain, the methyl groups connected to the tertiary carbon atoms have different arrangements. According to the different positions of the methyl groups, PP can be divided into isotactic polypropylene (iPP), syndiotactic polypropylene (sPP), and atactic polypropylene (aPP). Studies showed that due to the spatial regularity of CH_3 groups, iPP was easier to crystallize and had a high melting point, excellent processing properties, and good electrical and mechanical properties compared with aPP and sPP [70]. Some scholars also compared the molecular structure and electrical properties of sPP and iPP, finding that the particle size of spherulites formed by sPP at high temperature was 20–30 times smaller than that of iPP. Therefore, sPP was believed to possess better and thermal stability. However, because of its high price, sPP was difficult to put into use in large quantities [71]. Studies showed that the structures of the molecular chain affected the aggregation state, leading to the change of macroscopic properties. Table 1.2 shows the crystallinity of iPP/sPP/aPP blends. The crystallinity of the sPP5, sPP15, and sPP30 samples increased sequentially because sPP played the role of heterogeneous nucleation during the crystallization process, which promoted the crystallization of iPP. However, when the addition of sPP exceeded 45 wt%, with the addition of macromolecular segments, the stacking and diffusion of

Table 1.2 Crystallinity of iPP/sPP/aPP blends.

Index	iPP content (wt%)	sPP content (wt%)	aPP content (wt%)	Crystallinity (%)
sPP0	95	0	5	39.69
sPP5	90	5	5	37.45
sPP15	80	15	5	38.60
sPP30	65	30	5	41.27
sPP45	50	45	5	41.04

iPP were hindered, resulting in a decrease in the crystallinity. As the crystallinity of the iPP/sPP/aPP blend increased, the amorphous region in the matrix decreased, and the free volume available for the movement of molecular chains decreased. The entangled molecular chains were more difficult to solve because there was not enough space for movement. The charges trapped by the localized state in the molecular chain were more difficult escape, which in turn led to the formation of more deep traps [72].

Generally, iPP is a linear polymer. Since the first successful development of long-chain-branched PP through electron beam radiation by Denicola and others of Himont in the early 1990s, the development and research of long-chain-branched PP (LCBPP) have continued to deepen and achieved commercialization production [73]. The LCBPP is a kind of PP with long-chain branches, and its microscopic molecular structure is different from linear PP. LCBPP had high melting point with high price and was not suitable for large-scale insulation drawing and processing. The long-chain branching modification of PP was mainly used to improve the melt strength, inducing strain hardening in the molten state, which could effectively inhibit the increase of viscosity and improve the yield strength, flexural modulus, elastic modulus crystallization temperature, thermal denaturation temperature, and melting point. Thereby, the scope of application was broadened, especially in the field of foaming and thermoforming. Studies showed that the development of PP with high melt strength by sensitizing radiation method not only improved the melt strength but also greatly elevated the tensile strength and impact strength. LCBPP had a wide processing temperature range, which could also act as a nucleating agent to refine the nucleus. Its thermal properties were excellent and no harmful substances were produced during use [74]. The initial crystallization temperature and nucleation rate of LCBPP were higher than linear PP due to changes in the molecular chain structures [75]. As for micro-morphology, linear PP had large and few spherulites, and the boundaries between the boundaries were clear, showing a typical homogeneous nucleation behavior. The spherulites of pure LCBPP were very small, with large numbers and fuzzy boundaries [76]. It can be concluded that the branched structures affect both the aggregation state and thermal properties and thus may alter the electrical properties. The method based on molecular chain structure design needs further experiments and simulation studies to explore its possibility in the modification of PP insulations.

1.3.2 Effect of Aggregate Structures

During the crystallization process, PP molecular chains fold to form ordered lamellae, which are then piled up into spherulites. Researchers have carried out a lot of work on the relationship between aggregated structures and macro properties. After 626 hours of accelerated electrothermal aging treatment, the breakdown strength of the insulation with high crystallinity decreased by 4.1% while that of the insulation with low crystallinity decreased by 11.2%. It was indicated that dense and compact microstructures helped to enhance the performance of the insulation [77]. When a high electric field is applied to semicrystalline polymer, there is a difference in conduction mechanisms of the electron migration in the amorphous and crystalline regions of the polymer. The molecular chain structure in the crystal region is tightly arranged, and electrons are not easily transported in the crystal region. However, the molecular chain in the amorphous region is arranged loosely and disorderly, which facilitates the electron transport process [78].

In addition, it is found that the crystal form has an important influence on the growth mode of spherulites, crystal morphology, and mechanical, thermal, and electrical properties of PP materials. Figure 1.4 shows the schematic diagram of the interior and interface of α- and β-crystals. The α-crystals are interspersed by radial and tangential lamellae, forming perfect and dense network structures interlaced with each other. Such structure shows a strong stability and is not prone to slip deformation, leading to high strength and modulus. Different from α-crystal, β-crystal has bundle-like structures formed by the growth of side-by-side bundled platelets or spherulite structures formed by the collision of crystal bundles. With layered structures and lower bulk density, it is easier to induce plastic deformation and shear yielding, and effectively dissipate impact energy, thus causing stronger toughness. The γ-crystal form is relatively rare in PP materials. It generally exists in PP with low molecular weight or can be obtained by crystallization under high pressure. Although with a high degree of order, γ-crystal has almost no clear application value currently. The δ-crystal form exists only in aPP and sPP. Pseudohexagonal crystals belong to the thermodynamically unstable crystal form, which is a transition state structure between amorphous and crystalline states. Therefore, most studies in recent years have focused on α- and β-crystal

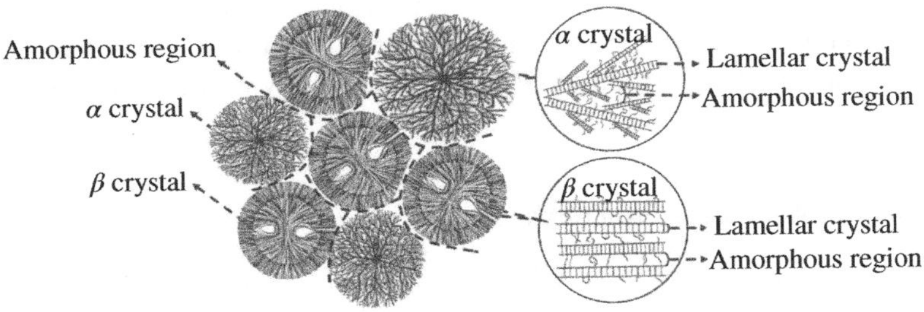

Figure 1.4 Schematic diagram of multiscale aggregation structure of PP insulation.

forms. The conduction currents of the three groups of samples are in the order of α-PP > PP > β-PP [79]. The analysis proved that the α-crystal caused a large number of shallow traps in PP, while the β-crystal resulted in deeper trap sites, which had the ability to inhibit carrier migration. The interface between the α-crystals was obvious, and the carriers were easy to migrate along the spherulite interface or the gap between the spherulites. More molecular chains were introduced in interfaces of β-crystals, resulting in deep trap sites.

However, most of the existing studies on the relationships between microstructure and macroscopic properties focus on a single applied voltage form. The effects of molecular chains and crystal morphology on electrical properties are insufficiently studied under complex electrical and thermal stresses. Moreover, the correlation between various microscopic parameters of PP insulations, such as crystallinity, spherulite size, and crystallization process, is complex and requires attention.

1.4 Effect of Operating Conditions on Dielectric Properties

1.4.1 Effect of Aging Treatment

Under the combined effect of electric field and high temperature for a long time, there would be new defects due to insulation aging [80]. The aging degradation of PP was divided into the induction and the acceleration stages. During the former period, PP only underwent relatively weak degradation. When the accumulated energy reached a certain level, under the effect of electric field and heat, the C—H bond or C—C on the molecular chain was broken to form macromolecular alkyl radicals. In the accelerated degradation stage, the aging rate of PP was significantly accelerated. At the same time, along with the aging and degradation of PP, the chain break was further induced, which reduced its relative molecular weight and weakened the intermolecular force [81, 82]. Various defects inside the polymer caused discontinuities in the energy band, forming charge-trapping centers and affecting charge transport. Unaged PP was nonpolar material and only CH_2 had a weak dipole moment. In addition, its permittivity did not change with the frequency of the electric field. However, as the heat aging progresses, it contained increasingly more polar radical carbonyl groups. The electron distribution of the carbonyl group was represented by the electrons on the double-stranded bands of two atoms with different electronegativity. The electrons were not shared equally, so the moving π electron orbital was strongly drawn toward the oxygen atom with higher electronegativity. Therefore, inherently polar free radicals were generated. Studies showed that after thermal aging treatment, small molecular groups inside the PP were formed and participated in polarization, resulting in increased dielectric loss [83, 84].

In [85], the breakdown strength of the insulation after aging can be divided into two parts. In the low breakdown strength area, the breakdown strength of the insulation decreases by 31% after aging. While in the high breakdown area, the negative effect of aging is only about 1.4%. The reason is that the weak points of the insulation are mainly located in the amorphous area. It is proved that at an ambient temperature of 100 °C, the aged PP sample after 30 days of heat treatment becomes very fragile and easily broken [86].

1.4.2 Effect of Thermal Stress

The high temperature also affected the charge transport and insulation microstructures. Therefore, the effect of thermal field was one of the main factors that caused serious degradation and failures [87]. According to the Schottky theory, high ambient temperature could promote the charge injection from the electrodes into the PP insulation, thereby increasing dielectric loss. Due to the low thermal conductivity of PP, it was hard to dissipate heat, causing the rapid internal temperature rise. It was found that the conductivity was elevated from roughly 4.4×10^{-17} S/m to 6.3×10^{-16} S/m as the testing temperature increased from 30 to 100 °C [88].

The temperature increase provided thermal activation energy for inert reactions. The excess activation energy stimulated the degradation of molecular chains and free radicals were formed, which increased molecular instability and led to the formation of low-density regions [89]. Under the effect of high temperature, the configuration of PP molecular segment was changed locally. Partially isolated free volume was connected and expanded, at the same time, the average free path of electrons and the number of high-energy carriers increased. Finally, impact ionization and PD were promoted [90, 91]. Studies showed that high temperature would cause insulation deterioration, and a temperature increase of 8 °C reduced the lifetime by about 50% [1, 76]. The characteristic 63.2% breakdown strength decreased from 684 kV/mm at room temperature to 602 kV/mm at 100 °C [92].

1.4.3 Effect of Voltage Stress

Within polypropylene materials, a spectrum of structural irregularities was evident, stemming from the inherent intricacies of the manufacturing process. These irregularities encompassed impurities, voids, amorphous domains, linear configurations, filament entanglement, fiber agglomerations, chain terminations, dislocation sites, and residual catalyst remnants. Notably, these minuscule imperfections served as the primary etiology of PDs, their inability to maintain electrical neutrality under the influence of a specific electric field intensity, thereby instigating the genesis of localized space charge trapping phenomena. Under long-term voltage stress, the insulation underwent irreversible insulation aging and partial breakdown. In [93], the space charge characteristics of PP insulations under different DC electric fields were studied. It was found that only negative space charge accumulation exists in PP insulations at 70 °C and 150 MV/m, and the maximum distortion of the local electric field was as high as 80% of the applied electric field. Studies showed that with the increase of the AC content in the DC superimposed AC voltage, the dielectric constant of the aged PP insulation increased. As the DC voltage was reduced to 37.5%, the value of the dielectric constant reached the maximum value of 3.2 [94]. This phenomenon was ascribed to the oxidation of the PP insulation. The elevation in the content of alternating current (AC) resulted in oxidative cleavage and degradation of the molecular chains. Notably, characteristic peaks indicative of hydroxyl (—OH) and carbonyl (C=O) functionalities were discernible within the sample. Concomitantly, the polarity of the molecules exhibited a progressive augmentation, further fortifying the orientation polarization.

The PP insulation also bears the transient overvoltage in actual operation. The data showed that considering the cumulative effect of 600 times of 15 kV standard lightning wave (1.2/50 μs), the molecular chain was broken and polar groups were generated in the PP insulation with a thickness of 12 μm, which increased the dielectric loss by 1.651 times [95]. The cumulative failure characteristics of PP insulations under impulse voltage were studied, and the influences of the cumulative effects on the dielectric properties were studied in [96]. The study found that with the decrease of the wave head time or the increase of the wave tail time of the impulse voltage, the number of times that the PP insulation could withstand the cumulative effect of impulse voltage gradually decreased. The cumulative effect of the impulse voltage would cause cumulative damage to the polymer insulations, resulting in an increase in its relative permittivity and dielectric loss angle tan δ. In summary, according to existing research, there are various factors for insulation failure under operating conditions. There is still a lack of testing experiments for PP insulations comprehensively considering different temperatures and voltage forms. And, the evolution law of the microstructure and dielectric properties of PP insulations has not been obtained. Therefore, it is urgent to further carry out the evaluation and assessment of the insulation performance, so as to guide the methods of improving the properties of the PP insulations.

1.5 Content of This Book

The book is organized as follows. This book includes Part I of polypropylene insulation for DC cables and Part II of polypropylene insulation for AC cables.

Part I: Polypropylene insulation for DC cables includes Chapters 2, 3, and 4. In Chapter 2, the space charge and dielectric breakdown of PP insulation are discussed:

- This chapter focuses on the space charge characteristics of polypropylene-based insulation. The effect of the elastomer blending and microstructure of PP composites on the space charge accumulation and trap level distribution are reviewed and analyzed. More importantly, the modification methods, including nanofillers and voltage stabilizer additives, are employed to further suppress the space charge accumulation of polypropylene cable insulation.

In Chapter 3, the electrical treeing phenomenon under DC operating condition is illustrated:

- This chapter focuses on the electrical treeing degradation and breakdown characteristics of polypropylene-based insulation. The effects of the ambient temperature and superimposed impulse voltage on the electrical treeing are of great concern. In addition to DC breakdown, the breakdown under DC polarity reversal is analyzed. It is proposed that the space charge behaviors under the polarity reversal conditions will significantly distort the electric field distribution and adversely affect the breakdown properties of PP cable insulation.

In Chapter 4, the insulation thickness design and thinning for HVDC cables, considering the effect of space charge, are tackled:

- This paper focuses on insulation thickness thinning based on space charge and electric field simulation for HVDC extruded cables. The modified bipolar electronic–ionic charge transport (BEICT) model for full-size cables, which considers both space charges generated by interface injection and ion dissociation, is established based on the finite element analysis (FEA) method.

Part II: Polypropylene insulation for DC cables includes Chapters 5–7. In Chapter 5, the polarization and dielectric relaxation characteristics of PP insulation are discussed:

- This work focuses on the polarization and dielectric relaxation of polypropylene-based insulation. The relationship between dielectric relaxation and trap characteristics, as well as their effects on the breakdown properties of polypropylene insulation is investigated.

In Chapter 6, the AC electrical treeing and breakdown phenomenon are illustrated:

- This work focuses on the AC PD, electrical treeing, and breakdown characteristics of polypropylene-based insulation. The relationship between the crystalline morphology and electrical treeing morphology and growth are investigated. The crystallization temperature and post-treatment process are also important external factors affecting the crystallization characteristics and modifying the electrical properties. Furthermore, effective methods to modify the crystallization are put forward, in order to improve the AC electrical treeing resistance of polypropylene insulation.

In Chapter 7, the aging and lifetime modeling under the effect of single and multiple stress are tackled:

- This chapter tackles the subject of the aging and life modeling of PP-based cable insulation, a term that is fairly popular in the literature on dielectrics and electrical insulation. The thermal aging and electrical–thermal coupling aging are investigated, and some lifetime models are reported and analyzed. Models of life expectancy could be set up in principle by using some of the modelist tools discussed in this chapter.

References

1 Montanari, G.C., Seri, P., Lei, X. et al. (2018). Next generation polymeric high voltage direct current cables-a quantum leap needed? *IEEE Electrical Insulation Magazine* 34 (2): 24–31.

2 Mazzanti, G. (2017). Life and reliability models for high voltage DC extruded cables. *IEEE Electrical Insulation Magazine* 33 (4): 42–52.

3 Zhou, Y., Dang, B., Wang, H. et al. (2018). Polypropylene-based ternary nanocomposites for recyclable high-voltage direct-current cable insulation. *Composites Science and Technology* 165: 168–174.

4 D. Kim, K. Yoshino, T. Inoue, M. Abe, and N. Uchikawa, "Influence of morphology on electrical properties of syndiotactic polypropylene compared with those of isotactic polypropylene," *Japanese Journal of Applied Physics*, vol. 38, no. 6R, 1999, Art. no. 3580.

5 Zha, J., Wu, Y., Wang, S. et al. (2016). Improvement of space charge suppression of polypropylene for potential application in HVDC cables. *IEEE Transactions on Dielectrics and Electrical Insulation* 23 (4): 2337–2343.

6 Kurahashi, K., Matsuda, Y., Ueda, A., et al. (2002). The application of novel polypropylene to the insulation of electric power cable. *IEEE/PES Transmission and Distribution Conference and Exhibition*, Yokohama, Japan, 1278–1283.

7 Belli, S., Perego, G., Bareggi, A., et al. (2010). P-laser: Breakthrough in power cable systems. *IEEE International Symposium on Electrical Insulation*, San Diego, CA, USA, pp. 1–5.

8 Green, C.D., Vaughan, A.S., Stevens, G.C. et al. (2015). Thermoplastic cable insulation comprising a blend of isotactic polypropylene and a propylene-ethylene copolymer. *IEEE Transactions on Dielectrics and Electrical Insulation* 22 (2): 639–648.

9 Huang, X., Zhang, J., Jiang, P., and Tanaka, T. (2020). Material progress toward recyclable insulation of power cables part 2: Polypropylene-based thermoplastic materials. *IEEE Electrical Insulation Magazine* 36 (1): 8–18.

10 Du, B., Li, Z., Zhou, S., and Fan, M. (2021). Research progress and perspective of polypropylene-based insulation for HVDC cables. *Journal of Electrical Engineering* 16 (2): 2–11. (in Chinese with English abstract).

11 Dang, Y., Wang, Y., Deng, Y. et al. (2011). Enhanced dielectric properties of polypropylene based composite using Bi_2S_3 nanorod filler. *Progress in Natural Science: Materials International* 21 (3): 216–220.

12 Takala, M., Sonerud, B., Ranta, H., et al. (2010). Effect of low amount of nanosilica on dielectric properties of polypropylene. *Proceedings of IEEE Conference on Solid Dielectrics*, Potsdam, Germany, 2010, 1–5.

13 Li, Z.L., Fan, M.S., Zhou, S.F., and Du, B.X. (2021). BNNS encapsulated TiO_2 nanofillers endow polypropylene cable insulation with enhanced dielectric performance. *IEEE Transactions on Dielectrics and Electrical Insulation* 28 (4): 1238–1246.

14 Zhou, Y., Dang, B., Hu, J. et al. (2021). Performance control of nano-MgO particles on polypropylene. *Journal of Electrical Engineering* 36 (24): 1–8. (in Chinese with English abstract).

15 Jiang, P.K., Sun, X.J., Huang, Y. et al. (2017). Preparation and properties of nano-magnesium oxide polypropylene composite insulating materials. *High Voltage Technology* 43 (02): 355–366. (in Chinese with English abstract).

16 Rytöluoto, I., Ritamäki, M., Lahti, K. et al. (2018). Compounding, structure and dielectric properties of silica-BOPP nanocomposite films. *Proceedings of IEEE International Conference on Dielectrics*, Budapest, Hungary, 1–4.

17 Lahti, K., Rytluoto, I., Niittymki, M., et al. (2020). From laboratory to industrial scale: comparison of short-and long-term dielectric performance of silica-polypropylene capacitor films. *Proceedings of IEEE International Conference on Dielectrics*, Valencia, Spain, 1–4.

18 Tang, J., Wang, Y., Liu, H., and Belfiore, L.A. (2004). Effects of organic nucleating agents and zinc oxide nanoparticles on isotactic polypropylene crystallization. *Polymer* 45 (7): 2081–2091.

19 Chan, C., Wu, J., Li, J., and Cheung, Y. (2003). Polypropylene/calcium carbonate nanocomposites. *Polymer* 43 (10): 2981–2992.

20 Zou, H., Wu, S., and Shen, J. (2008). Polymer/silica nanocomposites: preparation, characterization, properties, and applications. *Chemical Reviews* 108 (9): 3893–3957.

21 Zhang, C., Shi, W., Wang, Q. et al. (2020). Effect of nano-clay filler on the thermal breakdown mechanism and lifespan of polypropylene film under AC fields. *IEEE Access* 8: 2169–3536.

22 Takala, M., Ranta, H., Nevalainen, P. et al. (2010). Dielectric properties and partial discharge endurance of polypropylene-silica nanocomposite. *IEEE Transactions on Dielectrics and Electrical Insulation* 17 (4): 1259–1267.

23 Yu, H.S., Wang, Y., and Deng, Y. (2020). Preparation of polypropylene/boron nitride nanosheet composite film and study on dielectric energy storage performance. *Functional Materials* 51 (7): 7141–7147. (in Chinese with English abstract).

24 Huang, X.Y., Li, Y., Liu, F. et al. (2014). Electrical properties of epoxy/POSS composites with homogeneous nanostructure. *IEEE Transactions on Dielectrics and Electrical Insulation* 21 (4): 1516–1528.

25 Du, B.X., Su, J.G., Tian, M. et al. (2018). Understanding trap effects on electrical treeing phenomena in EPDM/POSS composites. *Scientific Reports* 8 (8481): 1–11.

26 Streibl, M., Streibl, M., Werner, S. et al. (2020). The influence of nanoparticles and their functionalization on the dielectric properties of biaxially oriented polypropylene for power capacitors. *IEEE Transactions on Dielectrics and Electrical Insulation* 27 (2): 468–475.

27 Takala, M., Karttunen, M., Salovaara, P. et al. (2008). Dielectric properties of nanostructured polypropylene-polyhedral oligomeric silsesquioxane compounds. *IEEE Transactions on Dielectrics and Electrical Insulation* 15 (1): 40–51.

28 Green, C., Vaughan, A.S., Stevens, G.C. et al. (2015). Thermoplastic cable insulation comprising a blend of isotactic polypropylene and a propylene-ethylene copolymer. *IEEE Transactions on Dielectrics and Electrical Insulation* 22 (2): 639–648.

29 Zhou, Y., Dang, B., and He, J.L. (2016). Large improvement in trap level and space charge distribution of polypropylene by enhancing the crystalline-amorphous interface effect in blends. *Polymer International* 65 (4): 371–379.

30 Han, Y. (2017). *Research on the influence of POE, INFUSE, SEBS and LLDPE on the microstructure and properties of polypropylene and its transparency*. M.S. thesis. China: Beijing University of Chemical Technology, Beijing. (in Chinese with English abstract).

31 Hou, Z.H. (2020). Insulation property modification methods of polypropylene for HVDC cable. Ph.D. dissertation. Tianjin University, Tianjin, China. (in Chinese with English abstract)

32 Fan, B., Zhou, M., Zhang, C. et al. (2019). Improved dielectric properties in polypropylene/poly(vinylidene fluoride) binary blends containing boron nitride nanosheets (BNNS): Towards high voltage current application. *Journal of Physical Chemistry C* 123 (18): 11993–12000.

33 Kim, S.H., Kang, H.S., Sohn, E.H. et al. (2020). High discharge energy density and efficiency in newly designed PVDF@SiO_2-PVDF composites for energy capacitors. *ACS Applied Energy Materials* 3 (9): 8937–8945.

34 Hilczer, B., Smogor, H., Kutek, J., and Goslar, J. (2006). Effect of fast electron irradiation on the dielectric and pyroelectric response of radially-oriented PVDF film. *IEEE Transactions on Dielectrics and Electrical Insulation* 13 (5): 1155–1161.

35 Xu, H. (2017). Modification and space charge investigation of polypropylene composites for HVDC cables. Ph.D. dissertation. Tianjin University, Tianjin, China. (in Chinese with English abstract)

36 Huang, X.Y., Fan, Y.Y., Zhang, J., and Jiang, P.K. (2017). Polypropylene based thermoplastic polymers for potential recyclable HVDC cable insulation applications. *IEEE Transactions on Dielectrics and Electrical Insulation* 24 (3): 1446–1456.

37 Zhang, W., Xu, M., Chen, G.H. et al. (2017). Structure and properties of isotactic and block polypropylene. *High Voltage Technology* 43 (11): 3634–3644. (in Chinese with English abstract).

38 Hu, S.X., Zhang, Y.R., Shao, Q. et al. (2021). Comprehensive performance comparisons of polypropylene-based HVDC cable insulating materials adopting different modification technical routes. *Proceedings of the CSEE* 41 (18): 6245–6258. (in Chinese with English abstract).

39 Liu, W., Cheng, L., Liu, X. et al. (2017). Correlation between morphology and electrical breakdown strength of the polypropylene/maleic anhydride grafted polypropylene/nano-ZrO_2 ternary system. *Journal of Applied Polymer Science* 135 (43): 1–8.

40 Zhang, G., Nam, C., Chung, T.C. et al. (2017). Polypropylene copolymer containing cross-linkable antioxidant moieties with long-term stability under elevated temperature conditions. *Macromolecules* 50 (18): 7041–7051.

41 Guo, Q., Li, X., Li, W., and Yao, Z.H. (2018). The balanced insulating performance and mechanical property of PP by introducing PP-g-PS graft copolymer and SEBS elastomer. *Industrial & Engineering Chemistry Research* 57 (19): 6696–6704.

42 Du, B.X., Xu, R.R., Li, J., and Li, Z.L. (2018). Improved carrier mobility dependent surface charge and flashover voltage of polypropylene film under DC and pulse voltages. *IEEE Transactions on Dielectrics and Electrical Insulation* 25 (3): 1014–1021.

43 Liu, H.L., Du, B.X., Xiao, M., and Tanaka, T. (2021). High-temperature performance of dielectric breakdown in BOPP capacitor film based on surface grafting. *IEEE Transactions on Dielectrics and Electrical Insulation* 28 (4): 1264–1272.

44 Liu, H.L., Du, B.X., and Xiao, M. (2021). Improved energy density and charge discharge efficiency of polypropylene capacitor film based on surface grafting. *IEEE Transactions on Dielectrics and Electrical Insulation* 28 (5): 1539–1546.

45 Y. H. Wu, J. W. Zha, W. K. Li, S. J. Wang, and Z. M. Dang, "A remarkable suppression on space charge in isotatic polypropylene by inducing the β-crystal formation," *Applied Physics Letters*, vol. 107, no. 11, 2015, Art. no. 112901.

46 Du, B.X., Zhu, L.W., and Han, T. (2017). Effect of ambient temperature on electrical treeing and breakdown phenomenon of polypropylene with repetitive pulse voltage. *IEEE Transactions on Dielectrics and Electrical Insulation* 24 (4): 2216–2224.

47 Matsui, K., Tanaka, Y., Takada, T. et al. (2005). Space charge behavior in low-density polyethylene at pre-breakdown. *IEEE Transactions on Dielectrics and Electrical Insulation* 12 (3): 406–415.

48 Lotz, B., Wittmann, J.C., and Lovinger, A.J. (1996). Structure and morphology of poly(propylene): a molecular analysis. *Polymer* 37 (22): 4979–4992.

49 Bruckner, S., Meille, S.V., Petraccone, V., and Pirozzi, B. (1991). Polymorphism in isotactic polypropylene. *Progress in Polymer Science* 16 (2-3): 361–404.

50 Miller, R.L. (1960). On the existence of near-range order in isotactic polypropylenes. *Polymer* 1: 135–143.

51 W. J Zhang (2020). Study on crystallization behavior and mechanical properties of nucleated polypropylene random copolymer. M.S. thesis. Zhejiang University, Zhejiang, China,. (in Chinese with English abstract)

52 Marco, C., Ellis, G., Gómez, M.A., and Arribas, J.M. (2002). Comparative study of the nucleation activity of third-generation sorbitol-based nucleating agents for isotactic polypropylene. *Journal of Applied Polymer Science* 84 (13): 2440–2450.

53 Zhang, Y.F. and Zhou, T. (2010). Non-isothermal crystallization behaviors of isotactic polypropylene with and without 1,3:2,4-bis(3,4-dimethylbenzylidene) sorbitol as a nucleating agent. *International Journal of Polymer Analysis & Characterization* 15 (7): 450–459.

54 Schawe, J., Budde, F., and Alig, I. (2019). Non-isothermal crystallization of polypropylene with sorbitol-type nucleating agents at cooling rates used in processing. *Polymer International* 68 (2): 240–247.

55 Zhang, W., Xu, M., He, M.X. et al. (2017). Effect of the β-crystal formation in isotactic polypropylene used for eco-friendly insulating material. *Proceedings of IEEE International Conference on Electrical Materials and Power Equipment*, Xi'an, China, 419–423.

56 Feng, Y., Jin, X., and Hay, J.N. (2015). Effect of nucleating agent addition on crystallization of isotactic polypropylene. *Journal of Applied Polymer Science* 69 (10): 2089–2095.

57 Li, C., Tong, C.C., Meng, X. et al. (2019). Effect of nucleating agent supported on zeolite via the impregnation on the crystallization ability of isotactic polypropylene and its mechanism. *Polymers for Advanced Technologies* 30 (11): 2674–2685.

58 Gui, Q., Xin, Z., Zhu, W.P., and Dai, G. (2003). Effects of an organic phosphorus nucleating agent on crystallization behaviors and mechanical properties of poly(propylene). *Journal of Applied Polymer Science* 88 (2): 297–301.

59 Cho, K., Saheb, D.N., Yang, H. et al. (2003). Memory effect of locally ordered α-phase in the melting and phase transformation behavior of β-isotactic polypropylene. *Polymer* 44 (14): 4053–4059.

60 Fillon, B., Thierry, A., Wittmann, J., and Lotz, B. (1993). Self-nucleation and recrystallization of polymers. isotactic polypropylene, β phase: β-α conversion and β-α growth transitions. *Journal of Polymer Science Part B: Polymer Physics* 10 (31): 1407–1424.

61 Lovinger, A., Chua, J., and Gryte, C. (1977). Studies on the α and β forms of isotactic polypropylene by crystallization in a temperature gradient. *Journal of Polymer Science Part B: Polymer Physics* 4 (15): 641–656.

62 Varga, J. and Karger-Kocsis, J. (1996). Rules of supermolecular structure formation in sheared isotactic polypropylene melts. *Journal of Polymer Science Part B: Polymer Physics* 34 (4): 657–670.

63 C. Zhang, H. Hu, D. Wang, S. Yan, and C. C. Han, "In situ optical microscope study of the shear-induced crystallization of isotactic polypropylene," *Polymer*, vol. 46, no. 19, pp. 8157–8161, 2005.

64 Grein, C., Plummer, C., Kausch, H. et al. (2002). Influence of β nucleation on the mechanical properties of isotactic polypropylene and rubber modified isotactic polypropylene. *Polymer* 11 (43): 3279–3293.

65 Xu, R.R. (2020). *Failure mechanism and suppression method of dielectric insulation of DC-link capacitor in flexible DC converter valve.* Ph.D. dissertation, Tianjin University, Tianjin, China. (in Chinese with English abstract).

66 Zhang, W., Xu, M., Huang, K. et al. (2019). Effect of β-crystals on the mechanical and electrical properties of β-nucleated isotactic polypropylene. *IEEE Transactions on Dielectrics and Electrical Insulation* 26 (3): 714–721.

67 Zhou, S.F., Fan, M.S., Li, Z.L., et al. (2021). Effect of β-spherulite on electrical tree characteristics of isotactic polypropylene insulation. *Proceedings of IEEE International Conference on Electrical Materials and Power Equipment*, Chongqing, China, 1–4.

68 Xu, H., Du, B.X., and Su, J.G. (2017). Space charge and trap distribution characteristics of polypropylene/polyolefin elastomer blends under tension. *High Voltage Technology* 43 (2): 453–459. (in Chinese with English abstract).

69 Law, A., Simon, L., and Lee-Sullivan, P. (2010). Effects of thermal aging on isotactic polypropylene crystallinity. *Polymer Engineering & Science* 48 (4): 627–633.

70 Kahouli, A., Gallotlavallée, O., Rain, P., et al. (2013). Relationship between structure and dielectric properties of bi-oriented isotactic polypropylene films for capacitors. *Proceedings of IEEE International Conference on Solid Dielectrics*, Bologna, Italy, pp. 1068–1071.

71 He, J.L., Peng, L., and Zhou, Y. (2017). Research progress of environment-friendly HVDC power cable insulation materials. *High Voltage Technology* 43 (02): 337–343. (in Chinese with English abstract).

72 Z. L. Li, Z. Y. Zhong, and B. X. Du, "Dielectric relaxation and trap-modulated DC breakdown of polypropylene blend insulation," *Polymer*, vol 185, no. 17, 2019, Art. no. 121935.

73 Zhao, W., Huang, Y., Liao, X., and Yang, Q. (2013). The molecular structure characteristics of long chain branched polypropylene and its effects on non-isothermal crystallization and mechanical properties. *Polymer* 54 (4): 1455–1462.

74 Ji, X.F. (2016). Preparation of long-chain branched polypropylene and control of its reaction selectivity. M.S. thesis, Harbin Engineering University, Heilongjiang, China. (in Chinese with English abstract).

75 Zeng, W., Liu, J.C., Zhou, J.F. et al. (2008). A comparison study on the melt crystallization kinetics of long chain branched and linear isotactic polypropylenes. *Science Bulletin* 53 (2): 188–197.

76 Zhou, S., Wang, W.X., Zhong, X. et al. (2016). Relationship between molecular structure, crystallization behavior, and mechanical properties of long chain branching polypropylene. *Journal of Materials Science* 51: 5598–5608.

77 Juraj, K., Michal, K., and Blazena, V. (2017). Breakdown voltage of polypropylene film during DC and thermal ageing. *Proceedings of IEEE International Conference on Electric Power Engineering*, Kouty nad Desnou, Czech Republic, 1–4.

78 Blaise, G. (2001). Charge localization and transport in disordered dielectric materials. *Journal of Electrostatics* 50 (2): 69–89.

79 Teng, C.Y., Zhang, L., Zhou, Y.X. et al. (2018). The effect of crystal form on the electrical conductivity and space charge characteristics of isotactic polypropylene. *High Voltage Technology* 44 (8): 2679–2686. (in Chinese with English abstract).

80 Yoshida, Y. and Nishimatsu, M. (1986). Power capacitors. *IEEE Transactions on Electrical Insulation* 21 (6): 963–972.

81 K. Maito, R. Watanabe, H. Hagihara, J. Mizukado, M. Tokita, and H. Shinzawa, "Accelerated aging-induced variation of polypropylene (PP) structure studied by two-dimensional (2D) small-angle X-ray scattering (SAXS) correlation spectroscopy," *Journal of Molecular Structure*, vol. 1207, no. 5, 2020, Art. no. 127764.

82 F. Zhang, Y. Cao, X. Liu, H. Xu, D. Lu, and R. Yang, "How small molecules affect the thermo-oxidative aging mechanism of polypropylene: a reactive molecular dynamics study," *Polymers*, vol. 13, no. 8, 2021, Art. no. 1243.

83 Umran, H.M., Wang, F., and He, Y. (2020). Ageing: causes and effects on the reliability of polypropylene film used for HVDC capacitor. *IEEE Access* 99: 40413–40430.

84 Ye, R.F., Pei, J.Y., Zheng, M.S., and Dang, Z.M. (2020). Anti-aging characteristics of dielectric and energy storage properties of high dielectric polypropylene-based nanocomposite films. *Journal of Electrotechnical Technology* 35 (16): 165–174. (in Chinese with English abstract).

85 Cheng, L., Qi, L.J., Yao, C. et al. (2015). The influence of electrothermal aging on the morphology and breakdown characteristics of polypropylene film. *Power*

Capacitors and Reactive Power Compensation 36 (6): 69–72. (in Chinese with English abstract).

86 Chang, T., Zhang, X., Guo, Q. et al. (2020). Terahertz dielectric spectroscopy based thermal aging analysis of polypropylene. *IEEE Transactions on Terahertz Science and Technology* 10 (4): 363–369.

87 Cheng, L., Liu, W., Liu, X. et al. (2019). Online degradation of biaxial-orientated polypropylene film from HVDC filter capacitors. *IEEE Transactions on Dielectrics and Electrical Insulation* 26 (1): 26–33.

88 Rytöluoto, I., Ritamäki, M., and Lahti, K. (2018). Short-term dielectric performance assessment of BOPP capacitor films: A baseline study. *Proceedings of IEEE International Conference on the Properties and Applications of Dielectric Materials*, Xi'an, China, 289–292.

89 Li, J., You, H.P., Li, J. et al. (2020). Application of quartz sand dielectric in heat dissipation improvement of self-healing low-voltage capacitors. *Power Capacitors and Reactive Power Compensation* 41 (4): 98–101.

90 Du, B.X., Su, J.G., Li, J., and Han, T. (2017). Effects of mechanical stress on treeing growth characteristics in HTV silicone rubber. *IEEE Transactions on Dielectrics and Electrical Insulation* 24 (3): 1547–1556.

91 Nie, H.Y., Wei, X.L., Yao, Y.H. et al. (2020). The influence of thermal aging on the overvoltage withstand characteristics of the inter-turn insulation of dry-type air-core power reactors. *Transactions of China Electrotechnical Society* 35 (24): 70–78. (in Chinese with English abstract).

92 Ritamäki, M., Rytöluoto, I., and Lahti, K. (2017). Temperature effect on breakdown performance of insulating polymer thin films. in *Proceedings of Nordic Insulation Symposium on Materials*, 75–79.

93 Zheng, F.H., Lei, Q.Q., Zhang, Y.W. et al. (2016). Space charge characterization in biaxially oriented polypropylene films. *IEEE Transactions on Dielectrics And Electrical Insulation* 23 (5): 3102–3107.

94 W. Zhao, W. H. Siew, M. J. Given, et al. (2016). Aging behaviour of polypropylene under various voltage stresses. *Proceedings IEEE Conference on Electrical Insulation Dielectric Phenomena*, Toronto, ON, Canada, 903–906.

95 Sima, W.X., Hua, J.F., Sun, P.T., and Chen, Q.L. (2018). Cumulative failure characteristics of polypropylene film used in shock capacitors under impulse voltage. *High Voltage Technology* 44 (2): 358–365. (in Chinese with English abstract).

96 Hua, J.F. (2017). Study on the cumulative effect of impulse voltage on polypropylene film for capacitors. M.S. thesis, Chongqing University, Chongqing, China. (in Chinese with English abstract)

Part I

Polypropylene Insulation for HVDC Cables

2

Space Charge and Dielectric Breakdown

2.1 Introduction

With the development of new energy technology in recent years, there has been a desire for more long-distance power transmission. In this context, high-voltage direct current (HVDC) transmission, which has the characteristics of high transmission power, low line cost, and narrow transmission corridors, has been widely developed [1, 2]. Thermoplastic polypropylene (PP) has drawn much attention as a potential HVDC cable insulation material due to its excellent dielectric properties, non-crosslinking, and recyclable properties [3]. Investigations have been performed to estimate the material property of PP as the primary insulation of DC cable, and it has been confirmed that both syndiotactic PP (s-PP) and isotactic PP (i-PP) exhibit the potential of being such insulation [4–8]. Recently, i-PP drew increasing attention as being the DC cable insulation due to its higher breakdown strength and lower manufacture cost [4–6]. However, an important problem that may influence the industry application of i-PP as cable insulation is the much higher brittleness, lower toughness, and higher glass transition temperature of about 0 °C of virgin i-PP than that of XLPE. So it must be modified to satisfy the demands for high-voltage cable transmission. The common method of modifying insulation is to blend PP with polyolefin elastomers or rubber materials, for example, styrene–ethylene–butylene styrene block copolymer (SEBS), polyolefin elastomer (POE), and propylene–ethylene copolymer [9]. Although this method improves the mechanical properties of PP, it also introduces a large number of interfaces and reduces its electrical properties, which may influence the space charge behaviors and DC breakdown strength of cable insulation. Currently, the balance of mechanical and electrical properties of the PP blend insulation is constantly desirable.

Polypropylene Cable Insulation, First Edition. Boxue Du and Zhonglei Li.

Published 2025 by John Wiley & Sons, Inc.

2.2 Effect of Elastomer on Space Charge and Breakdown Characteristics

In previous studies, many types of materials were selected as toughening agents to improve the flexibility of virgin PP. C.D. Green et al. [10] reported on optimization in the composition of a propylene homopolymer/propylene–ethylene copolymer blend which had similar electrical and mechanical performances to those of XLPE. But, the blend still held a higher modulus than XLPE with a flatter temperature profile up to 120–130 °C. Caramitu et al. [11] investigated the mechanical properties of PP modified by various rubber elastomers, and it was pointed out that the compatibility of PP and elastomers was bad. Zha et al. [12] demonstrated the improved mechanical property of PP blended with styrene–ethylene–butylene SEBS that had a high melting point and excellent compatibility. However, the introduction of the elastomer resulted in the accumulation of space charge obviously. Zhou et al. [13] pointed out that the PP/POE blend with proper POE content could possess better toughness as compared with the virgin i-PP and exhibits higher DC breakdown strength than XLPE. However, as revealed in our previous work, due to bad compatibility between PP and POE, the POE acted as "separated spots" within the PP matrix, which may result in the formation of physical defects that reduced the DC breakdown strength of PP [14, 15]. Accordingly, in order to obtain excellent electrical properties of PP/elastomer blend for DC cable insulation, the compatibility between PP and elastomer should be taken into account.

A new type of propylene-based elastomer (PBE), which has better compatibility with PP, has been proposed in this subsection. The electrical and mechanical properties of the PP/PBE blend have been estimated and compared to those of PP/POE blend with various elastomer contents. The PBE elastomer is composed of repetitive monomers of isotactic propylene (85 wt%) and randomly distributed ethylene (15 wt%). The PBE was prepared by means of a proprietary metallocene catalyst method invented by Exxon Mobil Corporation [16]. The POE elastomer is a type of thermoplastic elastomer consisting of ethylene–octane copolymer, and the proportion of ethylene is 75 wt%. In this work, the content of elastomer was selected at 10, 20, and 30 wt%. The base material and the elastomer were first cleaned with ethyl alcohol and were dried at 80 °C in a vacuum chamber. Then, the mixture was placed into an internal mixer with a temperature of 180 °C at a speed of 30 r/min for 5 minutes. The obtained mixtures were hot pressed under 16 MPa at 190 °C for 5 minutes, by which the blend samples with dimensions of 90 mm × 90 mm could be obtained. In order to easily index the blends with various types and contents of elastomers, the abbreviations of the samples are listed in detail in Table 2.1.

Table 2.1 Index for polymer samples investigated in this subsection. *Source:* Gao et al. [17]/with permission from IEEE.

Index	Elastomer content (wt%)	Elastomer type	ε_r
PP	0	—	2.18
PB10	10	PBE	2.19
PB20	20	PBE	2.20
PB30	30	PBE	2.20
PBE	100	PBE	2.39
PO10	10	POE	2.22
PO20	20	POE	2.23
PO30	30	POE	2.25
POE	100	POE	2.28

In order to better understand the blending status between the PP base material and the elastomers, scanning electron microscope (SEM) measurements were performed to observe the micromorphology of samples. The samples were initially immersed in liquid nitrogen, after which they were mechanically bent; hence, the cross section of the sample was obtained. The smooth section was then coated with Aurum for SEM measurement. The SEM used in this work was FEI Nanosem430 (FEI Company, USA). Thermal electrons were accelerated under a voltage of 10 kV, the energy of the electron beam was 10 keV with beam spot 3, and the magnification was 10,000. The microstructures of the elastomer dispersion manners in PP are illustrated by SEM images in Figure 2.1. In PP/POE blend, the POE is distributed in a dispersed way within the base material. A number of isolated spots exist in the PP, and with the increase of the POE content, the size of isolated spot appears to be larger, as can be found in Figure 2.1a–c. This may reveal that the PP and the POE have bad compatibility, which should be attributed to the fact that the two materials have different chemical groups [13]. It has been generally accepted that Van der Waals' force between organics with the same or similar groups is higher than that with different groups [18], so the compatibility would be better for the polymer matrix and the elastomer with the same or similar groups. In this work, Van der Waals' force between POE molecular chains would be expected to be higher than that between PP and POE molecular chains. Accordingly, as can be seen in Figure 2.1a–c, the POE elastomers are gathered and act as "separated spots" within the base PP matrix due to the separation of molecular chains microscopically. As the content of POE increases, the size of the "separated spots" appears to grow, which implies the enhancement in incompatibility between the base material and the elastomer. It could be expected that the

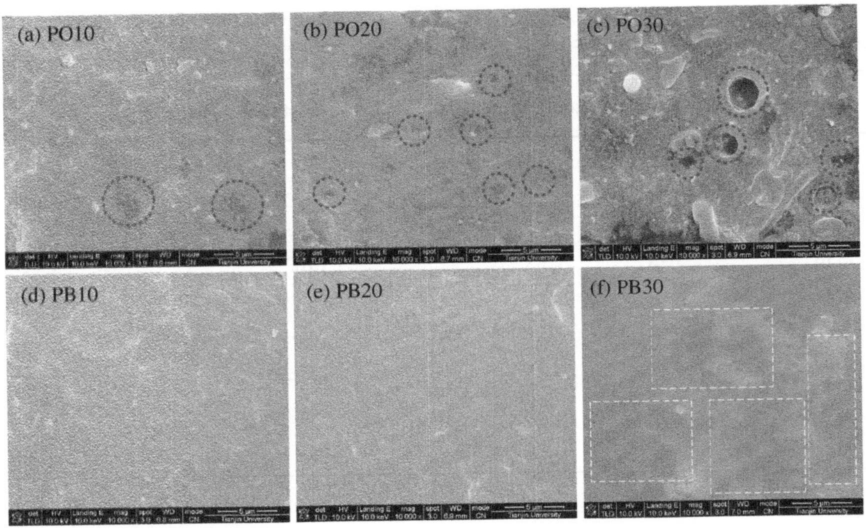

Figure 2.1 SEM photos for the PP/POE and the PP/PBE blends with various elastomer contents. (a) PO10. (b) PO20. (c) PO30. (d) PB10. (e) PB20. (f) PB30. *Source:* Gao et al. [17]/ with permission from IEEE.

introduction of POE into PP leads to the formation of physical defects at the interface between PP and POE due to the bad compatibility.

However, no such "separated spots" could be found in the PP/PBE blend, as shown in Figure 2.1d–f. The micromorphology of the cross section for PB10 and PB20 is rather smooth without visible distinct boundaries between the PP and the PBE, which reveals the excellent compatibility between the matrix and the elastomer. As for the PB30, the boundaries between the PP and the PBE are obvious, which are more likely as the features of "coastline," as indicated by the yellow dash line in Figure 2.1f. As compared with the PO30, the PB30 still presents better blending performance where no abrupt interfaces between the two materials are likely to be observed. The fact that PBE possesses better compatibility with PP could be ascribed to its nature in chemical composition. The PBE elastomer is a type of mixture that contains isotactic propylene and ethylene. Van der Waals' force between PP chains and isotactic propylene molecules of PBE is considered to be stronger. Thus, the PP/PBE blend exhibits better compatibility as compared with PP/POE. In addition, the ethylene groups of PBE have weak interaction with the base PP matrix, and they are more easily combined together. On the basis of SEM photos shown in Figure 2.1, it is assumed that the combination between the isotactic propylene components of PBE and the PP matrix is tight due to the higher content of the isotactic propylene in PBE (85 wt%). Thus, the boundaries between PP and PBE could not be observed at low content of the elastomer. As the PBE of 30 wt% is blended into PP, the boundary profile could be clearly seen since the gathering of ethylene monomers is considerable.

Carrier trap distribution is the key factor for the space charge behaviors, which could be derived from the ISPD measurement [19, 20]. The surface potential of the polymer sample is established through DC corona charging and then is allowed to decay under isothermal condition, associated with the charge de-trapping behavior closely [20–22]. Accordingly, the trap distribution could be derived from the ISPD measurement. To interpret the charge transportation behavior in detail, negative charge trapping and de-trapping process in energy band is shown in Figure 2.2a.

The typical surface potential decay behavior is depicted in Figure 2.2b and c, both the cases for positive and negative corona charging are presented. It can be observed that the potential decay exhibits nonlinear manner where a fast decay occurs at the initial stage that is then followed by a slow decay. Since the electric field generated by the implanted charges is as low as 5–7 kV/mm, injection of charge with the opposite sign from the grounded electrode should be negligible, and carrier recombination is not taken into consideration in this chapter. With the increase of elastomer content, the decay rate tends to increase for both the positive and the negative potentials. It has been demonstrated that the decay is associated with charge trapping and de-trapping processes [23]. Electric charges generated during the corona discharge process are driven by the field and migrate to the sample surface. As they arrive at the test specimen, charge transfer is considered to occur and the charge within the sample bulk becomes electronic in nature [24, 25]. Due to the presence of carrier traps as stated earlier in this chapter, the charges could be captured first and are then thermally activated to de-trap and transport to the grounded electrode, leading to the decrease of surface potential [26]. It has been suggested by G. Chen et al. that both deep and shallow traps are likely to contribute to the de-trapping process [27]. The initial fast decay of the potential should be ascribed to charge de-trapping from shallow traps, while the following slower decay of the potential is due to charge de-trapping from deep traps. As can be found in Figure 2.5a and b, the test data are in good agreement with the fitting curve as a bi-exponential function. The motivation that a bi-exponential function is employed to fit the test data is based not only on our experience but also on the basis of physical origin. It has been claimed that the thermal de-trapping probability of trapped charge P_{de} can be expressed by [24],

$$P_{\mathrm{de}} = \nu \exp\left(-\frac{\Delta E}{kT}\right) \tag{2.1}$$

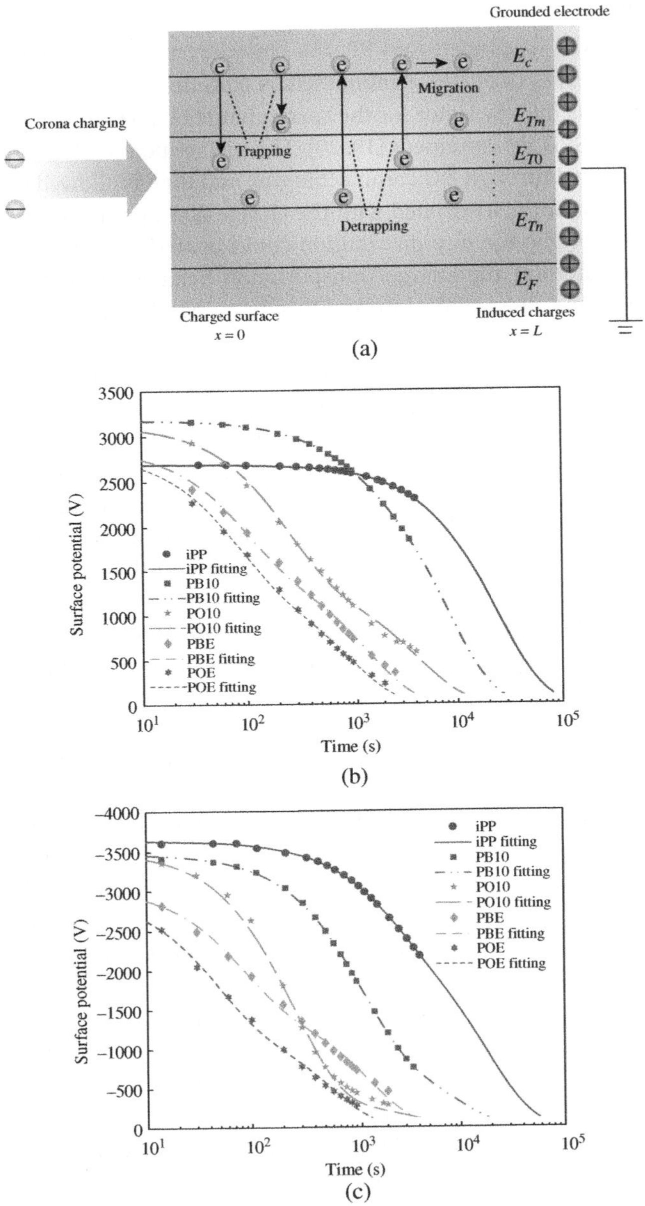

Figure 2.2 (a) Scheme of charge trapping and de-trapping process in energy band after negative corona charging. E_c is the bottom of the conduction band, E_{Tn}, E_{T0}, and E_{Tm} are traps of different energy levels, and E_F is the Fermi level. (b) Typical positive surface potential decay curves. (c) Typical negative surface potential decay curves. *Source:* Gao et al. [17]/ with permission from IEEE.

Accordingly, by assuming that the dissipation of trapped charge is induced by the de-trapping from both shallow and deep traps, the density of trapped charge can be expressed by [27],

$$N(t) = N_s \exp(-k_s t) + N_d \exp(-k_d t) \tag{2.2}$$

where $N(t)$ is the density of total trapped charge, N_s is the density of charge captured by shallow trap, N_d is the density of charge captured by deep trap, and k_s and k_d are de-trapping rates for shallow and deep traps, respectively. Since the decay of surface potential is closely related to the release of charge from both shallow and deep traps, the variation in surface potential with time is proposed as [24],

$$U_s = A \exp(\alpha t) + B \exp(\beta t) \tag{2.3}$$

where A and B are fitting parameters related to the potentials excited, respectively, with charges captured by shallow and deep traps, and α, β are the parameters with respect to the de-trapping rates of charges captured by shallow and deep traps.

Deep trap distributions for electrons for both PP/PBE and PP/POE blends with various elastomer contents are shown in Figure 2.3. As regard the PP/PBE blend, as shown in Figure 2.3a, with the increase of the PBE content, the trap distribution curve shifts toward the left. It can be seen that the virgin PP covers the deep trap range from ~0.76 to ~1.00 eV, while the PB30 exhibits a trap distribution range from ~0.73 to ~0.94 eV, which is remarkably lower than that for the virgin PP. As a comparison, the deep trap distribution for virgin PBE is also presented, the trap distribution starts from 0.71 to 0.94 eV. Furthermore, as the PBE is added, the density at the trap center tends to reduce. It can be found that the density at the trap center for PB10, PB20, and PB30 is in the range from 2.23×10^{19} to $3.72 \times 10^{19}\ \mathrm{m^{-3}/eV}$, which is only one-third to half the value of the virgin PP ($6.71 \times 10^{19}\ \mathrm{m^{-3}/eV}$). As for the PP/POE blend, the trap depth appears to be shallower with the growth of elastomer content as well. Another important fact is that the density at the trap center for the PP/POE blend becomes reduced as compared with the virgin PP by a factor of ~6, which is more significant than that in the PP/PBE blend. The variation of POE content from 10 to 30 wt% does not induce obvious change in the density at the trap center, which is close to $1.0 \times 10^{19}\ \mathrm{m^{-3}/eV}$. It is then concluded from Figure 2.3a and b that the presence of elastomers leads to the reduction in both trap depth and trap density for electrons in the virgin PP, and such a reduction is more obvious in PP/POE than in the PP/PBE blend.

Figure 2.4 shows the distributions of shallow electron trap for both PP/PBE and PP/POE blends with various elastomer contents. The distribution of shallow electron trap for the PP/PBE blend is shown in Figure 2.4a. With the increase of the elastomer content, the trap depth tends to be shallower obviously. The trap center is reduced from 0.889 eV for virgin PP to 0.824 eV for PB30. The virgin PBE has an even shallower trap center at 0.803 eV. As for the density at the trap center, it is

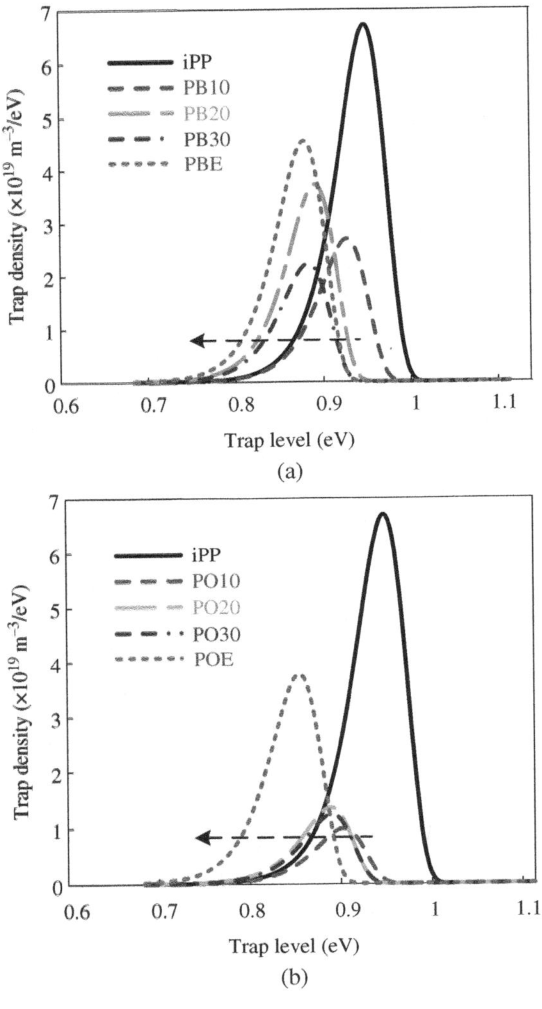

Figure 2.3 Deep trap distribution for electron for the two types of blends. (a) PP/PBE blend. (b) PP/POE blend. *Source:* Gao et al. [17]/ with permission from IEEE.

noticed that an enhancement of the density occurs with the addition of PBE elastomer by a factor of ~2 to the virgin PP. The trap distribution for the PP/POE blend is illustrated in Figure 2.4b. The presence of POE elastomer leads to the shallower electron trap as well. On the other hand, the density at the trap center becomes increased with the presence of POE elastomer. The PP/POE blend has a density at the trap center in the range of 7×10^{19}~ 9×10^{19} m^{-3}/eV, which is higher than that for the virgin PP (2.86×10^{19} m^{-3}/eV) by a factor of 2–3. By comparing the electron trap distribution curves presented in Figure 2.4a and b, it is assumed that

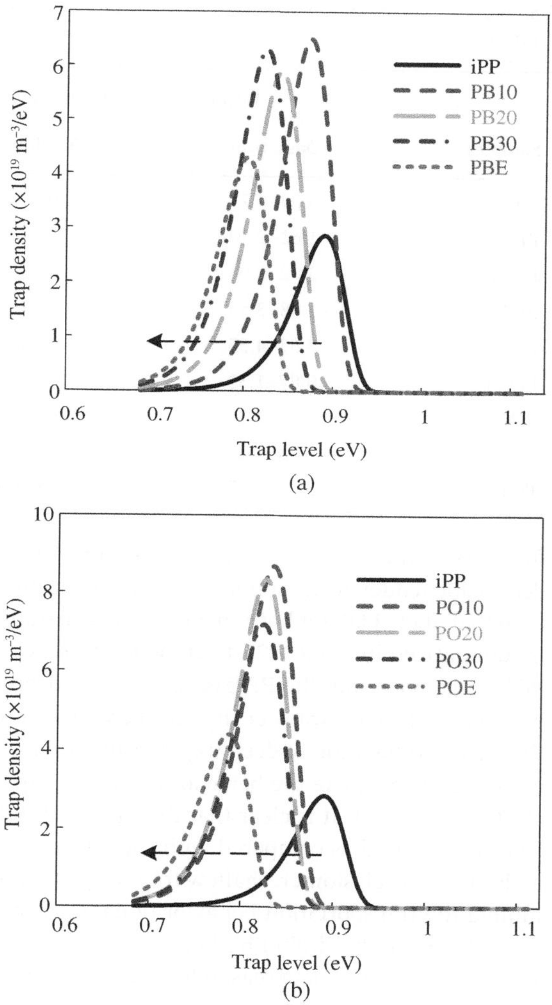

Figure 2.4 Shallow trap distribution for electron for the two types of blends. (a) PP/PBE blend. (b) PP/POE blend. *Source:* Gao et al. [17]/ with permission from IEEE.

the addition of POE leads to a more remarkable increase in the density of shallow trap than that of PBE.

It must be argued that due to the randomness of corona discharge [24], the imperfections of polymer samples which are originated from the semi-crystal morphology of PP and the nonideally uniform dispersion of elastomer, it is almost impossible to allow the samples to be charged under completely identical conditions. The deviations in test data caused by the inhomogeneity in the polymer sample and the discharge randomness are therefore unavoidable. As the content grows

Table 2.2 Density at electron trap centers. *Source:* Gao et al. [17]/with permission from IEEE.

Samples	Trap density ($\times 10^{19}\ m^{-3}/eV$)		Typical $k_{d/s}$
	Deep trap	Shallow trap	
iPP	6.09~6.71	2.86~3.26	2.346
PB10	1.62~2.70	6.04~6.50	0.415
PB20	3.02~3.94	4.69~5.89	0.632
PB30	1.98~3.55	5.88~6.39	0.354
PBE	3.67~4.55	2.96~4.29	1.061
PO10	0.87~1.32	7.98~8.66	0.114
PO20	1.05~1.36	8.03~8.49	0.163
PO30	0.78~1.24	7.17~8.60	0.173
POE	3.78~5.27	2.73~4.39	0.861

from 0 (virgin PP) to 100 wt% (virgin PBE or POE), the trap center exhibits a clear decreasing tendency. It is noticed that the trap center of PP/PBE is basically higher than that of PP/POE for both shallow and deep traps. The trap density at the center is summarized in Table 2.2. It can be found that as the elastomer, either PBE or POE, is blended into the PP base matrix, the deep trap density becomes reduced whereas the shallow trap density gets increased for both electron and hole. The $k_{d/s}$ that refers to the ratio of deep trap density to shallow trap density at the center is calculated to estimate the behavior of carrier trap in response to the variation in elastomer content. It is clear that the $k_{d/s}$ for PP/PBE or PP/POE blend is significantly decreased as compared with the virgin PP, which indicates that with the addition of the elastomer, shallow trap would play a more important role in determining the transportation behavior of charge carriers. It is also noticed that the virgin POE and the virgin PBE hold lower $k_{d/s}$ than that of virgin PP, which is considered as one candidate reason for the relatively low $k_{d/s}$ in the blends. Another interesting feature that could be derived from Tables 2.2 and 2.3 is that for both electron and hole traps, the virgin PBE possesses higher $k_{d/s}$ than the virgin POE, and the PP/PBE blend exhibits higher $k_{d/s}$ than the PP/POE blend. This reveals that the introduction of PBE into the PP matrix would be more beneficial from deep trap than that of POE. In addition, the total integrated trap density is not calculated and compared. Since the trap center plays a dominant role in determining the charge transportation behavior in polymer, attentions are usually paid to the energy level and density at the trap center [17, 28].

DC breakdown strength is an important parameter to estimate the insulation performance of polymers. It has been pointed out that the interfacial region

Table 2.3 Index for polymer samples investigated in this subsection. *Source:* Gao et al. [17]/with permission from IEEE.

Index	PP content (wt%)	Elastomer content (wt%)	ZnO (wt%)
P100	100	0	0
P60E40	60	40	0
P60E40N0.5	60	40	0.5
P60E40N2.0	60	40	2.0

between the polymer base material and the fillers usually plays a key role in dielectric breakdown behavior [29], and deep traps have remarkable influence on the short-term breakdown behavior [30]. As mentioned in the above section, the density of deep trap decreases by the introduction of elastomers, while on the other hand the density of shallow trap increases. The estimation of DC breakdown strength of the blends is expected to be carried out to reveal the influence of trap on the insulation performance. Figure 2.5 shows the relationship between the DC breakdown strength and the elastomer content for both PP/PBE and PP/POE blends. It can be observed that the breakdown strength of virgin PP is ~358 kV/mm, which is higher than that of XLPE (~320 kV/mm) [30]. As the elastomers are blended in the virgin PP, the breakdown strength is obviously reduced, and, with the increase of the elastomer content, the DC breakdown strength tends to decrease monotonically. As the elastomer content reaches 30 wt%, the breakdown strengths of PP/PBE and PP/POE are, ~300 and ~280 kV/mm, respectively, obviously lower than that of XLPE. It indicates that the addition of elastomer results in unacceptable DC breakdown strength for cable insulation if the elastomer content is too high. It is also noticed that the breakdown strength of PP/PBE is higher than that of PP/POE. In particular, the breakdown strength of PB10 is ~348 kV/mm, which is over that of XLPE. This suggests that the addition of PBE elastomer with low content (10 wt%) could meet the basic requirement in electrical performance for DC cable insulation.

In order to estimate the effect of elastomers on mechanical properties of blends, elongation at break and tensile strength of the samples have been measured. Such two parameters are selected to show the toughness modification performance, which have been widely employed in previous works [13, 31–35]. It should be mentioned that a thorough examination of mechanical property is certainly needed for industry application of PP/elastomer blends; however, since the very issue discussed in this chapter is the electrical property, other mechanical properties are not taken into consideration and further work in this respect will be performed in the future.

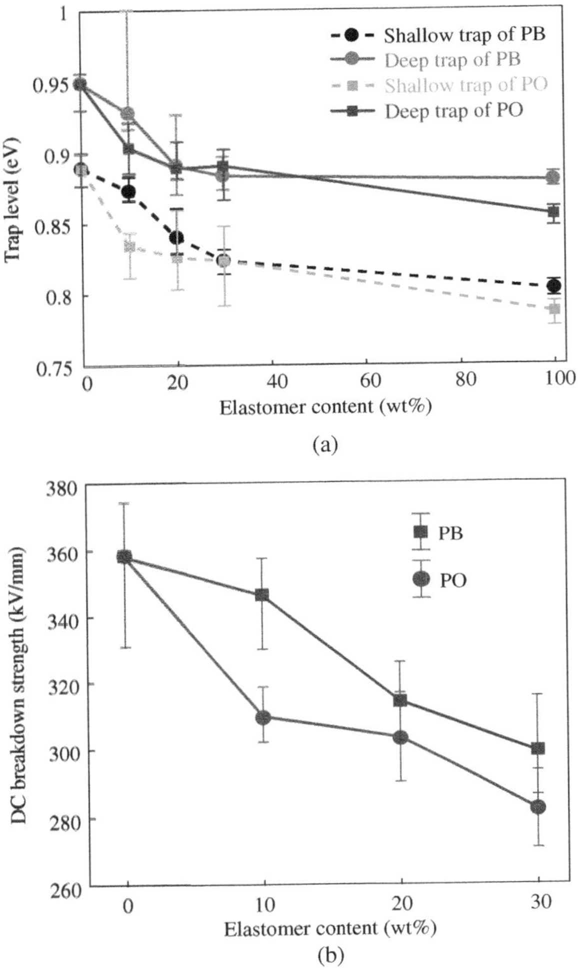

Figure 2.5 Dependence of the trap level distribution and DC breakdown strength upon the elastomer content for both PP/PBE and PP/POE blends. (a) Trap level distribution. (b) DC breakdown strength. *Source:* Gao et al. [17]/with permission from IEEE.

The relationship between the elongation at break and the elastomer content for both PP/PBE and PP/POE blends is illustrated in Figure 2.6a. With the growth of the elastomer content from 0 to 30 wt%, the elongation at break increases elastomer content for both PP/PBE and PP/POE blends significantly from ~40% to ~1200%. It reveals that the flexibility of virgin PP is remarkably improved. The

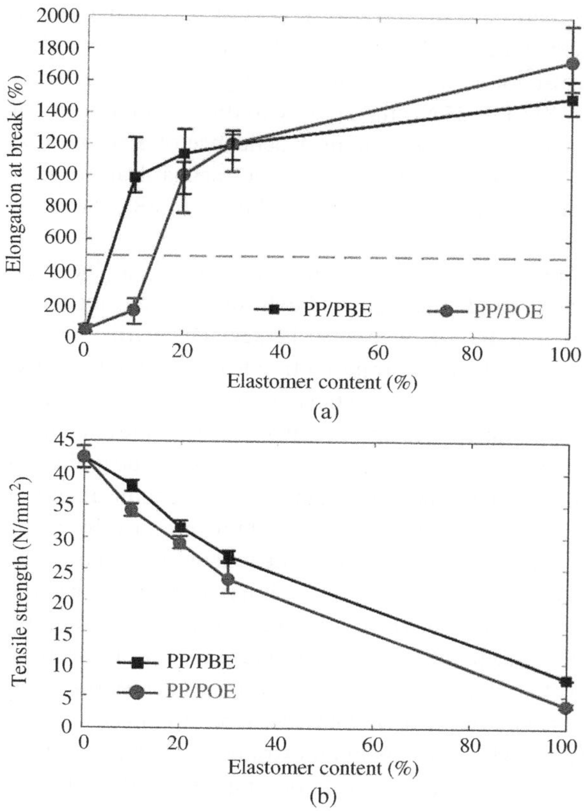

Figure 2.6 Relationship between the elongation at break, tensile strength, and the elastomer content. (a) Elongation at break. (b) Tensile strength. *Source:* Gao et al. [17]/ with permission from IEEE.

addition of PBE possesses better performance in flexibility modification than that of POE, especially at low elastomer content. However, the elongation at the break of POE is higher than that of PBE. That may be due to the fact that PP and PBE have similar structures and the compatibility of them is better. PP and PBE molecules bond closely, so the toughening effect is more significant. It can be observed that the addition of PBE with 10 wt% contributes to the elongation at a break of ~988.5%, which is higher than that of XLPE insulation (~500%) [36]. This suggests that the addition of PBE elastomer with low content could lead to remarkable improvement in the flexibility of virgin PP. Figure 2.6b shows the influence of elastomer content on tensile strength of PP/PBE and PP/POE samples. The tensile strength decreases obviously as the elastomer content increases from

0 to 30 wt%, and that of pure elastomers is the lowest. PP/PBE has higher tensile strength than PP/POE, which indicates that the PP/PBE has better mechanical property than PP/POE. It is also found that the tensile strengths for both PP/PBE and PP/POE are higher than XLPE (~25 N/mm^2) [35] with the elastomer content no higher than 20 wt%. By considering the content dependence of elongation at break and tensile strength, it is proposed that acceptable mechanical property for recyclable cable insulation could be achieved by the addition of PBE elastomer with low content.

Carrier traps in polymer insulating materials which are known as localized states presented at the forbidden band are considered to be originated from chemical and/or physical defects within the polymer matrix [21, 22]. One particular type of defect that plays an important role in determining trap feature is the interface. The interface in the polymer is usually referred to the boundary between two parts with different properties, for example, the interface between crystalline and amorphous regions in a semi-crystalline polymer like polyethylene [37]. Due to the disappearance of continuity in molecular chain arrangement, some chemical and/or physical disorder in chain alignment would be presented at the interface, by which the carrier traps are formed. It has been revealed that the fold of molecular chain, chain branch, and cavity or void could be found at the interface between crystalline and amorphous regions in XLPE [38]. The carrier trap formed at the interface could be understood from the viewpoint of energy structure as well. It has been reported by previous studies that when charge carriers transfer across the interface, the energy band bending and drifting occur because of the difference in Fermi levels at the different materials, which results in an extra potential barrier at the interface. Such a potential barrier could lead to charge trapping at the interface [39]. In short, the presence of interface between the PP matrix and the elastomers, shown in SEM images, contributes to the formation of carrier traps, which would allow charge trapping and de-trapping dynamics, thereby influencing the electrical performance of polymer insulation.

In order to better elucidate the possible mechanism of such interfaces in varying the trap characteristics, a schematic diagram is presented in Figure 2.7. It is considered that the carrier traps within the PP blend should be originated from three routes, that is, PP matrix, elastomer, and interface between them. Since both the PBE and the POE elastomers have shallower traps as compared with the virgin PP, it is reasonable that the blending of elastomer results in shallower trap in the PP/elastomer blend. Furthermore, since the PBE and the POE have different compatibilities with PP matrix, the interfaces exhibit different features that remarkably affect the trap characteristics. As for the PP/PBE blend, since both PP and PBE molecular chains contain propylene groups, Van der Waals' forces between them are strong. Thus, the interaction of the molecular chains at the interface between PP and PBE is tight. Accordingly, as carriers are transported across the interface, they are likely to be captured by the traps formed due to molecular chains, as

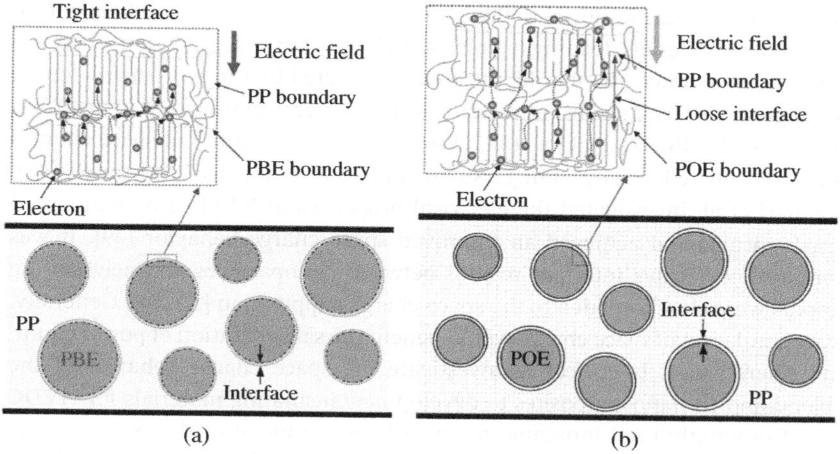

Figure 2.7 Schematic diagram of microscopic morphology at the interface between the PP and the elastomers. (a) PP/PBE blend. (b) PP/POE blend. *Source:* Gao et al. [17]/ with permission from IEEE.

depicted in Figure 2.7a, taking electrons as an example. Since the disorder of molecular chains usually acts as a chemical defect with the deep trap center, the interface at PP/PBE exhibits deep trap. As a comparison, the trap characteristics at the PP/POE interface should be different because of the bad compatibility. As can be seen in Figure 2.7b, the molecular chains of PP and POE have weak interactions, the interface between them is therefore considered to be a loose one. This is confirmed by the SEM photo shown in Figure 2.1c. Accordingly, as carriers are transported across the interface, the loosely bonded interface would allow the migration of charges easily. Such a loose interface should be recognized as physical defect that leads to the formation of shallow trap. Consequently, the PP/POE possesses shallower trap for both electron and hole than PP/PBE. It is, therefore, proposed that the deeper trap in the PP/PBE blend should be attributed to the better compatibility between PP and PBE.

2.3 Effect of Inorganic Nanofiller on Space Charge and Dielectric Breakdown

2.3.1 Metal Oxide Nanoparticles

Blending PP with polyolefin elastomer is an effective way to modify its stiffness and brittleness at room temperature. However, the elastomer blending also weakens the space charge behavior and breakdown strength of PP-based insulation, as a great concern under dc stress. In the operation of DC cables, space charge accumulation will cause electric field distortion and increase the local field within the insulating materials, leading to partial discharge, faster degradation, and even premature breakdown [40–45]. Many studies indicated that nanoparticles doping showed great performance in suppressing space charge accumulation in dielectric materials [6, 29, 46, 47]. Dakka et al. observed that the quantity of space charges was remarkably reduced by adding a small amount of nanoparticles to the PP [48]. Montanari et al. investigated the electrical properties of PP filled by nano-sized layered silicates and achieved an improved space charge behavior [49]. It was reported the extensive interface regions between nanoparticles and neighboring materials were closely related to the space charge suppression [50–54]. Generally, less accumulation of space charge would benefit the safe operation of power equipment. Therefore, it is critical to investigate the space charge behavior of the PP blend and its nanocomposites to develop new insulating materials for HVDC cables. Commonly used inorganic nanoparticles are metal oxides, such as zinc oxide (ZnO) and magnesium oxide (MgO). In this subsection, the ZnO nanofiller, with an average size of ~40 nm, is employed to improve the electrical properties of the PP blend insulation. For the convenience of description, the designation, composition, and weight ratio of each type of specimen are given in Table 2.3.

Space charge measurements were performed by the PEA method at room temperature. The experiment arrangement is shown in Figure 2.8. The

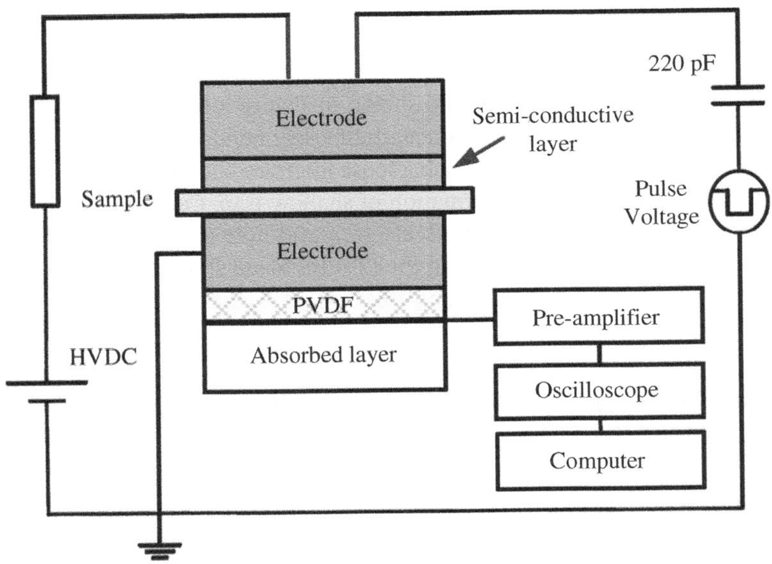

Figure 2.8 Schematic illustration of the experimental setup for PEA. *Source:* Du et al. [33]/ with permission from IEEE.

semi-conductive material was used as an anode (the upper electrode) and aluminum was used as a cathode (the lower electrode). Each specimen with a thickness of 250 μm was stressed under a positive DC voltage with an electric field of 70 kV/mm. The DC poling for all specimens was performed for 3600 seconds, and the measurement continued for 600 seconds after poling. The space charge behavior of each group was confirmed by repeating the tests five times. All the specimens were short-circuited and thermally treated at 80 °C for 48 hours and then cooled to room temperature in a vacuum oven before each measurement. Figure 2.9 shows the dynamic space charge distributions of different specimens under a DC electric field of 70 kV/mm for 3600 seconds. In order to represent the results clearly, space charge distributions in the thickness range of 35–215 μm are also shown in the partly enlarged graphs. For referencing purpose, Figure 2.9a shows the space charge behavior of pure PP. Due to the semicrystalline nature of PP, the interfaces between the crystalline and amorphous phases can introduce trapping sites, which would capture the injected charges [55]. With increasing the poling time, homo-charges injected from the cathode increase, and hetero-charges also appear in the vicinity of the anode. It is obvious that the space charge development in the pure PP bulk is minimal.

By comparing the space charge accumulation results in Figure 2.9a and b, it is found that much more charges are injected from the cathode in the PP blend insulation. With the lapse of time, the injected charges increase and advance deeper into the middle of the specimen, which is related to the microstructure of the PP blend insulation. With the addition of elastomer, the interfaces between elastomer and PP introduce many trapping sites, which can easily capture injected charges during the polarization process [13].

Just by loading the PP blend insulation with a small amount of nano-ZnO, the space charge behavior is improved, as shown in Figure 2.9c and d. The charge accumulation in the vicinity of the cathode is obviously suppressed, and the injected charges move much more slowly compared to the PP blend insulation and accumulate in the vicinity of electrodes. Obvious positive charge injection from the anode can be seen in the case of P60E40N2, with low charge density and deep injection distance of about 125 μm. On the whole, the space charge behaviors of nanocomposites are dramatically improved with the addition of nano-ZnO compared to that of blend insulation.

The total absolute charge amount for each specimen was calculated according to the following equation:

$$Q(t) = \int_0^L |\rho(x,t)| S dx \tag{2.4}$$

where $\rho_f(x, t)$ is the charge density in depth x (distance from the cathode) at poling time t. S is the electrode area. L is the thickness of the specimen.

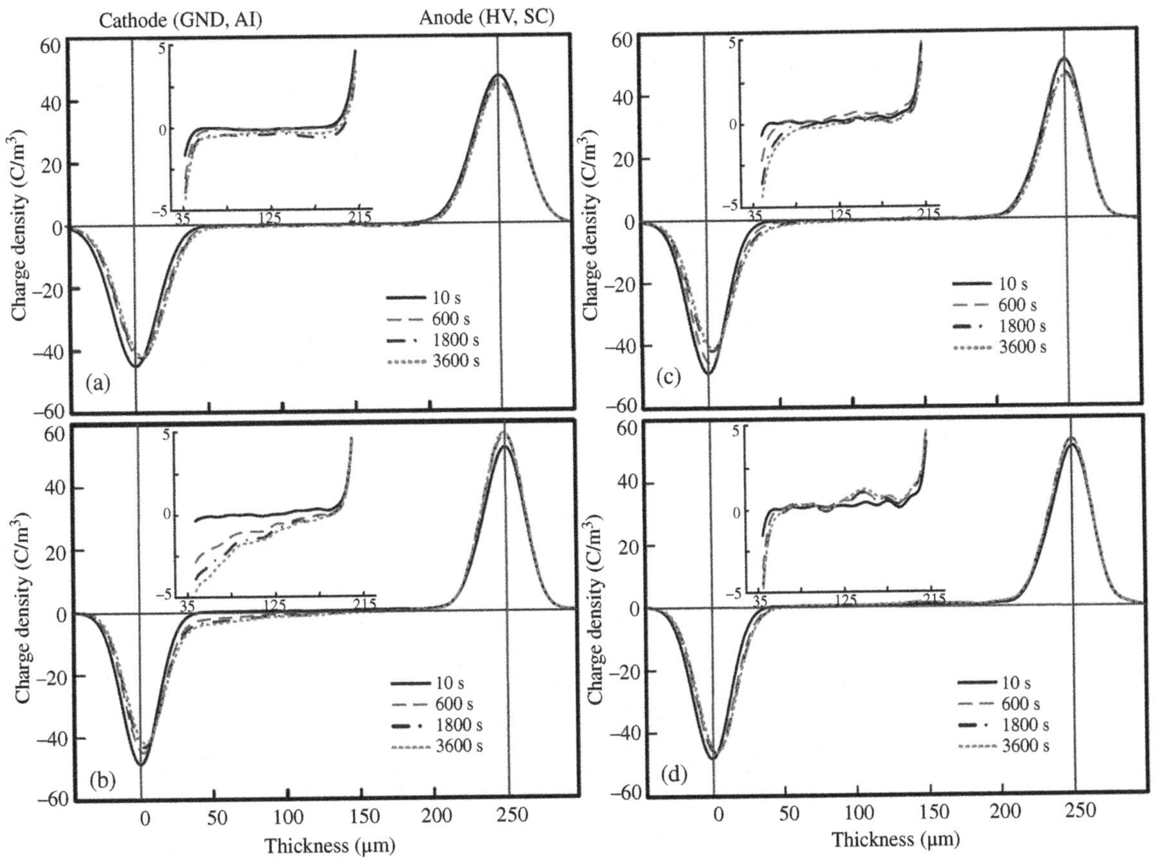

Figure 2.9 Effect of the nanofiller content on space charge distribution at 70 kV/mm DC field. (a) P100. (b) P60E40. (c) P60E40N0.5. (d) P60E40N2. *Source:* Du et al. [33]/with permission from IEEE.

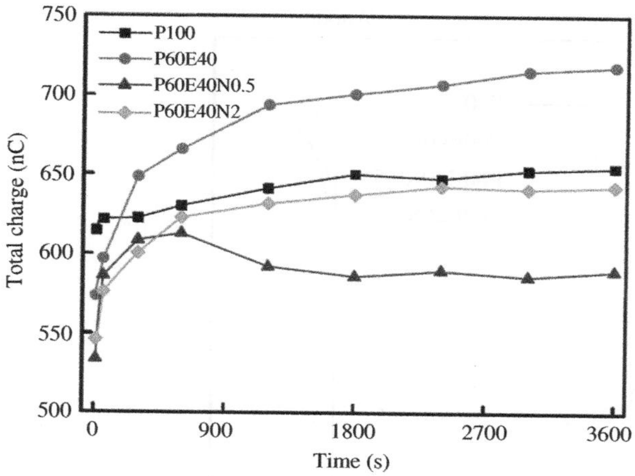

Figure 2.10 Relationship between the total charge and the polarization time. *Source:* Du et al. [33]/with permission from IEEE.

The calculated total charge of each specimen is shown in Figure 2.10. With the same applied stress, the total charge of pure PP shows the highest start point and is relatively stable. For PP blend insulation, the total charge amount is the highest and keeps increasing with poling time. The total charge amounts of nanocomposites increase at the initial period and then become stable with poling time. The total charge amount of different specimens after poling for one hour is in this order: P60E40 > P100 > P60E40N2 > P60E40N0.5.

As the electric field distortion due to space charge formation under DC stress is a great concern, the electric field along the specimen was calculated according to the following equation:

$$E(x) = \frac{1}{\varepsilon_0 \varepsilon_\mathrm{r}} \int_0^L \rho(x)\mathrm{d}x \tag{2.5}$$

where $\rho_f(x, t)$ is the charge density, ε_0 is the vacuum permittivity, and ε_r is the relative permittivity of the specimen.

The electric field distortion results at the polarization time 3600 seconds are shown in Figure 2.11a. It can be seen that the electric field is greatly distorted in PP/elastomer blend bulk compared to that of pure PP. For the PP/elastomer blend, the electric field is inhibited in the vicinity of the cathode, while it is enhanced in the vicinity of the anode, which is attributed to the large number of homo-charges injected from the cathode. Compared to the electric field distribution of PP/elastomer blend, electric field distortion in 0.5 and 2 wt% nanocomposites gets better. It is observed that the electric field of P60E40N0.5 is in close

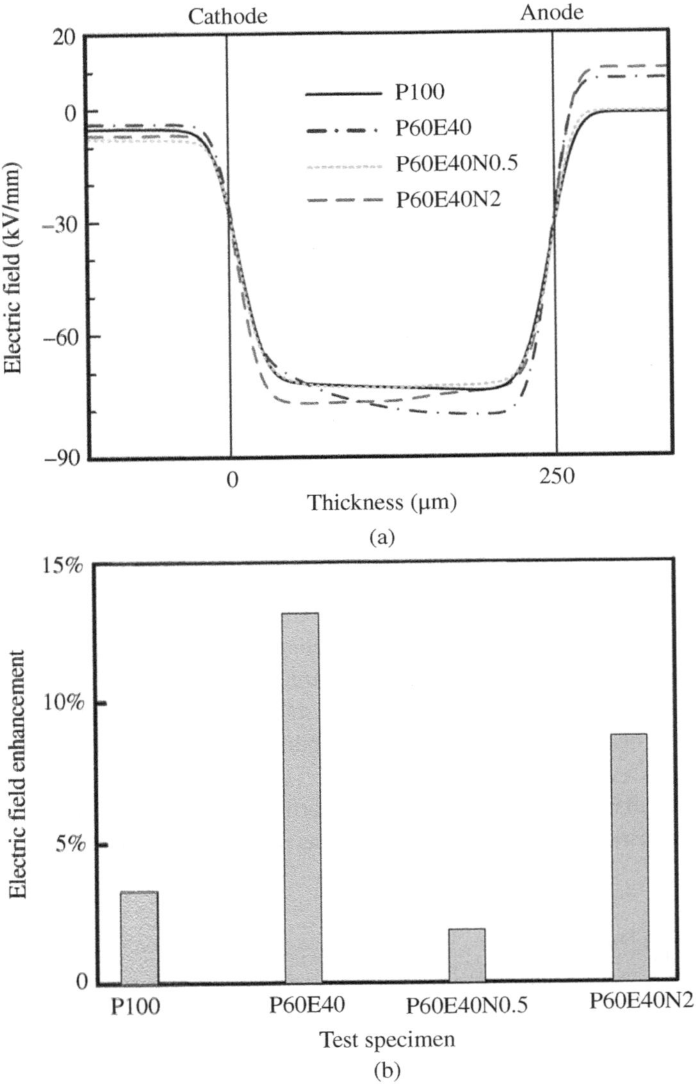

Figure 2.11 Effect of the nanofiller content on the electric field at 70 kV/mm DC field. (a) Electric filed distributions. (b) Electric filed enhancements. *Source:* Du et al. [33]/with permission from IEEE.

proximity to the ideal electric field of 70 kV/mm, indicating that the electric field distribution is rather uniform in the polymer. In the case of 2 wt% nanocomposites, the electric field in the vicinity of the cathode gets enhanced, which is related to the homo-charges injection from the anode.

The distortion factor of the electric field ΔE can be obtained by the following equation:

$$\Delta E = \frac{E_{\max} - E_{\mathrm{av}}}{E_{\mathrm{av}}} \tag{2.6}$$

where the $E_{\max}$ is the highest value of the electric field in the specimen bulk after DC poling of one hour. E_{av} is the ideal electric field in the absence of space charge. The results are shown in Figure 2.11b. It is observed that the PP/elastomer blend shows the highest distortion factor, which is increased by four times as compared with pure PP. The severe electric field distortion induced by space charge formation may lead to partial discharge in the micro defects, thus accelerating the insulation aging. Nanoparticles doping can greatly mitigate the electrical field distortion. The distortion factor of specimen with 0.5 wt% nanoparticles is much lower than those of the other specimens, indicating uniform electric field in the polymer.

Compared to the results of pure PP, the addition of elastomer into PP matrix results in more space charge accumulation in the vicinity of the cathode and higher electric field distortion factor, which is unexpected for insulating materials in HVDC cable application. While all the nanocomposites show lower quantity of space charges and lower distortion factor than those of PP/elastomer blend. It is proved that the nano-ZnO doping with a content of 0.5 wt% can effectively suppress the space charge accumulation and electric field distortion in polymer, which makes the nanocomposites more suitable for HVDC cable application than PP blend insulation.

The Weibull plots of dc breakdown strength of different specimens are shown in Figure 2.12. It indicates the PP/elastomer blend has a lower breakdown strength than pure PP. It may result from the extensive interfaces between elastomer and PP phase, which can trap charge carriers and distort electric field. Besides, it was reported that the crystallization temperature and crystallinity of PP blend insulation were lower than pure PP, therefore, the lamellae thickness and free volume in PP/elastomer blend changed, resulting a decrease in dc breakdown strength [13]. It also shows that nano-ZnO doping enhances the DC breakdown strength. The literature has suggested that the enhancement of breakdown strength in nanocomposite is closely related to the space charge and traps in the interfacial regions [51, 56]. It is confirmed that the incorporation of ZnO particles can suppress space charge accumulation and inhibit local electrical field distortion. All these factors together give rise to an increase in breakdown strength of PP-based nanocomposites.

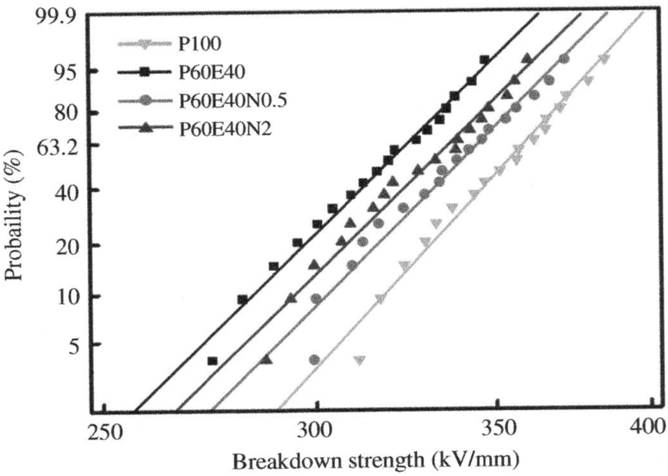

Figure 2.12 DC breakdown strength of different specimens. *Source:* Du et al. [33]/with permission from IEEE.

The volume resistivity was also measured to evaluate the insulating properties as well as the DC breakdown strength [57, 58]. The charging currents of the specimens under 30 kV/mm DC electric field are shown in Figure 2.13a, which all decay quickly at the initial period and then reach the stable stage. The current values are averaged over the period from 750 to 800 seconds, respectively, to calculate the volume resistivity, and the results are shown in Figure 2.13b. The volume resistivity of PP/elastomer blend is higher than that of pure PP. This may be attributed to the numerous interfaces between elastomer and PP.

In conclusion, the introduction of nano-ZnO into PP/POE blend introduces the deep trap sites at the interfacial regions between the nanofillers and polymer dielectrics, which can lead to the reduction in charge injection, thus, suppressing space charge accumulation and electric field distortion in the nanocomposites bulk. Besides, the nanocomposites show great performance in permittivity, breakdown strength, and volume resistivity. With the improved mechanical properties and excellent electrical properties, the PP-based nanocomposites have much potential for HVDC cable application.

2.3.2 Nanoplatelets

The last subsection reports that the accumulation of space charge in polymer insulation can be suppressed by doping nanoparticles, which is attributed to the interfacial regions between the nanofillers and polymer dielectrics, which can capture the mobile charges and restrict the mobility of carriers, thus suppressing the

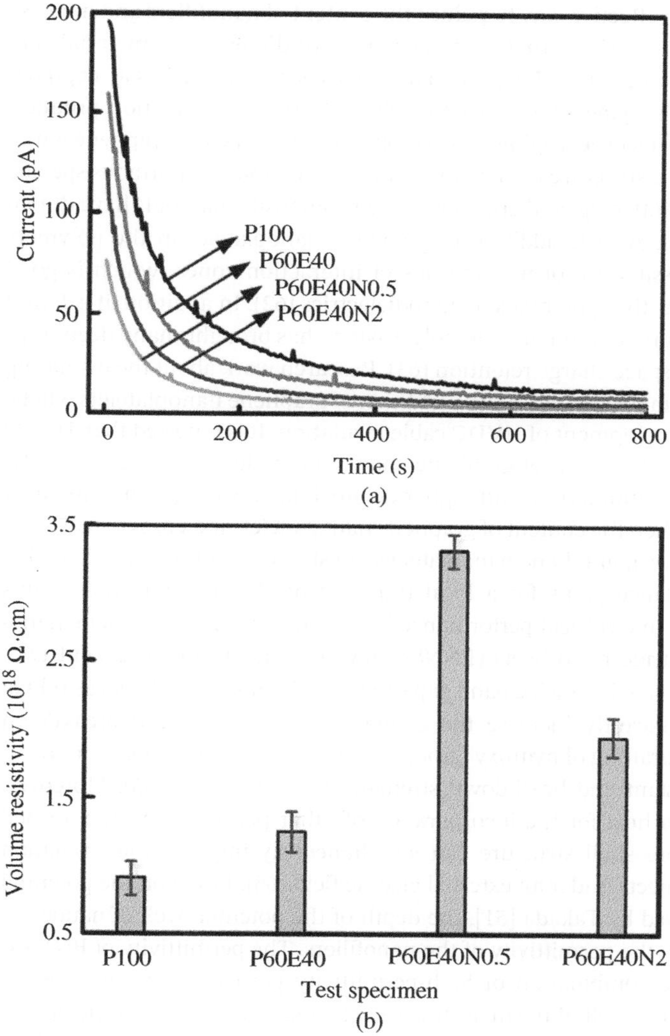

Figure 2.13 (a) The charging current and (b) volume resistivity of different specimens. *Source:* Du et al. [33]/with permission from IEEE.

injection and transport of space charges [33, 59]. Usually, it is difficult for the common nanofillers, such as ZnO and MgO, to obtain a specific surface area of 1000 m^2/g in theory. In order to achieve a desired effect, a high content of nanofillers should be doped, which can bring in particle agglomeration or even involve defects in the polymer matrix [6]. Increasing the specific surface area of nanomaterials is an effective way to reduce the content of nanofillers and increase the interface area, so mesoporous materials and two-dimensional materials are introduced. Mesoporous silica MCM-41 nanoparticles can be used as a "degradation inhibitor" for polymer dielectrics to isolate electrical degradation products and then inhibit the electrical aging process [60]. Recent years, the graphene nanoplatelet with specific structure of one atomic thick layer and an enormous specific surface area over 2000 m^2/g, and graphene-based composite materials have drawn much concern [61]. A little addition of graphene nanoplatelet in the polymers would make it possible to form numerous of interaction zones, which is good for exploiting the further potentials of nanodielectrics [62]. In addition, it is found that the introduction of graphene into polyethylene has brought many deep traps and restrained the space charge retention [63]. Research work about the insulating properties of PP composites with a little addition of graphene nanoplatelets will be beneficial to the development of HVDC cable insulation. It is reported that The PP blends with 0.01 wt% content of graphene show a much lower DC conductivity, less space charge accumulation, and higher dc breakdown strength than the neat PP samples. But, when the content of graphene nanoplatelets increases to 0.05 wt%, more contacts between graphene nanoplatelets are structured in the matrix, which provides low-resistance paths for a local transport of electrons and holes, thus causing a decrease in electrical performance [59]. Compared with graphene nanoplatelets, boron nitride nanosheets (BNNSs) have become a more excellent two-dimensional material with a wide band gap of 5.9 eV. For instance, BNNS as robust scaffolds can significantly increase the energy storage density and breakdown strength [64]. The grafting of hydroxyl groups onto BNNS simultaneously improves the dielectric constant and breakdown strength of the dielectric [65]. Moreover, BNNS can act as a host for the incorporation of other particles, and the formed hybrids with a core–shell structure can simultaneously improve the insulation and polarization effects under an external electric field [66]. Based on the potential well theory proposed by Takada [51], the depth of the potential well of nanocomposites is related to the permittivity of the nanofillers. The permittivity of BNNS is only ~4.5, and the combination of high-permittivity particles may increase the depth of the potential well of the dielectric, thereby increasing its dielectric performance. This subsection will report novel TiO_2@BNNS nanofillers fabricated by a high-temperature calcination method.

Figure 2.14a illustrates the preparation process of TiO_2@BNNS nanofillers. BNNS and TiO_2 are previously dried in a hot oven at 90 °C for 12 hours. One gram of BNNS and 200 ml of absolute ethanol are added to the beaker, and the suspension is magnetic stirred at 2000 rpm for 5 minutes. Then, 0.2 g of TiO_2 is added thereto. The mixture is dispersed by using mechanical stirring and ultrasonic vibration for 120 minutes. The ethanol is evaporated to dryness at 90 °C, to obtain the BNNS/TiO_2 composite fillers with a BNNS and TiO_2 mass ratio of 5:1.

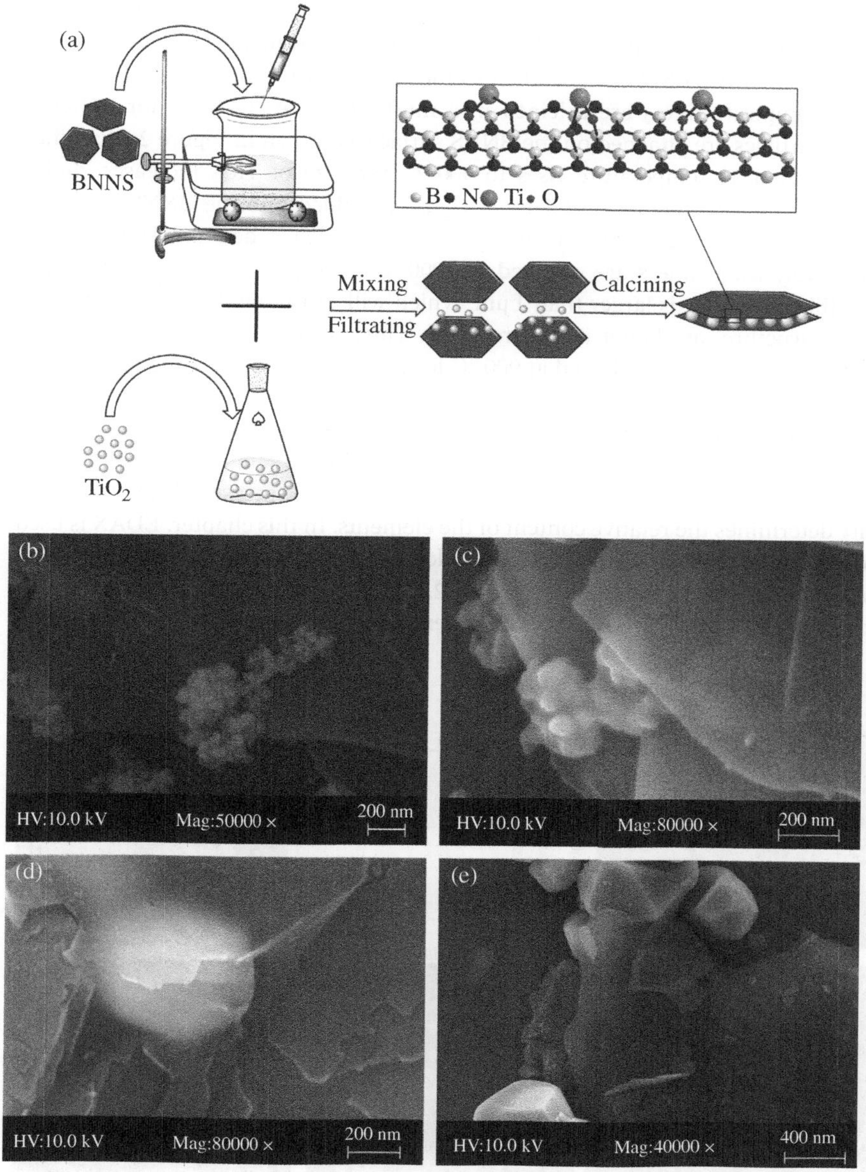

Figure 2.14 (a) Illustration for the preparation process of TiO_2@BNNS nanofillers. SEM images of TiO_2/BNNS, (b) without calcination, (c) calculated at 800 °C, (d) calculated at 900 °C, and (e) calculated at 1000 °C. *Source:* Li et al. [67]/with permission from IEEE.

Figure 2.14b presents the scanning electron microscope (SEM) image of BNNS/TiO_2 nanofillers without calcination, showing the TiO_2 nanoparticles are mostly free between the BNNS or adsorbed to the surface of the BNNS. As the BNNS/TiO_2 composite fillers are calcined in a muffle furnace at 800 °C for 20 hours, TiO_2 particles are attached on the BNNS surface as shown in Figure 2.14c. When the calcination temperature rises to 900 °C, the size of TiO_2 increases, and it reduces the agglomeration of TiO_2. TiO_2 is embedded into the BNNS to form a core–shell structure, as shown in Figure 2.14d, which is denoted TiO_2@BNNS. However, when fillers are calcined at 1000 °C, TiO_2 grain further growth, and the filler diameter is larger than 1 μm, which will cause the destruction of core–shell structure, as shown in Figure 2.14e. Thus, TiO_2@BNNS nanofillers used in this subsection are calcined at 900 °C for 20 hours, with a heating and cooling rate of 3.65 °C/min.

The energy dispersive X-ray analysis (EDAX) is realized by analyzing the X-ray wavelength and intensity of the elements emitted by the sample. According to the wavelength, the elements contained in the sample can be measured, and the intensity determines the relative content of the elements. In this chapter, EDAX is used to determine the type and content of the elements and characterize the distribution of N, B, Ti, and O elements in Figure 2.15a. As shown in Figure 2.15b–d, the content of B and N elements is higher, which represents BNNS substance,

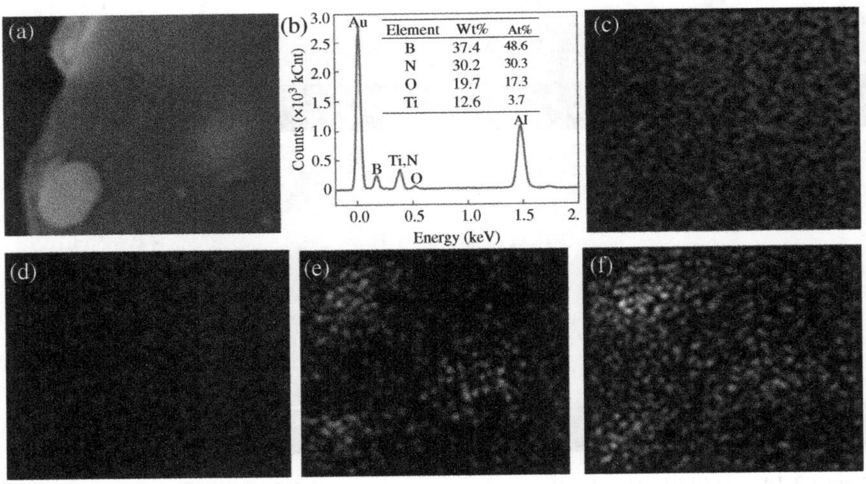

Figure 2.15 EDAX results of TiO_2@BNNS nanofillers calculated at 900 °C. (a) SEM images of TiO_2@BNNS. (b) Mass fraction of each element. The distributions of (c) N, (d) B, (e) Ti, and (f) O elements are presented by the spots. *Source:* Li et al. [67]/with permission from IEEE.

corresponding to the gray area in Figure 2.15a. The white spherical particles corresponding to the Ti and O elements described in Figure 2.15e and f are TiO_2. The comparison shows that the N and B elements are similarly distributed, and the corresponding substance is BNNSs, but the distributions of the O and Ti elements are not exactly the same. It is deduced that the substance containing the O element is not only TiO_2 but the B element or the N element may also be chemically bonded to the O element during the calcination process, thereby further reacting with TiO_2.

The X-ray diffraction (XRD) results of BNNS and TiO_2 are shown in Figure 2.16, the intensity of BNNS diffraction peak remains basically unchanged after calcination at different temperatures, while the peaks corresponding to other crystal faces increase. A new diffraction peak appears in TiO_2@BNNS after calcination, and the diffraction peak corresponds to a chemical formula of TiB_mO_n, that is, boron-doped rutile-type titanium dioxide, corresponding to PDF#87-0921. As the calcination temperature increases from 800 to 900 °C, the intensity and the width of the TiB_mO_n diffraction peak increase, which proves that the combined area of BNNS and TiO_2 increases, and the content of TiO_2@BNNS with core–shell structure increases significantly. In addition, the intensity of the TiO_2 diffraction peak is

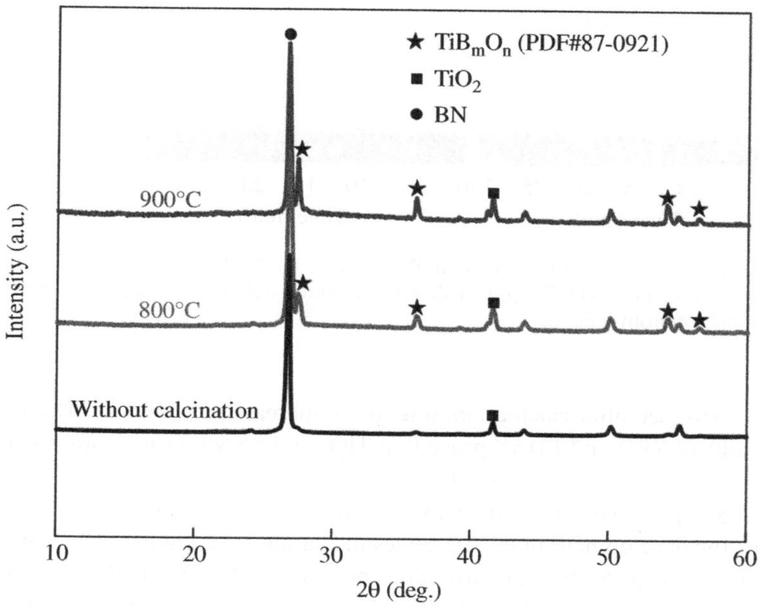

Figure 2.16 XRD patterns of BNNS and TiO_2 at room temperature, and calcination at 800 and 900 °C. *Source:* Li et al. [67]/with permission from IEEE.

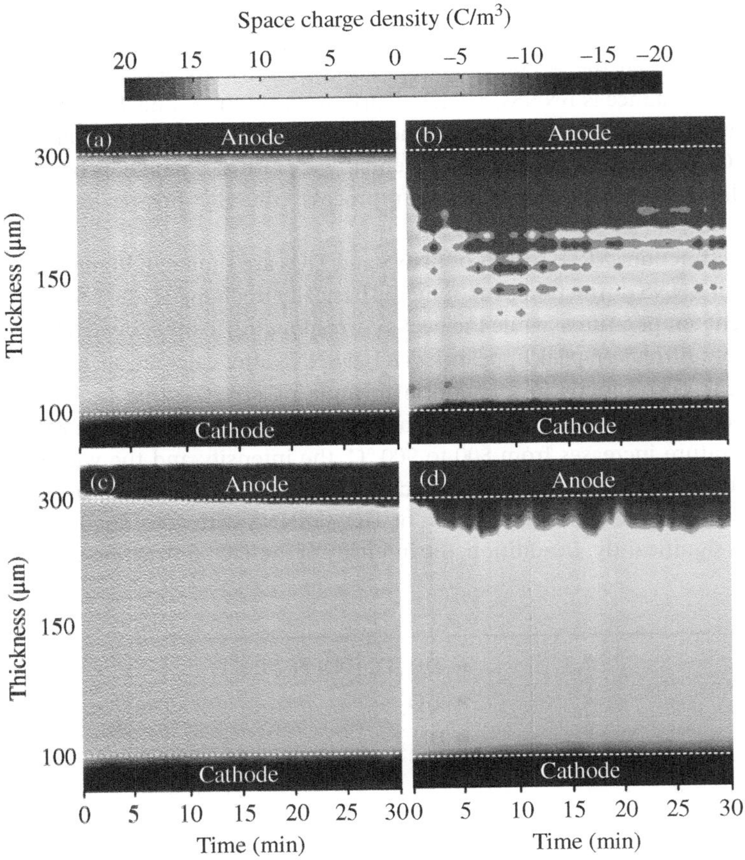

Figure 2.17 Space charge distributions of (a) PP at 50 °C, (b) PP at 90 °C, (c) PP/TiO_2@BNNS at 50 °C, and (d) PP/TiO_2@BNNS at 90 °C during polarization. *Source:* Li et al. [67]/with permission from IEEE.

also slightly increased after calcination, due to the increase in grain size. Therefore, the results of XRD and EDAX prove that TiO_2 and BNNS form a core–shell structure in the form of chemical bonds.

Space charge injection behavior at an operational temperature of up to 90 °C is a key issue for the development of HVDC cables insulation [68]. Figure 2.17 shows the time-dependent space charge distributions of PP and PP/TiO_2@BNNS during the polarization process with an applied field strength of 40 kV/mm at 50 and 90 °C. The horizontal axis and the vertical axis represent the polarization time and the thickness of the sample, respectively. PP has a small amount of homopolar

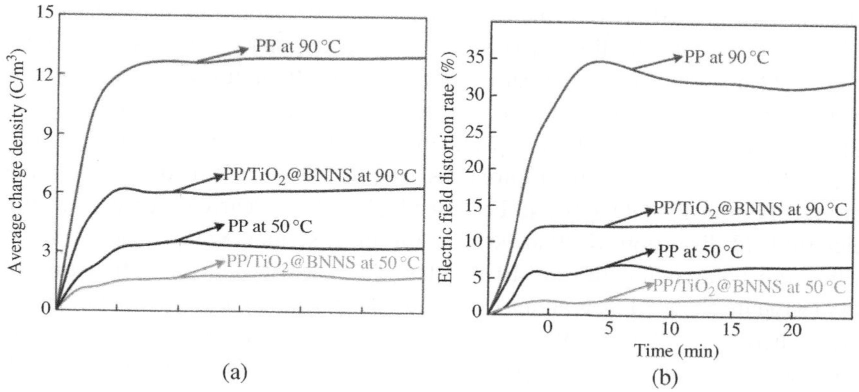

Figure 2.18 (a) Average charge density and (b) electric field distortion rate of PP and PP/TiO_2@BNNS at 50 and 90 °C. *Source:* Li et al. [67]/with permission from IEEE.

space charge injected near the anode at 50 °C, and when the temperature is raised to 90 °C, a large amount of homopolar space charges are injected from the anode into the PP matrix. For the PP/TiO_2@BNNS nanocomposite, almost no space charge injection is observed at 50 °C, and only a small amount of space charge accumulates near the anode at 90 °C, which inhibits space charge injection into the dielectric.

The average charge density results of PP and PP/TiO_2@BNNS are shown in Figure 2.18a. It is obvious that the space charge injection is mainly concentrated in the first 5 minutes of polarization. The average charge densities of PP and PP/TiO_2@BNNS are 3 C/m^3 and 1.8 C/m^3, respectively, and most of their space charges accumulate near the anode. When the temperature rises to 90 °C, a large amount of homopolar space charge is injected into PP, and PP/TiO_2@BNNS has only a small amount of homopolar space charge at the anode, and the injection depth is significantly less than that for PP. When the polarization time is 30 minutes, the average charge density of PP and PP/TiO_2@BNNS at 90 °C is 12.9 and 6.3 C/m^3, respectively. It is indicated that TiO_2@BNNS with a two-dimensional structure can provide a large number of local states to capture the charges and effectively block the injection of charges into the bulk especially under high electric field and high temperature, which provides a new idea for the development of HVDC cables to higher temperatures.

The injection of homopolar space charge can weaken the electric field near the anode, and the electric field inside the insulating material will be greatly enhanced. The distortion of the electric field greatly affects the insulation performance of the dielectric, thus the electric field distortion rate is employed to characterize the time-dependent electric field distribution inside the sample. The electric field distortion rate increases the fastest in 5 minutes of polarization, which corresponds to the injection of homopolar space charges, as shown Figure 2.18b. When the polarization time is up to 10 minutes, its electric field distortion rate reaches 36.3%. The electric field distortion rate of PP and PP/TiO_2@BNNS is 2.1 and 7%, respectively, when the polarization time is 30 minutes at 50 °C. When the temperature rises to 90 °C, the electric field distortion rate of both PP and PP/TiO_2@BNNS becomes higher. It is clearly observed that the electric field distortion of PP at 90 °C is the most serious, which reaches 36.3% when the polarization time is 10 minutes. The electric field distortion rate of PP/TiO_2@BNNS at 90 °C is about 13%, when the polarization time exceeds 5 minutes. Therefore, PP/TiO_2@BNNS can effectively suppress the distortion of the internal electric field of the sample under high electric field and high temperature.

Figure 2.19 shows the two-parameter Weibull distribution of the DC breakdown strength at 30, 50, 70, and 90 °C for the five samples. E_{bd} is the measured DC field strength at which there is a 63.2% probability for the specimens to breakdown, which are marked in the figures. PP/TiO_2@BNNS nanocomposite has the highest DC breakdown strength, which is 336.2, 284.5, 268.9, and 255.0 kV/mm at 30, 50,

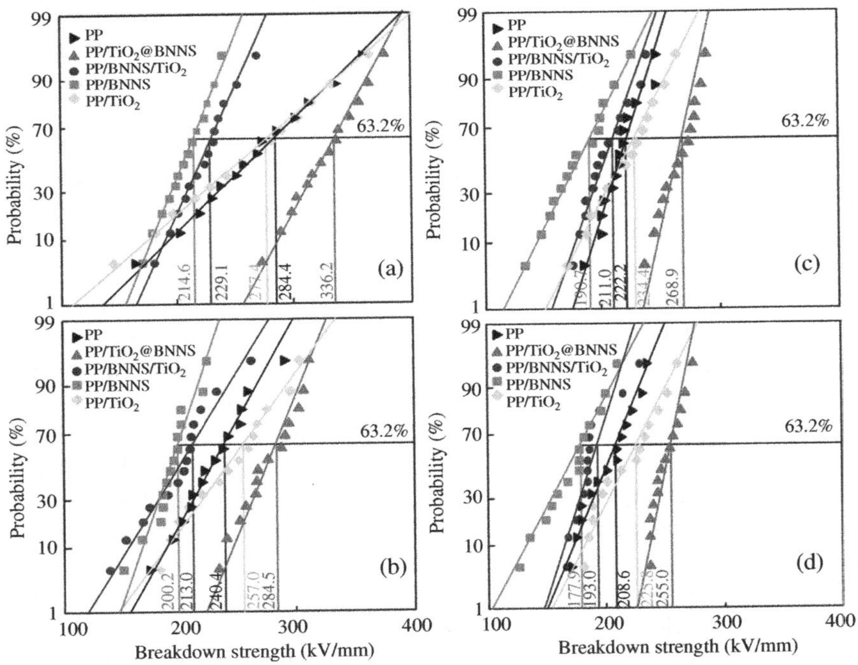

Figure 2.19 DC breakdown strength of PP composites at (a) 30 °C, (b) 50 °C, (c) 70 °C, and (d) 90 °C. *Source:* Li et al. [67]/with permission from IEEE.

70, and 90 °C, respectively. The breakdown strength of PP/TiO_2@BNNS samples increases by 18.2, 18.3, 21, and 22.2% compared with that of PP varying from 30 to 90 °C. In addition, the breakdown strength of PP/TiO_2@BNNS shows higher shape parameters at various temperatures, indicating a more concentrated breakdown strength distribution. Therefore, the doping of TiO_2@BNNS also optimizes the internal microstructure of PP, thereby reducing the dispersion of the breakdown strength of the sample. Moreover, the agglomeration of a small part of TiO_2 particles will cause their loose layer regions to overlap, and the carriers are prone to de-trapping in this region to participate in the charge transport. This process results in the breakdown strength of the PP/TiO_2 local area at a lower probability even lower than that of PP. Therefore, the shape parameter of PP/TiO_2 is always lower than that of PP, and the breakdown strength of PP/TiO_2 at low probability (such as 0.1%) is sometimes lower than that of PP. The outstanding breakdown strength of the PP/TiO_2@BNNS nanocomposites can be partially attributed to unique core–shell structure of TiO_2@BNNS as shown in Figure 2.14d, which prevents agglomeration, and the high-polarity TiO_2 strengthens the interface effect between nanofillers and PP. Thus, the nanofillers of two-dimensional layered BNNS encapsulated high-polarity TiO_2 can effectively reduce the probability of the formation of hot electrons to increase the breakdown strength of PP/TiO_2@BNNS.

The breakdown strength of PP/TiO_2 nanocomposites increases by 6.9, 5.5, and 8.2 compared with PP at 50, 70, and 90 °C, respectively. It is deduced that more traps are introduced at the interface area between TiO_2 and PP, and the interface area between PP and TiO_2 strengthens the binding effect on carriers, which decrease the mobility of carriers. Moreover, the TiO_2 nanoparticles can also be considered as scattering sources to reduce the energy of carriers, thereby preventing the electrical treeing propagation [69].

PP/BNNS and PP/BNNS/TiO_2 nanocomposites show worse breakdown performance than PP, which is related to its micro-morphology as shown in Figure 2.14b. BNNS has a very large specific surface area, and there is a large adsorption force between BNNS and TiO_2, which will cause agglomeration. There are lots of interface areas between nanoparticles and PP, and the agglomeration will cause the interface area to overlap, so it is easier for carriers to pass through the overlap area. The overlapping area weakens the ability to confine the carriers and accelerates the process of impact ionization, thus, it is not conducive to the improvement of breakdown performance.

The trap level distributions are considered to be associated with the dielectric performance of PP nanocomposites from the microscopic point of view. The ISPD curves of PP composites at 90 °C are in Figure 2.20a; the decay rate of PP/TiO_2@BNNS and PP/TiO_2 surface potential is significantly slower than the other three samples. According to the ISPD curves, the trap level distribution

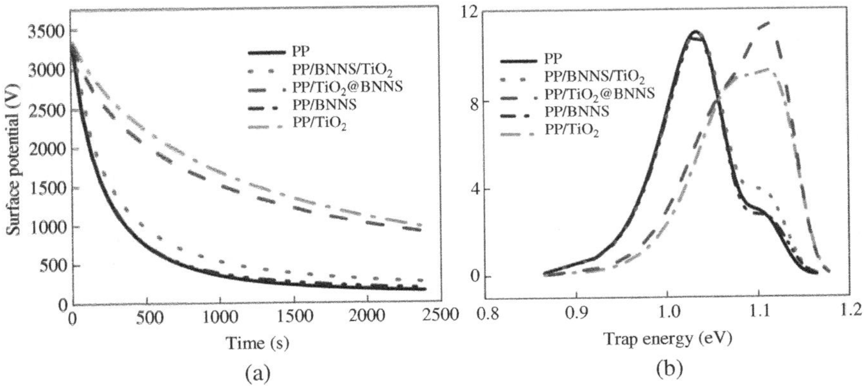

Figure 2.20 (a) Surface potential decay and (b) trap level distribution of PP nanocomposites. *Source:* Li et al. [67]/with permission from IEEE.

characteristics of each sample can be calculated based on the method introduced in Section 2.2 and are shown in Figure 2.20b [68]. The trap level centers of PP/TiO_2@BNNS and PP/TiO_2 (1.12 eV) are significantly higher than those of the other three samples (1.03 eV), and the deep trap density at 1.10–1.15 eV of the PP/TiO_2@BNNS sample is higher than that of the PP/TiO_2 sample. The unique core–shell structure of TiO_2@BNNS interacts with the PP matrix to form an interaction zone structure, introducing a lot of potential wells [29, 66]. Carriers in PP/TiO_2@BNNS are easily blocked by BNNS during the transportation process, so the energy obtained from the electric field when migrating in the free volume is reduced, it is not easy to cross the potential wells. In addition, the increase in the energy level and the density of the potential wells further strengthen the capturing effect [70], so that collision ionization is not easy to occur. The potential wells can weaken the equivalent electric field of charge injection at the interface between the electrode and the dielectric after trapping the charge, thereby suppressing the charge injection and weakening the distortion of the electric field. Hence, PP/TiO_2@BNNS exhibits higher DC breakdown strength and less space charge accumulation.

In addition to the above experimental result analysis, theoretical simulation analysis is also employed to investigate the influence of TiO_2@BNNS nanofillers on the dielectric performance. The dielectric constants of PP, BNNS, and TiO_2 in this model are set to 2.3, 4.5, and 114, respectively [71]. The large and small rectangles represent PP and BNNS, respectively, and the electric field applied to PP matrix is 1 kV/mm as depicted in Figure 2.21.

According to the electrical potential well theory proposed by T. Takada [51], free carrier charges are more likely to accumulate near the fillers and captured by

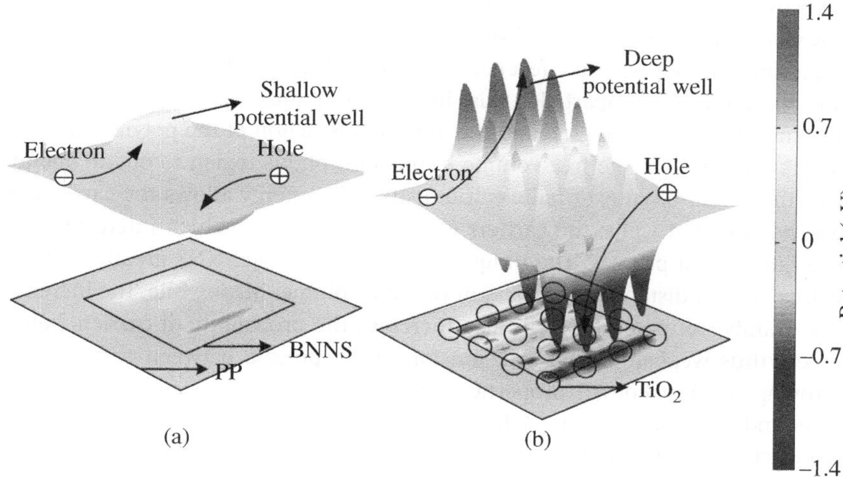

Figure 2.21 Three-dimensional electrical potential distributions of (a) PP/BNNS and (b) PP/TiO_2@BNNS. *Source:* Li et al. [67]/with permission from IEEE.

potential wells at the interface between the fillers and the PP matrix. Additionally, Figure 2.21a and b shows three-dimensional views of the potential well distribution of PP/BNNS and PP/TiO_2@BNNS. It can be clearly seen that the potential well depth and density of PP/BNNS are significantly less than PP/TiO_2@BNNS. Previous work [72] showed that when there is a large difference in the dielectric constant values of the fillers and the polymer matrix, the polymer composites will show an electric field concentration phenomenon. The two-dimensional structure of BNNS and its wide band gap feature make it have a charge-blocking effect, so it is difficult for hot electrons to pass into the conduction band. Compared with the PP/BNNS, the electric field is more concentrated near the TiO_2@BNNS fillers in the PP composite under the same applied electric field. The synergistic effect of BNNS with a wide band gap and high-polarity TiO_2 enables TiO_2@BNNS nanoparticles to provide more potential wells for PP and further enhance the charge-trapping ability [50]. When the charge drifts along the electric field near the potential well, the charge will be blocked by the BNNS with a two-dimensional slice structure and then captured by the potential well produced by TiO_2 nanoparticles, thereby preventing further injection of homopolar space charge. Therefore, the synergistic effect of BNNS and TiO_2 endows PP for better insulation performance based on the core–shell structure.

Moreover, lots of deep potential wells are introduced into PP by doping the TiO_2@BNNS nanoparticles, thus the transport characteristics of carrier in PP matrix are limited at high temperature, which causes a reduction in the conductivity activation energy of PP/TiO_2@BNNS. At the same time, deep potential wells in PP/TiO_2@BNNS also suppress the injection of homopolar space charge, and the suppression effect becomes more obvious with the higher temperature. In addition, nanoparticles exist in the amorphous region or the interface between the crystalline region and the amorphous region of the polymer matrix. Carriers accelerate in the free volume in the amorphous region to obtain energy under the action of an external electric field. If the energy allows the carriers to cross the potential wells, the carriers continue to migrate, which intensifies the impact ionization process. The unique structure of TiO_2@BNNS greatly reduces the acceleration distance of carriers in the free volume. Meanwhile, the increase in the number of potential wells also increases the probability of carriers being trapped, thus weakening the collision ionization process between carriers and improving the breakdown properties. Thus, the theoretical simulation results further indicate that the TiO_2@BNNS nanofillers have a profound impact on the dielectric properties of PP composites.

2.4 Effect of Organic Compounds on Space Charge and Dielectric Breakdown

2.4.1 Introduction

Inorganic nanofillers still face the problem of difficulty in dispersing in polymer matrices during large-scale manufacturing processes, but organic molecular additives can effectively avoid this problem. Organic molecular monomers are mixed into the PP matrix through melt blending or solution blending methods, and the high polarity of the molecules is utilized to change the local state structure of the PP matrix, introducing trap energy levels and capturing migrating charge carriers. This method has significant modification effects and is suitable for large-scale industrial production. At present, the mainstream molecular monomers are divided into three categories: polar molecules, voltage stabilizers, and antioxidants. Polar molecules are generally suitable for various working conditions and voltage stabilizers are mainly aimed at improving the electrical tree characteristics of PP, while antioxidants effectively inhibit the aging of PP cables, while enhancing their insulation strength. The specific situation is shown in Table 2.4.

2.4.2 Voltage Stabilizer

Voltage stabilizers are put forward to solve the problem of XLPE insulation once again in recent years, which can attenuate the energy of high-energy electrons that injected into the materials from electrodes under the condition of high electric

Table 2.4 Effect of different grafted molecules on the dielectric properties of PP or XLPE.

Molecule type	Grafted monomer	Conductivity	Space charge	Breakdown strength	Electrical tree	Optimal content
Ordinary polar molecule	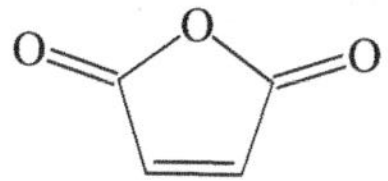 (MAH)	Improves high-temperature stability	Inhibits space charge injection and accumulation, and the maximum electric field distortion is 20% of PP	DC breakdown strength is 13.5% higher than PP	/	2.0 wt%
	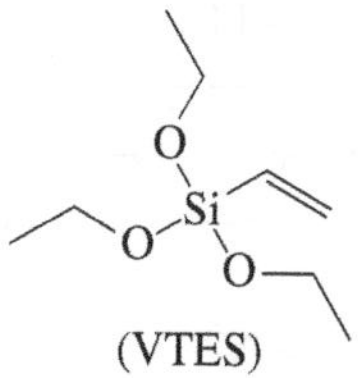 (VTES)	One order of magnitude lower than PP	Inhibits space charge injection and accumulation, and the electric field distortion rate is 36.4% of PP	DC breakdown strength is 18.8% higher than PP	/	2.5%
	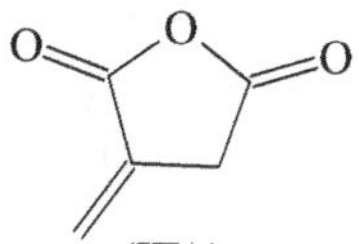 (ITA)	One order of magnitude lower than PP	/	DC breakdown strength is 42.9% higher than PP	Inhibits the growth of the electrical tree and reduces the fractal dimension of the electrical tree	5 wt%
	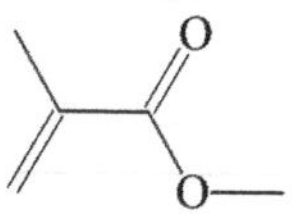 (MMA)	/	/	/	Inhibits the initiation and growth of electrical tree, and the length of electrical branches is 11% of PP	10 wt%

(*Continued*)

Table 2.4 (Continued)

Molecule type	Grafted monomer	Conductivity	Space charge	Breakdown strength	Electrical tree	Optimal content
	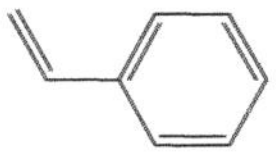 (St)	Two order of magnitude lower than PP	Inhibits space charge injection and accumulation, and the electric field distortion rate is 16.7% of PP	DC breakdown strength is 36.0% higher than PP	/	5 wt%
Functionalized grafted molecule—voltage stabilizer	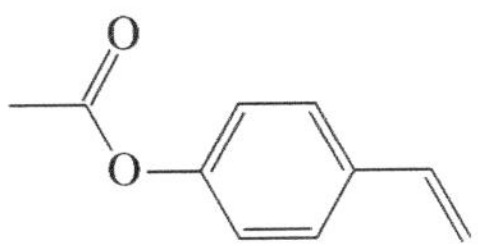 (AOS)	74.4% lower than PP	Inhibits space charge injection and accumulation	DC breakdown strength is 53.0% higher than PP	/	/
	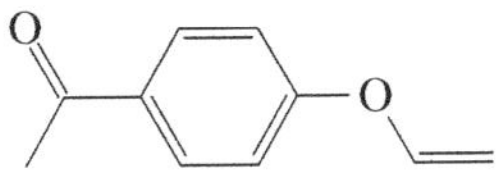 (VPE)	/	/	AC breakdown strength is 15.5% higher than XLPE	The starting voltage of the electrical tree is 39.6% higher than XLPE	1 wt%
	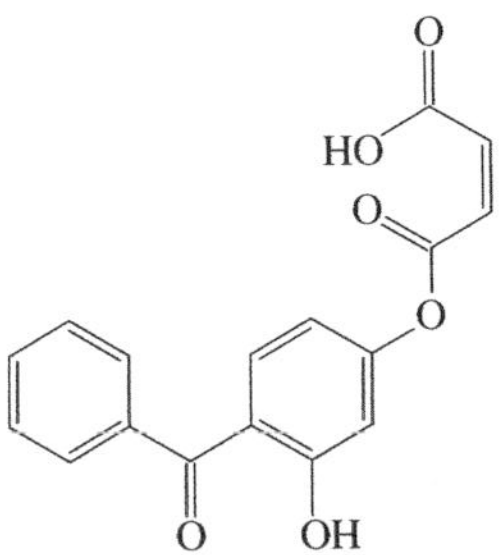 (MDVS)	Two order of magnitude lower than XLPE	Inhibits space charge injection, and the maximum space charge density is 43.8% of XLPE	DC breakdown strength is 14.8% higher than XLPE	/	1.2 wt%

	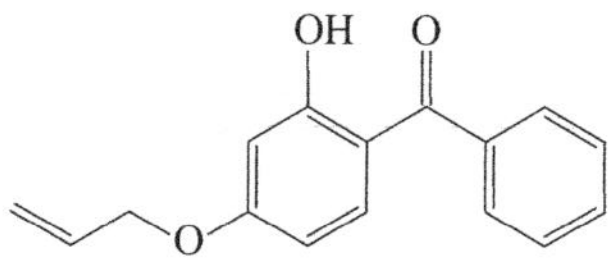 (AOHBP)	/	/	AC breakdown strength is 9.7% higher than XLPE	Improves the starting voltage of the electrical tree and inhibits its growth	0.8 phr
Functionalized grafted molecule—antioxidant	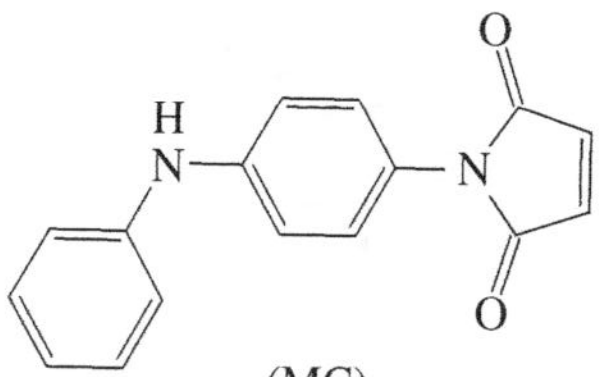 (MC)	/	Inhibits space charge accumulation and the electric field distortion rate is 10.2% of XLPE	DC breakdown strength is 11.8% higher than XLPE	/	/
	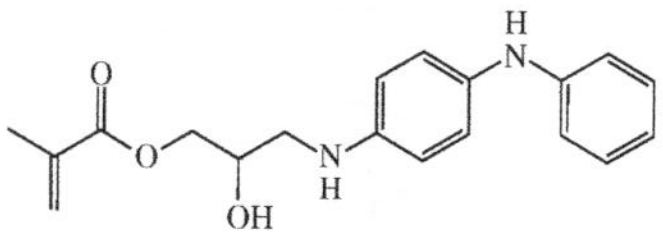 (GA)	One order of magnitude lower than XLPE	/	DC breakdown strength is 15.3% higher than XLPE	/	/
	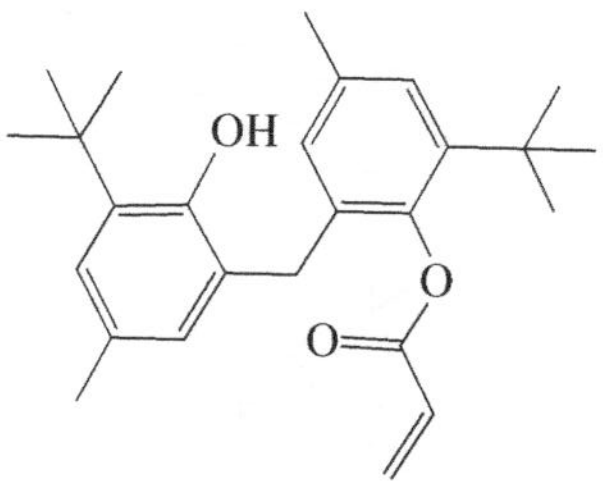 (AO3052)	One order of magnitude lower than PP	Inhibits space charge accumulation and the electric field distortion rate is 44.2% of PP	DC breakdown strength is 32.2% higher than PP	/	/

field, prevent the energy from these electrons being transferred to the polyethylene chains and leave the polymer matrix unharmed [73].

The related researches on voltage stabilizers have lasted for a long time, and it is a few decades earlier than the research on nanocomposites. While due to its poor compatibility with polymeric matrix, it did not cause much attention and application until the voltage stabilizer with higher solubility was obtained by attaching alkyl side chains to an aromatic core recently [74]. Several new types of voltage stabilizers with excellent compatibility have been proposed and its stabilizing effect on inhibiting the growth of electrical tree and improving the breakdown strength were reported, which may provide an alternative solution for high-performance HVDC cables [75].

In this subsection, the ultralow density polyethylene (ULDPE) was blended with PP as an effective impact modifier and three different voltage stabilizers were introduced for insulating property improvement. The breakdown strength and space charge behavior of PP composites were researched under 30, 60, and 90 °C. The three different voltage stabilizers were purchased from BASF Group, called as Tinuvin 622 (T622 for short), Tinuvin 770 (T770), and Chimassorb 944 (C944). The chemical structural formulas of the voltage stabilizers are shown in Figure 2.22. A certain proportion of PP, ULDPE, and voltage stabilizer were mixed by melt blending at 190 °C. Details are listed in Table 2.5.

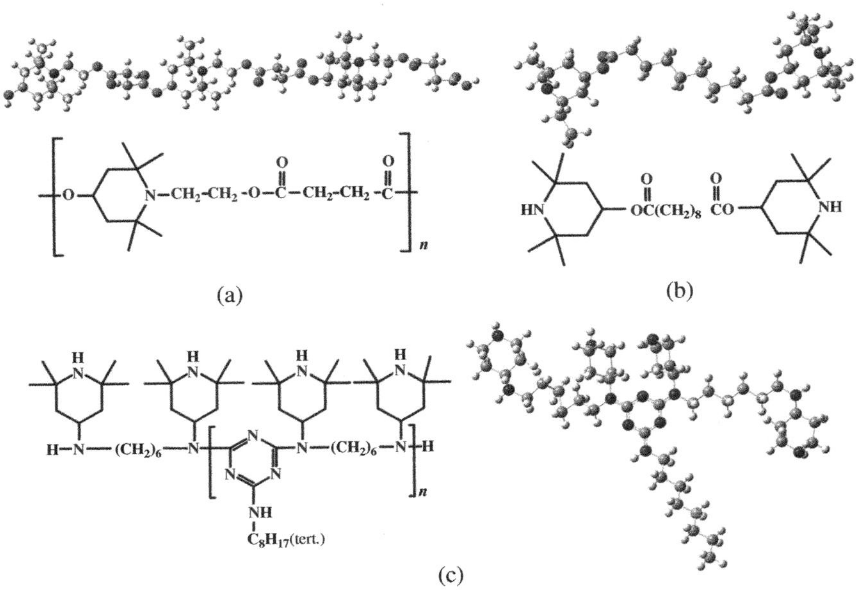

Figure 2.22 Chemical formula of the voltage stabilizers. (a) Tinuvin 622. (b) Tinuvin 770. (c) Chimassorb 944. *Source:* Hou et al. [76]/with permission from IEEE.

Table 2.5 Index for polymer samples investigated in this subsection. *Source:* Hou et al. [76]/with permission from IEEE.

Index	PP content (wt%)	ULDPE (wt%)	T622 (wt%)	T770 (wt%)	C944 (wt%)
PU	85	0	0	0	0
PU622	85	40	0.3	0	0
PU770	85	40	0	0.3	0
PU944	85	40	0	0	0.3

The chemical structure of PU blends, PU622, PU770, and PU944 were studied by FT-IR. As shown in Figure 2.23, PU presents typical characteristic peaks of PP and ULDPE mixtures. The wide absorption bands near 2962 and 2872 cm^{-1} are related to the stretching vibration of methyl ($-CH_3$). The peaks near 2926 and 2853 cm^{-1} were assigned to methylene ($-CH_2-$) stretching. The band with a center of 1242 cm^{-1} is caused by C–C stretching vibration. Because of the low content of voltage stabilizer additives, the characteristic peaks of infrared spectra are not obvious. Compared with the spectra of the original PU blends, a new ester group

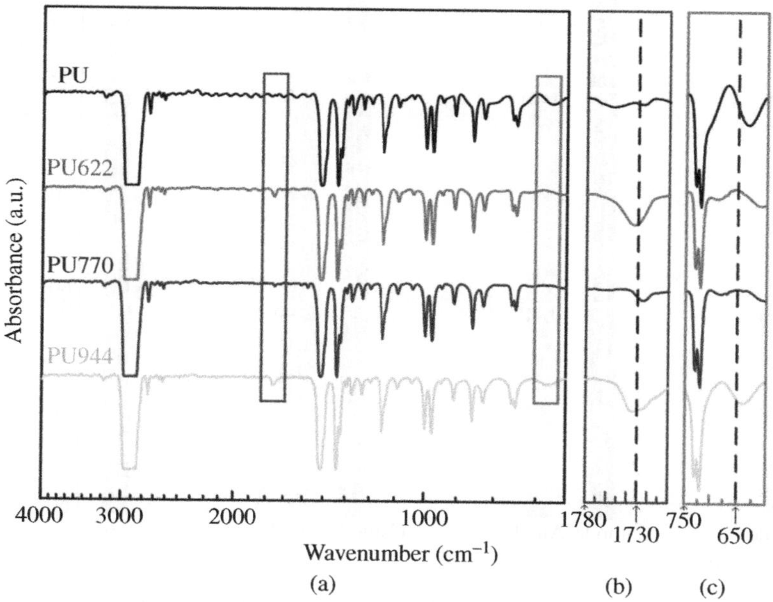

Figure 2.23 FT-IR spectra: (a) PU, PU622, PU770, and PU944 composites; (b) and (c) partial enlargement of the FT-IR spectra. *Source:* Hou et al. [76]/with permission from IEEE.

(R–COO–R_1) stretching band peak appeared at 1730 cm^{-1} in PU622 and PU770. The characteristic peak at about 1530 cm^{-1} is caused by striazine derivatives of PU944 [77].

As mentioned above, space charge performance is an important aspect of HVDC-extruded cable insulation, which needs to be taken seriously. Figure 2.24 shows the space charge polarization distribution of PP/ULDPE blends

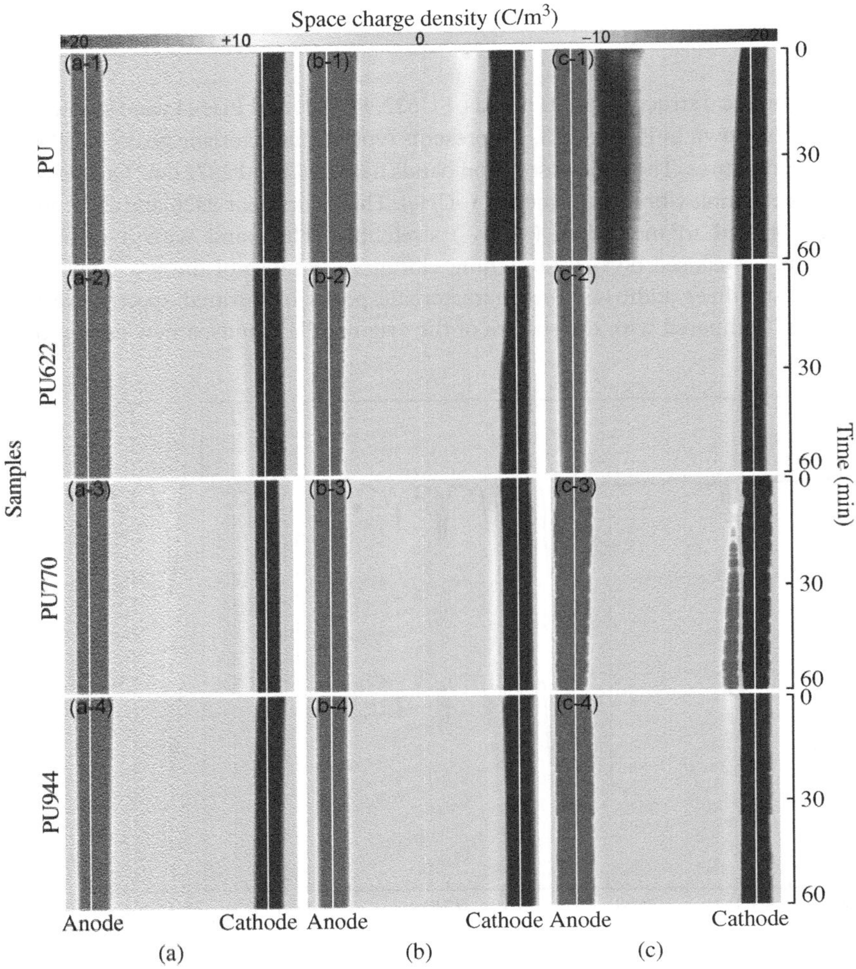

Figure 2.24 Space charge distribution in PU and its voltage stabilizer composites under 60 kV/mm for 60 min at different temperatures: (a) 30 °C, (b) 60 °C, and (c) 90 °C. *Source:* Hou et al. [76]/with permission from IEEE.

and the voltage stabilizer-modified composites (i.e. PU, PU622, PU770, and PU944) during the polarization time for 60 minutes and 60 kV/mm. Figure 2.25 shows the charge in depolarization distribution of different samples after 60 kV/mm polarization for 60 minutes. Both the voltage-on and short-circuit tests were tested at 30, 60, and 90 °C.

For the space charge behavior of PU during polarization at 30 °C, the homo-charges injected from the cathode increase obviously as shown in Figure 2.24a-1. During the polarization time, the injected charges move to the anode gently, leading to an obvious accumulation of homo-charges in the bulk of PU sample, especially near the cathode. For PU at 60 °C, the heterogeneous charge appears near the two electrodes. Figure 2.24b-1 shows that the negative charge accumulated near the anode during the applied voltage is larger. The negative charge dominates near the anode, and only a small positive charge accumulates near the cathode at 90 °C, as shown in Figure 2.24c-1.

For PU622 at 30 °C, with the increase of polarization time, a small amount of hetero-charge appears near the cathode of PU622 as shown in Figure 2.24a-2. When the test temperature is 60 °C in Figure 2.24b-2, the homo-charge injected from the two electrodes increases with time. When the temperature reaches 90°, a certain number of positive heteropolar charges appear near the cathode of the sample. When the temperature reaches 90°, a certain number of positive heteropolar charges appear near the cathode of the sample. Moreover, the accumulated positive polar charges accumulated rapidly in the initial few minutes, and then gradually decreased with time.

When it comes to PU770, a very small positive charge was injected into the PU770 and accumulated near the cathode at 30 °C, as shown in Figure 2.24a-3. For PU770 at 60 °C in Figure 2.24b-3, with the passage of time, the accumulation of hetero-charges near the cathode in the sample increases, and more hetero-charges occur at the same time than at 30 °C. The accumulation of hetero-charges in the sample near the cathode is further intensified at 90 °C as depicted in Figure 2.24c-3.

Concerning PU944, a little negative charge accumulates near the cathode inside the specimen PU944 at 30 °C and a positive charge is injected into the position of the specimen near the anode as shown in Figure 2.24a-4. For the situation of 60 °C in Figure 2.24b-4, a small amount of hetero charge accumulated near the cathode in the sample increases slightly with time. In terms of PU944 at 90 °C, the homo-charge injection is enhanced and only a very small amount of negative space charge is accumulated in the sample. The hetero-charge accumulated near the cathode increases rapidly at first and then decreases lightly with time in PU944.

The space charge behavior of each sample during de-polarization at different temperatures can be seen in Figure 2.25. As shown in Figure 2.25a-1, the space charge dissipates quickly at the very beginning of the 30 seconds in PU near both

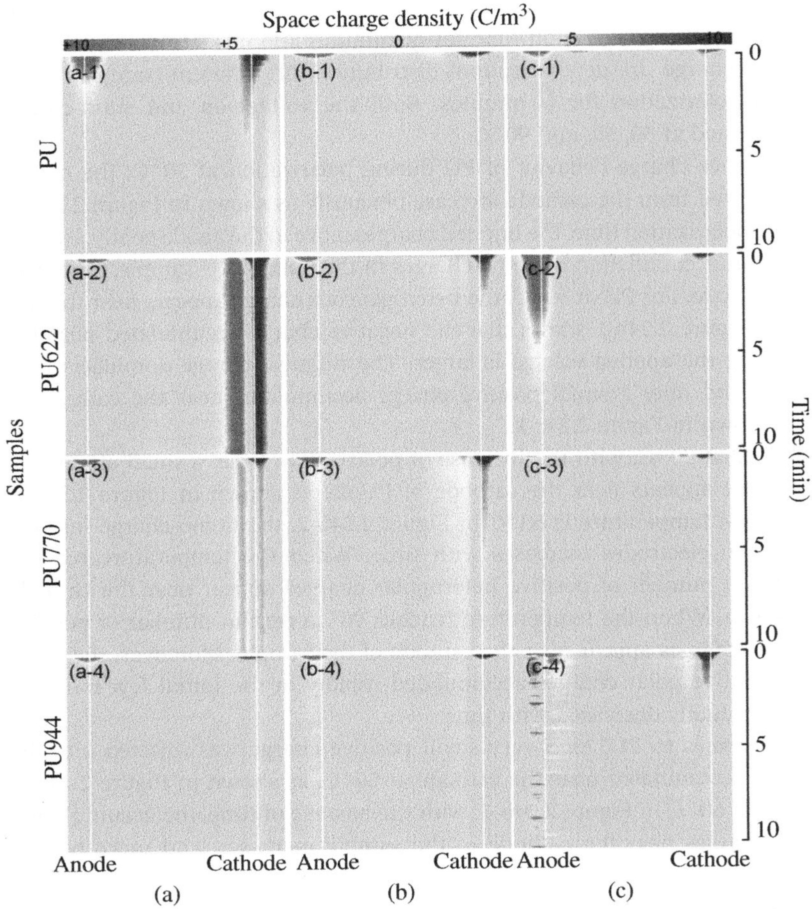

Figure 2.25 Space charge depolarization in PU and PP/ULDPE/voltage stabilizer composites at different temperatures: (a) 30 °C, (b) 60 °C, and (c) 90 °C. *Source:* Hou et al. [76]/with permission from IEEE.

the anode and the cathode and then dissipates slowly for the restriction of deep traps at 30 °C. A little hetero-charge can be found near both electrodes and almost disappears after one minute at 60 °C as depicted in Figure 2.25b-1. In Figure 2.25c-1, the negative charge appears near both the anode and cathode electrodes and dissipates slowly with the voltage-off time at 90 °C.

As depicted in Figure 2.25a-2, the space charge in PU622 at 30 °C dissipates very slowly and there is still a certain amount of positive charge accumulated near the cathode in the sample, after 10 minutes of depolarization. For the 60 °C, a small amount of negative charge accumulated near the positive electrode, while a little positive charge accumulated inside the sample, especially near the cathode as shown in Figure 2.25b-2. With the short-circuit time, the positive charges in the sample dissipate gradually. When the experimental temperature is 90 °C in Figure 2.25c-2, both the deeper homo-charge injected from the positive electrode during polarization and the hole traps introduced by T622 into the sample have made a lot of positive charges accumulated near the positive electrode and a very slow dissipation rate.

For PU770 at 30 °C, the space charge dissipation speed is also slow, but the charge accumulates little inside the specimen and near the cathode in Figure 2.25a-3. It can be seen from Figure 2.25b-3 that the homo-charge injected from the two electrodes dissipates rapidly within the first minute of depolarization at 60 °C. A little positive charge accumulates in the specimen near the cathode and dissipates very slowly. The depolarization space charge behavior of the sample at 90 °C is shown in Figure 2.25c-3. In the sample, only a little charge accumulated near the electrodes and dissipated very rapidly, almost disappearing at 10 minutes.

The space charge of PU944 dissipates rapidly in the initial 15 seconds, leaving only a small amount of homo-charges on both sides of the sample in Figure 2.25a-4 at 30 °C. As depicted in Figure 2.25b-4, the homo-charge injected from the cathode dissipates rapidly in the first few seconds of the short circuit at 60 °C. A small amount of positive charge accumulated near the cathode, but almost no charge accumulated in the sample. As the short circuit progresses, the positive charges near the cathode and the anode in the sample dissipate gradually. As shown in Figure 2.25c-4, a certain amount of positive charge accumulates in the sample at 90 °C. The homo-charge injected from the two electrodes dissipates slowly with the short circuit time. After 10 minutes of short circuit, there is still a little positive charge in the sample, especially near the cathode.

The absolute values of charge density after 60-minute polarization and 10-minute depolarization are taken as the reference for quantitative analysis, respectively, as shown in Figure 2.26. As depicted in Figure 2.26a, the mean volume charge density of all specimens with voltage stabilizers is smaller than that of pure PU specimens after 60 minutes of polarization at each test temperature. The average charge density of PU decreases slightly with the increase of temperature and then becomes larger. The mean charge density of PU622 increases first and then decreases. For PU770, the average charge density of decreases with temperature, while that of PU944 increases slightly. In addition, the average charge density of PU770 is slightly higher than that of PU622 and PU944 at each test temperature, and the difference between PU770 and PU622 or PU944 decreases with temperature.

As shown in Figure 2.26b, after 10 minutes of depolarization, the average charge densities of PU622 and PU770 are smaller than those of PU at 90 °C, and the

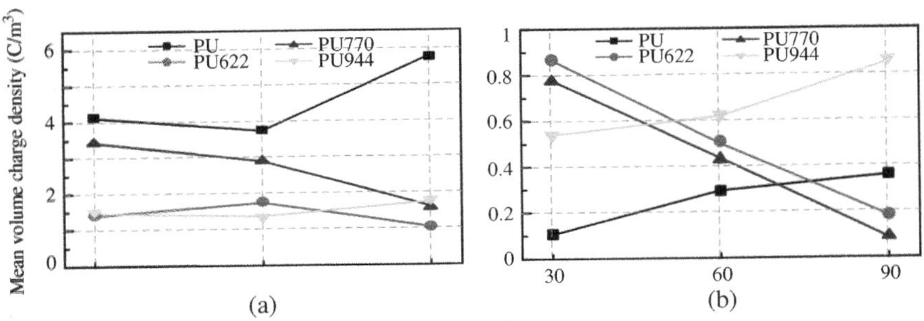

Figure 2.26 Relationship between mean volume charge density of different samples under 30, 60, and 90 °C: (a) polarization for 60 minutes and (b) de-polarization for 10 minutes. *Source:* Hou et al. [76]/with permission from IEEE.

average charge densities of other samples or temperatures are higher than those of PU. With the increase of temperature, the average charge densities of PU and PU944 increase gradually, while the average charge densities of PU622 and PU770 decrease gradually.

Figure 2.27 shows the DC breakdown strength of PP/ULDPE blends and their voltage stabilizer-modified composites in Weibull distribution. Different specimen show different distributions of breakdown strength at different temperatures. With the increase of temperature, the breakdown strength of all samples generally decreases. However, the breakdown strength of the sample with voltage stabilizers is always better than that of the pure PU blends, at different test temperatures. The breakdown strength of PU622 performs well at the three test temperatures. PU770 has the highest breakdown strength at 30 °C than other samples, but it is slightly lower than PU622 and PU944 at 60 and 90 °C. Compared with others, PU944 has the highest breakdown strength at 60 and 90 °C.

To further study the effect of voltage stabilizer on space charge behaviors and breakdown characteristics, it is necessary to find the trap position in the chemical structure from the calculated band. The electron iso-surface and the energy band are then calculated using DFT with the B3LYP hybrid function and the 6-31G basis function in Gaussian 09 in this chapter [38]. As shown in Figure 2.28, the 3D electrostatic potential distribution of the three different voltage stabilizers is depicted based on the quantum chemistry calculation [78]. The right end of the colorbar represents negative electrostatic potential, which is attractive to holes and transform to hole-type traps. The left side shows positive electrostatic potential, which can capture electrons and turn them into electron-type traps. In Figures 2.28a and 2.26b, for the different electronegativity of the nitrogen-hydrogen bond, the single bond and the double bond of carbon to oxygen in T622 and T770, the deviated orbital electron in molecular chains create a trap site by inherent dipole moments,

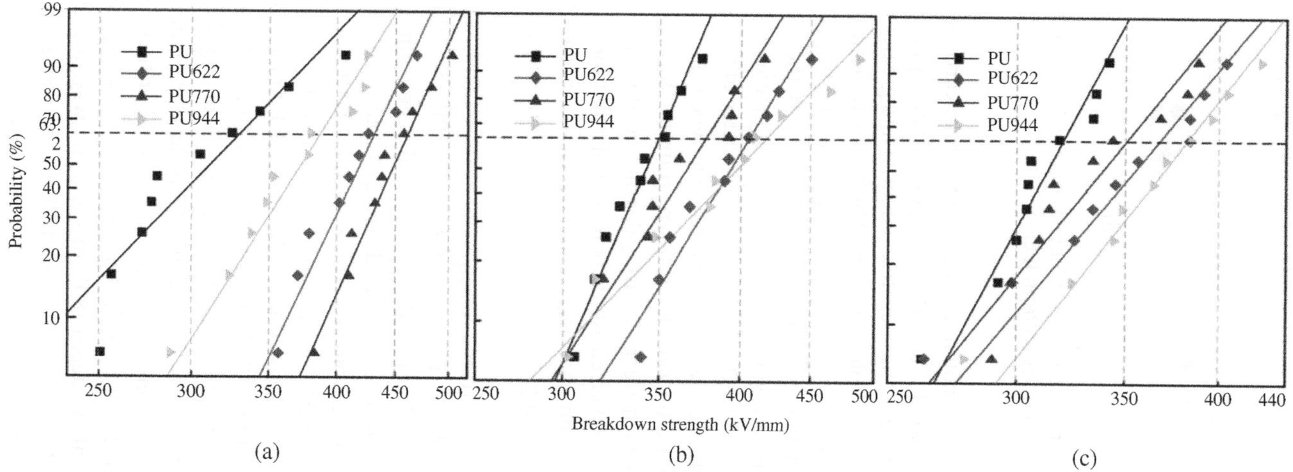

Figure 2.27 Weibull distribution of DC breakdown strength at different temperatures: (a) 30 °C, (b) 60 °C, and (c) 90 °C. *Source:* Hou et al. [76]/ with permission from IEEE.

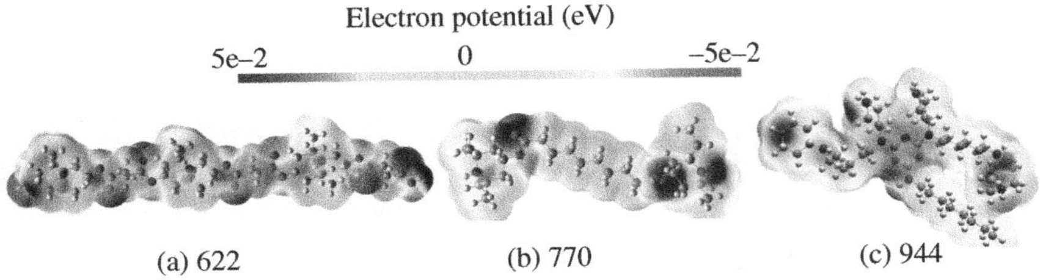

Figure 2.28 3D electrostatic potential distribution in (a) T622, (b) T770, and (c) C944. *Source:* Hou et al. [76]/with permission from IEEE.

which generated an electric potential to capture charge carriers. Compared with the nitrogen–hydrogen bond, the carbon–oxygen bonds, especially those in ester groups (R–COO–R1), have higher negative potential and greater attraction to positive charges, which become hole-type traps. Unlike the trap sites in the molecular chain of T622, which alternate along the main chain, the trap sites in each T770 molecular are almost symmetrically distributed. In T770, the electrostatic potential of methylene chain between two ester groups is higher and appears as an electron trap. As shown in Figure 2.28c, the trap sites located at the triazine groups in C944 are negatively charged, which generates hole-type traps. In addition to the triazine group, a large number of electron-type traps are formed along the nitrogen–hydrogen bonds which are periodically distributed along the molecular chain. It can be seen from the 3D electrostatic potential distribution in Figure 2.28, the trap sites formed by carbon–hydrogen bond and carbon–carbon bond are almost continuous and overlapping, which generates electron-type traps to capture electrons.

Figure 2.29 shows the electronic energy band in PP, ULDPE, and three different voltage stabilizers (i.e. T622, T770, and C944). For PP and ULDPE, the lowest unoccupied molecular orbital (LUMO) is 2.17 and 2.1 eV, the highest occupied molecular orbital (HOMO) is −7.54 and −5.82 eV, and the energy gap is 9.71 and 7.92 eV. For the three voltage stabilizers, the LUMO is −0.51, −0.1, and 0.59 eV; the HOMO is −5.51, −5.51, and −5.1 eV; and the energy gap is 5, 5.41, and 5.69 eV, respectively. Below the LUMO level of PP or ULDPE, there exists the energy gap above the LUMO level of T622, T770, and C944 as shown in Figure 2.29a–c, which usually introduce a localized energy state for electrons [38, 79]. Taking the gap of 1.49 eV above the LUMO level in Figure 2.29a, for example, electrons need more energy from the electric field or thermal source to hop over the energy level gap for conduction, which can be regarded as electron-trapping sites with shallow depth [80]. In the same way, the energy difference gap below the HOMO levels of PP or ULDPE in the voltage stabilizers, as the gap of 1.95 eV shown in Figure 2.29a, can introduce a trap site with deep depth

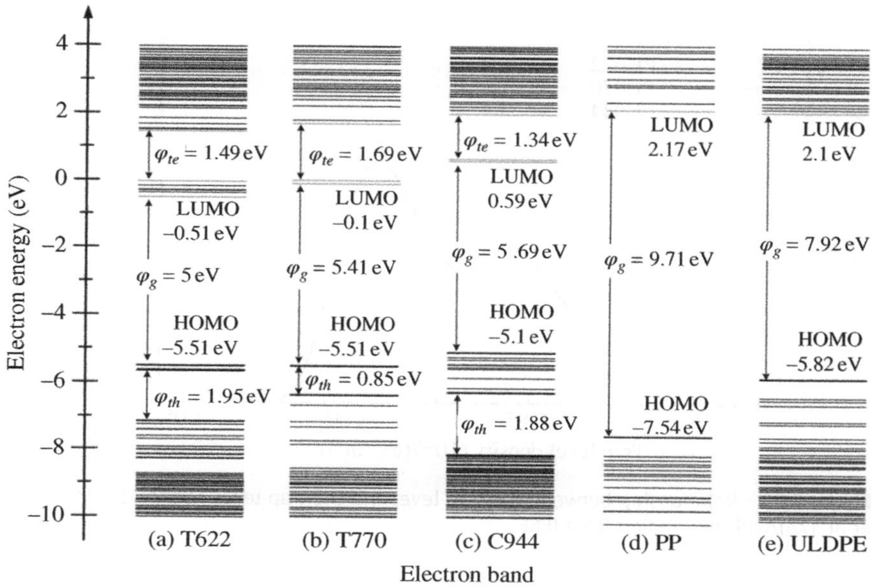

Figure 2.29 Electronic energy band in PP, ULDPE, and three different voltage stabilizers. (a) T622. (b) T770. (c) C944. (d) PP. (e) ULDPE. *Source:* Hou et al. [76]/with permission from IEEE.

for holes, by limiting its transport of hopping or tunneling process. Depending on the molecular structure, the electron or hole trapping sites can be ascribed to the carbon–oxygen, nitrogen–hydrogen, or triazine groups, as discussed based on its 3D electrostatic potential distribution. As depicted in Figure 2.29, different voltage stabilizers introduced the hole-type or electron-type traps with different energy levels and difference.

In order to further study the effect of different voltage stabilizers on the trap level distribution of PP/ULDPE blends, Figure 2.30 depicts the relationship between the trap level and the trap level density based on the results of IDC test [81]. As shown in the figure, the trap levels of the samples are all in the range of 0.7–1.05 eV, with two trap level density peaks. The first trap level density peak is located at 0.86 eV and the second density peak is about 0.95 eV. It can be seen from the results that compared with pure PU, all the samples with free radical traps introduced deep traps to varying degrees. Besides PU770, the shallow trap depth of the other two samples is also greater than that of pure PU. According to the results of quantum chemistry calculation, especially the difference of electron bands between PP, ULDPE, and three voltage stabilizers in Figure 2.29, the trap level depth of the three voltage stabilizers introduced is different according to their molecular structures.

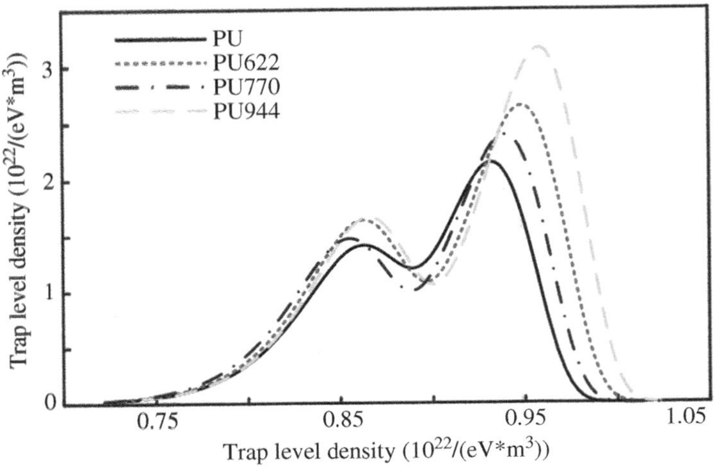

Figure 2.30 Relationship between the trap level and the trap level density. *Source:* Hou et al. [76]/with permission from IEEE.

The trap distribution in the composite and the additives' chemical structure characteristics can greatly influence the behavior of space charge and breakdown strength of polymer insulation. Traps can form localized states that have strong attraction for charges and then restrict its movement or capture charges in polymer during charge injection, migration, and extraction. Both the terminals, additives and interfaces, between crystalline and amorphous polymers can introduce traps. In the case of organic additives, such as voltage stabilizers, their effects on space charge and breakdown properties relate not only to the density and depth of the traps that their own structures can introduce but also to their chemical activity.

From the dynamic change of space charge during the polarization and depolarization process in Figures 2.24 and 2.25, the accumulation of hetero charges near the electrodes in all specimen increases with the temperature. Generally, there are two sources of hetero-charge under the polarization voltage [30, 82]. One is the dissociation of impurities in the sample, which usually occurs near the electrode at the very beginning of polarization measurement. Another reason is that, the injected holes or electrons accumulate near the opposite electrode due to the limited extraction. Generally, it takes some time for the charge injected from the electrode to reach the opposite electrode through the whole sample. It can be seen from Figure 2.24 that the hetero-charge accumulated near the electrode belongs to the latter situation. Compared with pure PU, the hetero-charge accumulation of the samples with voltage stabilizer was greatly improved. This relates to the traps introduced by the molecular structure of additives. From the experimental results, it can be seen that T622, T770, and T944 all introduce shallow traps and deep traps at the same time, but the ability of each voltage stabilizer to trap electrons and holes is different. In addition, as the test temperature increases, the charge injection, migration, and extraction all intensify. However, it should be noted that at different temperatures, the distribution of molecular chains and the local motion of branched chains of organic additives would change, which is different from inorganic additives.

Moreover, the experimental results show that the overall breakdown strength of the samples decreases with the rise of temperature. However, at each test temperature, the strength of the specimens with voltage stabilizer is all higher than that of pure PU specimens. There are two reasons for the increase of breakdown strength [30, 82]. On the one hand, when the charges injected or ionized under the applied electric field get in or escape from the traps introduced by additives, their migration behavior is limited. This weakens the energy carried by the charges, thereby weakening their ability to destroy molecular chains. On the other hand, the short-term high-intensity electric field, Joule heat, and mechanical stress can lead to the breaking of molecular chains and the generation of free radicals. All three voltage stabilizers can capture free radicals and inhibit the destruction of molecular chains [83], thus improving the breakdown strength. In addition, with the increase of temperature, the sample with the best breakdown strength changes. The highest 63.2% breakdown strength is PU770, PU944 (slightly higher than PU622), and PU944 at 30, 60, and 90 °C, respectively. This difference is related to the molecular structure of additives.

From the chemical structure characteristics of the additive itself, T770 is a single molecule additive, which has the fast migration and fast reaction time. Therefore, T770 has high breakdown strength at low temperature. However, T770 is also easy to migrate or even precipitate at high temperature, which is also the reason why the breakdown strength of PU770 decreases at high temperature. In addition, under the high temperature and electric field, T770 migrates to the surface near the cathode of the sample as a polarized charge and is not easy to extract from the interface, which can explain the rapid disappearance of the hetero-charge at the very beginning of the depolarization, which accumulated near the cathode of the sample PU770.

Although both T622 and C944 are polymer additives, T622 with short-branched chains at lower temperatures gives it better mobility and faster response in a local range, while C944 with long-branched chains has limited local mobility. However, at high temperatures, the space of long-branched chain of C944 becomes larger, and the space of chemical reaction becomes larger. Unlike the nearly linear distribution of T622, the connection between C944 and polymer is more similar to surface contact. This makes it possible to introduce traps with wide distribution and to react rapidly when polymer molecular chains are destroyed under high-temperature and high electric field. This is consistent with the results of slower charge dissipation in depolarization process and higher breakdown strength of PU944. Therefore, the change of temperature will affect not only the charge behavior but also the performance of the voltage stabilizer itself.

2.4.3 Antioxidant Additives

The free radicals (i.e. atoms, molecules, or ions with irrelevant valence electrons) produced by ionization, polarization, and degradation of polymer insulating materials under high-temperature and high electric field can play an important role in the formation of space charges [84]. These pose a potential threat to the aging of insulating materials and the safety of their dielectric properties [44, 85] In order to suppress the ageing process of cable insulation, antioxidants are usually used as the additives [86, 87]. In the past few decades, various types of antioxidants have been developed, in which the hindered phenolic antioxidants have become the main antioxidants commonly used in polyolefin materials, and its anti-oxidation mechanism and advantages have been fully studied [88]. Researchers found that nanosilica-immobilized antioxidant can significantly increase the oxidation induction time (OIT) of PP [89]. In addition to increasing the service life of PP, whether the antioxidants can affect the electrical properties of PP insulation is rarely studied by researchers. Therefore, in this subsection, two isomeric hindered phenolic antioxidants are selected to explore the effects of antioxidants on the electrical properties of PP insulation. However, the migration and precipitation of small organic molecules in the blend can change the microstructure of the sample and reduce the antioxidant capacity of the material [90]. Grafting modification techniques can graft monomers with target functional groups onto polymer molecular chains to effectively solve the problem of small molecule migration [5]. Studies have shown that grafting hindered amine antioxidant MC onto polyethylene (PE) can effectively improve trap density, space charge suppression, and DC breakdown strength [7]. In addition, grafting maleic anhydride (MAH) and styrene (St) onto PP can efficiently suppress space charge injection, enhance breakdown strength, and homogenize the electric field distribution [91, 92]. Meanwhile, grafting long molecular chains can also inhibit the slip of molecular chains, thus enhancing the mechanical properties of PP [93]. Furthermore, a study has investigated the effect of 4-allyloxy-2-hydroxybenzophenone (AHB) grafting rate on PP and concluded that a grafting rate of 0.73% resulted in the best improvement of electrical properties [94]. However, most small molecules used in grafting lack antioxidant function, which cannot limit the aging deterioration of polymer insulation caused by oxidation reactions and has limited improvement in electrical properties [95]. Therefore, further investigation in grafting hindered phenolic antioxidants onto PP is of great research value to solve physical blending disadvantages, inhibit the oxidation process, and improve the electrical performance of insulation materials. This subsection reports the effect of the grafting with hindered phenolic antioxidant (AO3052) on the electrical properties of PP, particularly DC and DC-prestressed breakdown strength.

In this subsection, the reactive hindered phenolic antioxidant used is acrylic acid 2-*tert*-butyl-6-(3-*tert*-butyl-2-hydroxy-5-methylbenzyl)-4-methylphenyl ester (AO3052) and the initiator is dicumyl peroxide (DCP). The melt-free radical grafting procedure is shown in Figure 2.31. Grafted samples (PP-g-AO) and blended samples (PP/AO) with a content of 0.5 wt% are prepared, taking the PP as a comparison. The samples are degassed in a vacuum drying oven for 24 hours to remove the residual air and volatile components. Further, to remove the unreacted antioxidants, a small amount of the PP-g-AO is put into xylene to be heated, dissolved, and refluxed, until it is completely dissolved. Next, it is poured into unheated acetone solution, left to precipitate, and filtered. At last, the grafted samples are put into the vacuum oven again and the sample preparation is completed.

The PP insulation structures have been characterized by Fourier transform infrared (FT-IR) spectra instrument, as shown in Figure 2.32a. The C=O stretching vibrational absorption peak is contained in AO3052 at around 1725 cm^{-1}. As can be seen, AO3052 showed an absorption peak at around 1725 cm^{-1}, demonstrating the successful grafting of AO3052 onto PP.

The oxidative induction time (OIT) is the time at which the catalyze oxidation reaction of a material begins under high temperature and oxygen condition,

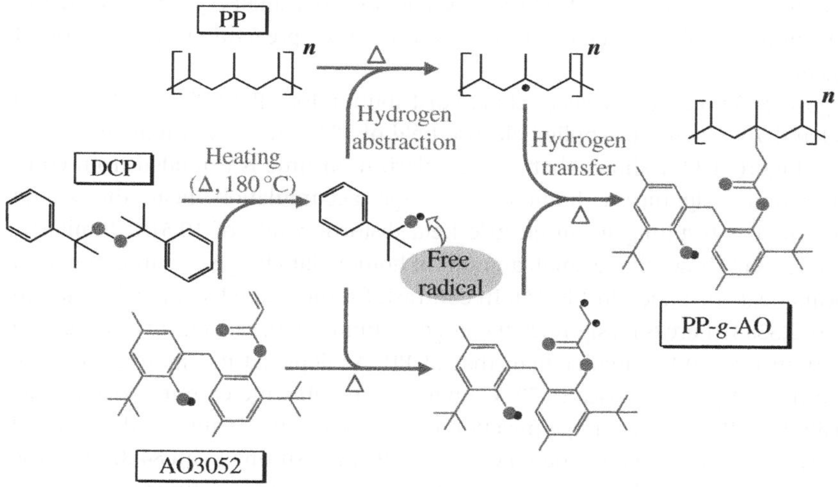

Figure 2.31 Scheme of melt-free radical grafting procedure. *Source:* [96]/Boxue Du et al./with permission from John Wiley & Sons.

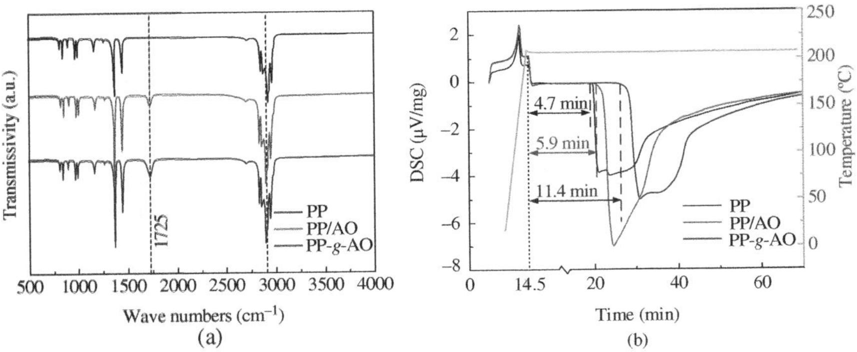

Figure 2.32 (a) FT-IR spectra of the samples. (b) OIT of the samples by DSC. *Source:* [96]/with permission from John Wiley & Sons.

characterizing the thermal stability of the material. Figure 2.32b shows the OIT for the samples based on the differential scanning calorimetry (DSC) method at 210 °C. The OIT of PP is the lowest among the three samples. However, after adding antioxidants, the OIT of PP/AO has been improved to 5.9 minutes, and the OIT of PP-g-AO has been further improved to 11.4 minutes. Thermal stability experiments further confirm that the antioxidants enhance the antioxidant performance of PP, with the grafted samples demonstrating superior effectiveness compared to the blended samples. It is very probable that the spatial blocking effect of grafted side chains, which indicates that the growth of spherical crystals will stop when they encounter grafted side chains, could form a small crystalline structure with complex molecular chain structures in PP, thus preventing the diffusion of oxygen.

Figure 2.33 presents the space charge distribution for PP, PP/AO, and PP-g-AO at 30 and 70 °C with an applied electric field of 40 kV/mm. At a temperature of 30 °C, Figure 2.33a1 shows that as the polarization time is extended from 60 to 1800 seconds, significant homopolar charge accumulation near the anode occurs in PP, resulting in an electric field distortion rate of 10.51%. Similarly, Figure 2.33a2 shows that a small amount of homopolar charge accumulation also appears near the anode in PP/AO. In contrast, for the grafted sample depicted in Figure 2.33a3, there is no significant charge accumulation, causing an electric field distortion rate 59.1% smaller than that of PP. As depicted in Figure 2.33b1–b3, when the temperature rises to 70 °C, there is a notable increase in space charge density for PP, and the charge injection depth is close to 150 μm. In the blended sample, the charge injection depth exceeds 100 μm, resulting in an 84.8% increase in the distortion rate. While in contrast, the grafted sample continues to show no significant space charge accumulation.

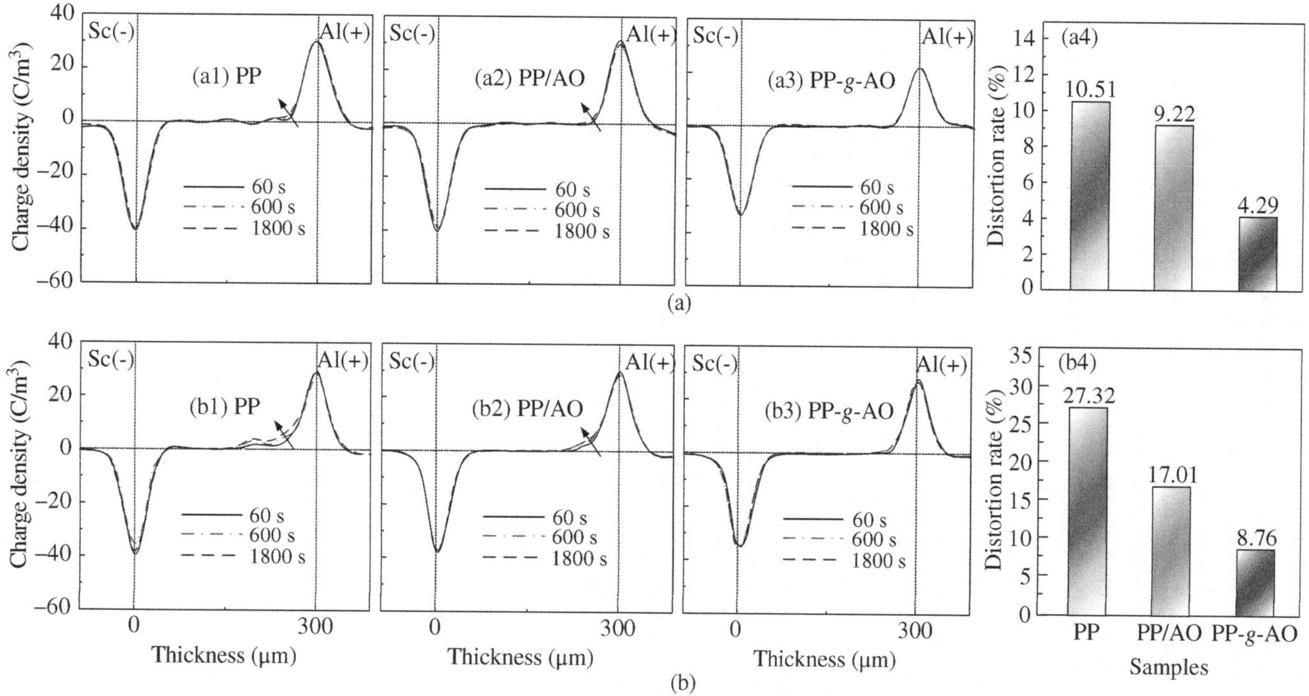

Figure 2.33 Space charge distributions and electric field distortion rate of the samples. (a) 30 °C. (b) 70 °C. *Source:* [96]/Boxue Du et al./ with permission from John Wiley & Sons.

It can be concluded that the injection of space charges leads to electric field distortion within the sample, and these two factors exhibit a positive correlation. The increase in temperature can exacerbate the accumulation of space charges and injection depth. Compared to PP and PP/AO, the grafted sample can effectively suppress the injection and accumulation of space charges in the PP insulation layer under both room-temperature and high-temperature conditions, consequently reducing the extent of electric field distortion.

Figure 2.34 depicts the DC breakdown strengths of PP, PP/AO, and PP-g-AO at 30, 50, 70, and 90 °C, respectively. The characteristic DC breakdown strength of PP, PP/AO, and PP-g-AO at 30 °C exhibits relatively close values of 380.7, 383.5, and 391.1 kV/mm, respectively. However, all three samples show a decreasing trend with increasing temperature. At 90 °C, the breakdown strength of PP decreases to 230.1 kV/mm, PP/AO decreases to 260.6 kV/mm, and PP-g-AO decreases to 302.1 kV/mm. In addition, the order of DC breakdown strength of the three samples at different temperatures is PP < PP/AO < PP-g-AO. The difference in the breakdown strengths among the samples gradually increases as the temperature increases. This indicates that the antioxidant-grafted sample has a more pronounced effect on the breakdown strength of PP, and this enhancement effect increases with the temperature. In particular, the antioxidant-grafted sample has a greater effect on improving the breakdown properties of PP. It is superior to the antioxidant blended sample of the same mass fraction.

To assess the effect of voltage polarity reversal, apply a prestressed electric field of −40 kV/mm to PP, PP/AO, and PP-g-AO for 60 minutes at 30, 50, 70, and 90 °C.

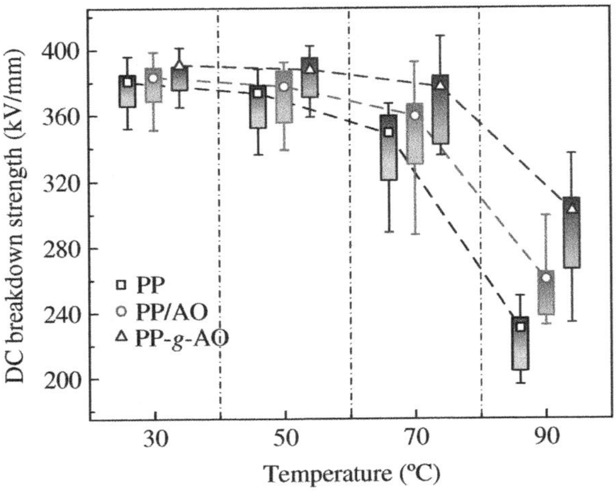

Figure 2.34 Effect of antioxidant grafting on DC breakdown strength. *Source:* [96]/Boxue Du et al./with permission from John Wiley & Sons.

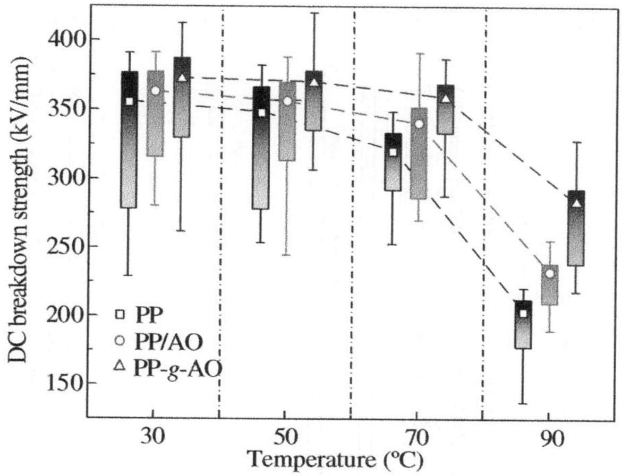

Figure 2.35 Relation between DC-prestressed breakdown strength and temperature with the different samples. *Source:* [96]/Boxue Du et al./with permission from John Wiley & Sons.

Then, the DC breakdown strength of the three samples at 40 kV/mm is measured, as shown in Figure 2.35. The breakdown strength of all three samples is significantly lower than that of the non-prestressed samples, indicating that polarity reversal has a detrimental effect on the DC breakdown characteristic of PP. The breakdown strength of PP, PP/AO, and PP-*g*-AO is 356, 363.8, and 372.9 kV/mm, respectively, at 30 °C. However, as the temperature increases to 90 °C, the breakdown strength of the samples decreases to 203.4, 232.6, and 284.1 kV/mm, respectively, representing a decrease of 42.9, 36.1, and 23.8% compared to the measurements at 30 °C. It indicates that the breakdown property of the PP-*g*-AO appears to be less affected by the test temperature.

To investigate the effect of hetero-polarity prestressed time on breakdown strength, the DC breakdown strength of the three samples is measured at prestressed times of 0, 10, 30, and 60 minutes at 30, 50, 70, and 90 °C, as shown in Figure 2.36. To analyze the effect of polarity reversal conditions better, the parity reversal breakdown coefficient k is defined in this chapter, as shown in Eq. (2.7),

$$k = \frac{E_{\text{prestressed-}t}}{E_0} \tag{2.7}$$

where $E_{\text{prestressed-}t}$ is the breakdown strength measured after hetero-polarity DC-prestressed, t is the time of prestressed, and E_0 is the breakdown strength without prestressed procedure. The coefficient k for each set of samples is calculated and shown in Figure 2.36.

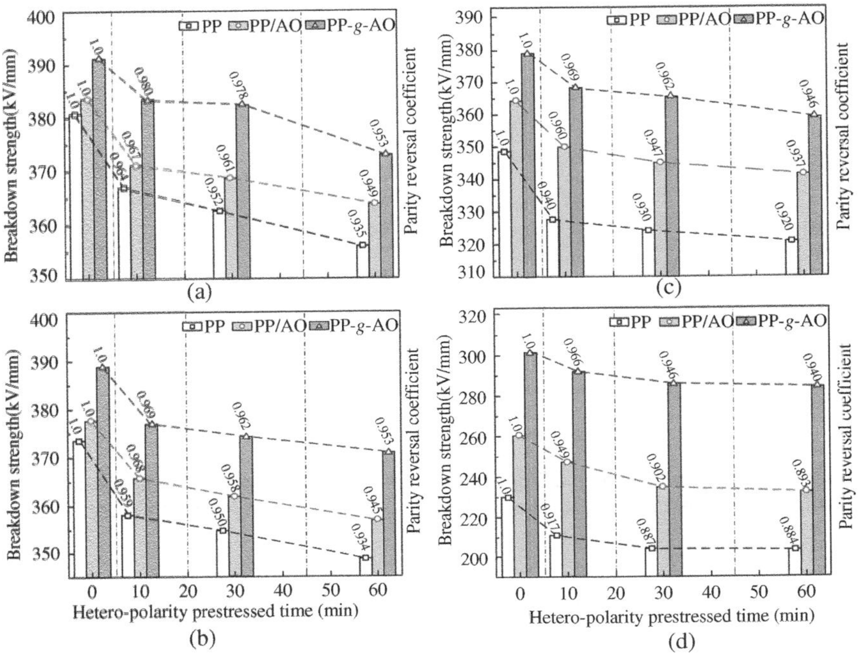

Figure 2.36 DC-prestressed breakdown strength after the hetero-polarity prestressed times at (a) 30 °C, (b) 50 °C, (c) 70 °C, and (d) 90 °C. *Source:* [96]/Boxue Du et al./with permission from John Wiley & Sons.

The polarity reversal breakdown strength of PP, PP/AO, and PP-g-AO at different temperatures shows a decreasing trend as the prestressed time increases. Notably, the fastest rate of decrease is observed within the first 10 minutes of prestressed time. The relationship between the breakdown strength of the three samples after the same prestressed time follows the order of PP < PP/AO < PP-g-AO. For instance, Figure 2.36a shows that compared to the non-prestressed sample at 30 °C, the breakdown strength of PP after 60 minutes DC-prestressed decreases from 380.7 to 356 kV/mm, and the coefficient k decreases from 1.0 to 0.935. Similarly, the k of PP/AO and PP-g-AO samples decreases to 0.949 and 0.953, respectively. In addition, the k of PP-g-AO is always higher than that of PP and PP/AO under the same experimental conditions. It can be inferred that PP-g-AO has a higher DC and DC-prestressed breakdown strength.

As shown in Figure 2.36a–d, the degree of reduction in breakdown strength due to prestressed increases for all three samples with the temperature increasing. The coefficient k after 60 minutes of prestressed decreases from 0.935 to 0.884 for PP, from 0.94 to 0.893 for PP/AO, and only from 0.953 to 0.940 for PP-g-AO. This is because the carrier mobility increases with temperature, which increases the injection and accumulation of space charge, and the breakdown strength of the sample decreases more significantly. It can be concluded that the hetero-polarity DC-prestressed breakdown performance of PP-g-AO has a high-temperature stability.

The electrostatic potential on the polar group of the antioxidant has a specific charge-trapping effect. To further analyze the impact, the wave function files of PP, AO3052, and PP-g-AO are further analyzed using Multiwfn. The electrostatic potential is calculated on a surface with an electron density of 0.001 with a lattice spacing of 0.25 Bohr. The surface electrostatic potentials of the samples are presented by VMD, as shown in Figure 2.37. From Figure 2.37a to c, it is evident that grafting AO3052 introduces both positive and negative potential traps into PP within the energy range of −2 to 2 eV. The high-energy electrons in PP-g-AO experience a Coulomb attraction effect from the positive electrostatic potential on the

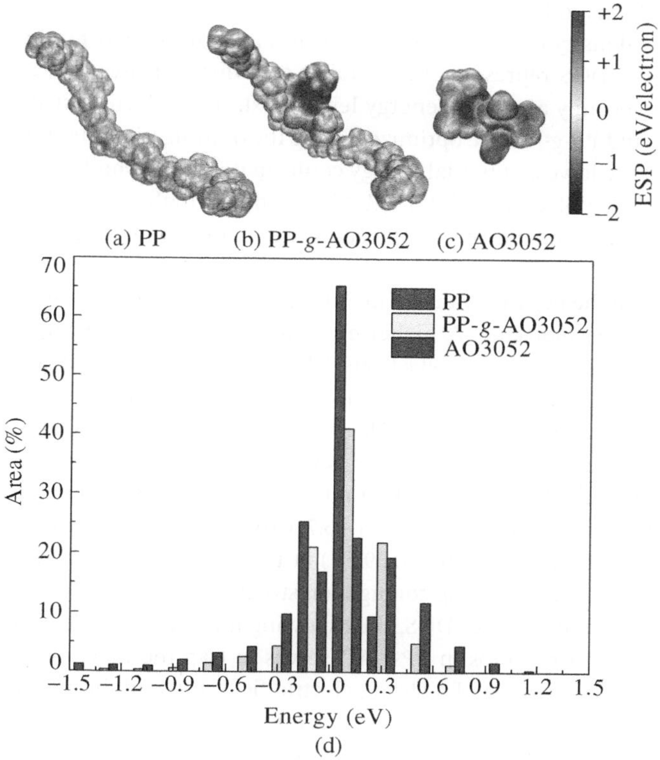

Figure 2.37 Surface electrostatic potential distribution of the samples. (a) PP. (b) PP-g-AO. (c) AO3052. (d) Electrostatic potential distribution. *Source:* [96]/Boxue Du et al./with permission from John Wiley & Sons.

molecular surface, while the holes experience a Coulomb attraction effect from the negative electrostatic potential. The charge carriers are trapped by the potential traps on the molecular surface, which suppresses the accumulation of space charge, thus reducing the degree of electric field distortion.

Figure 2.37d illustrates the surface electrostatic potential ratios of PP, AO, and PP-g-AO. Potential traps in PP are primarily distributed within the electrostatic potential range of −0.15 to 0.3 eV. In comparison, PP-g-AO demonstrates a significantly higher number of potential traps in the electrostatic potential range below −0.3 and above 0.3 eV. The surface area of the electrostatic potential region serves as an important parameter for evaluating the effectiveness of potential traps. Within a specific interval, a larger surface area of the electrostatic potential region could lead to a stronger effect of trapping charge carriers, which can reduce the mobility of the carriers and decrease the energy of the electron obtained from the acceleration process in the electric field. Consequently, this characteristic exerts a more pronounced inhibitory effect on the process of high-energy electrons breaking molecular chains to form low-density regions.

In Figure 2.38, the density-of-states (DOS) distribution is presented for PP, AO3052, and PP-g-AO. DOS represents the number of different states that an electron is allowed to occupy at a given energy level. Next, the molecular structures of PP, AO3052, and PP-g-AO are optimized using the optimization function in Gaussian to obtain the lowest potential energy conformation of the molecules. The lowest unoccupied molecular orbital (LUMO) and the highest occupied molecular orbital (HOMO) of the three molecules are extracted, as shown in Figure 2.38.

Research suggests that the negative value of LUMO energy level reflects the electron affinity energy of the molecule. A higher electron affinity energy means a stronger ability of the molecule to trap electrons. The negative value of HOMO energy level reflects the ionization energy of the molecule. A lower ionization energy implies a stronger ability to trap holes [79]. It can be noticed that the LUMO and HOMO of PP-g-AO are mainly occupied by benzene rings and C–O groups. The HOMO–LUMO gap of PP is 9.829 eV. Within the HOMO–LUMO gap range, grafting the antioxidant introduces a hole trap (−6.716 eV) at the highest energy level of −7.471 eV and an electron trap (−0.09 eV) at the lowest energy level of 2.353 eV, mainly due to the benzene ring conjugation structure. Notably, the electron traps of PP-g-AO exhibit a higher DOS, thus showing it forms more electron traps. The presence of electron traps in PP-g-AO serves to trap the high-energy electrons generated in the presence of a strong electric field, thus inhibiting the injection of space charge and the breakdown degradation process.

Figure 2.39a illustrates the energy level and corresponding molecular orbitals distribution for the sample/measurement electrode without the application of a polarization electric field. When the sample is not in contact with the measuring

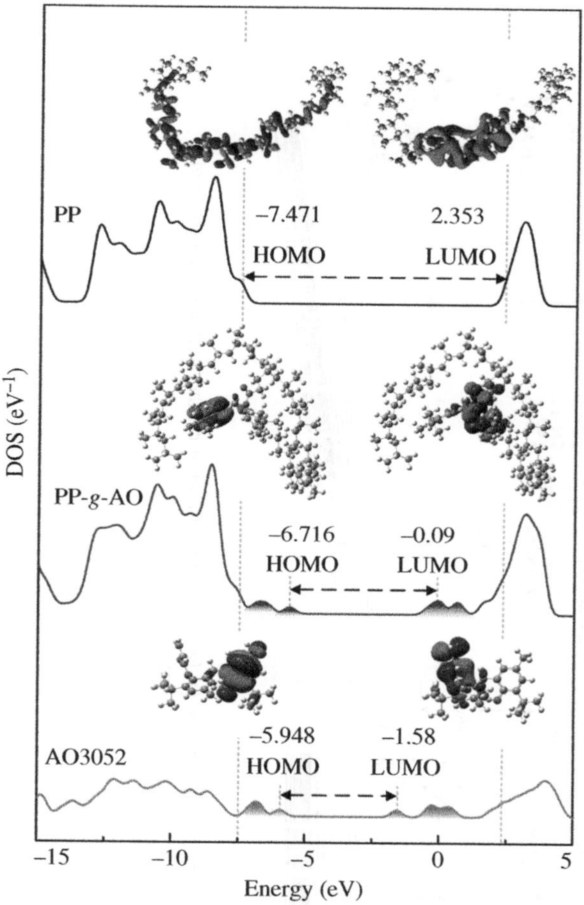

Figure 2.38 DOS and HOMO–LUMO orbitals of the samples. *Source:* [96]/Boxue Du et al./with permission from John Wiley & Sons.

electrode, the distance between them is set to be much greater than 10 Å [97]. From the intrinsic energy level distribution, the electron affinity (χ) of PP is −2.353 eV, the ionization energy (Φ_i) is 7.471 eV, and the HOMO–LUMO gap (Φ_g) is 9.824 eV. The electron injection barrier at the electrode is denoted as Φ_{Be}, and the hole injection barrier is denoted as Φ_{Bh}. The energy difference between the vacuum energy level (VL) and the Fermi energy level (E_F) of the electrode materials is defined as the work function (Φ_W) of the electrode. In this chapter, Φ_W(SC) and Φ_W(Al) are, respectively, set to 5.56 and 4.25 eV. The Fermi energy

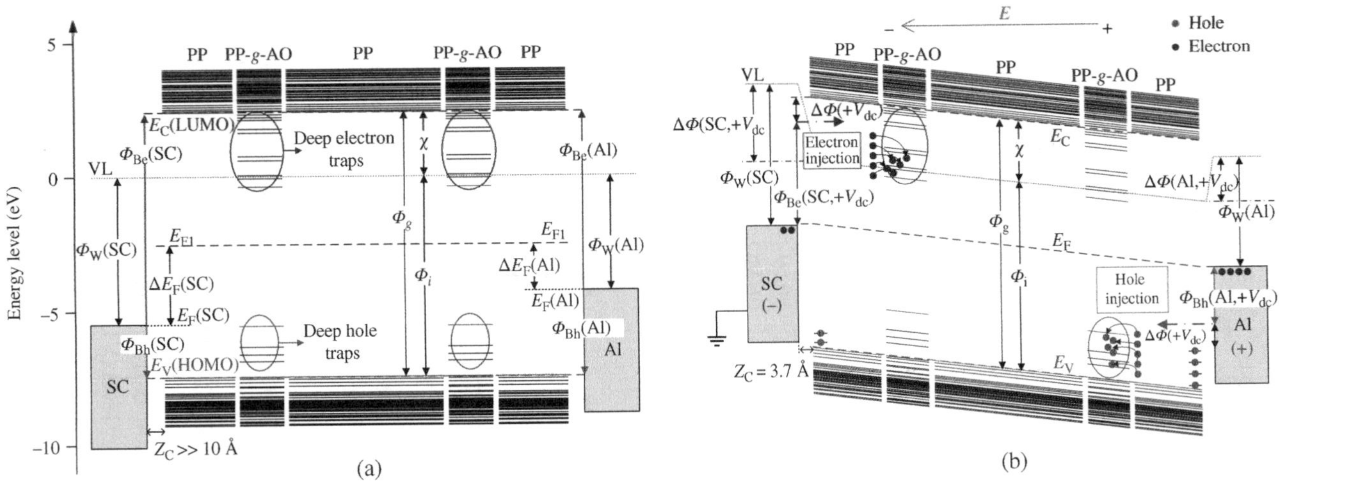

Figure 2.39 Energy level and corresponding molecular orbitals. (a) Before contact. (b) After contact and under electric field. *Source:* [96]/Boxue Du et al./with permission from John Wiley & Sons.

level differences between PP and the electrode material are denoted as ΔE_F(SC) and ΔE_F(Al).

After contact, the sample/measurement electrode forms a stable thermodynamic system where all the Fermi energy levels coincide, and the vacuum energy level (VL) undergoes changes. Figure 2.39b presents the energy level distribution of the sample/measurement electrode with a DC electric field. Upon contact, as the Fermi energy levels of the SC and Al electrodes are higher than the HOMO energy level of the PP, electronic charges are transferred from the PP material to the electrodes. Under the electric field ($+V_{dc}$), the migration of the charge generated by the energy level difference at the electrode interface will alter. Affected by the electric field polarization, the charge injection barriers in the sample/electrode system decrease, which is commonly governed by the Schottky emission effect, as expressed in Eq. (2.8),

$$\Delta\phi(+V_{dc}) = \sqrt{\frac{e^3 E}{(4\pi\varepsilon_0\varepsilon_r)}} \tag{2.8}$$

where e is the element charge of the electron (1.6×10^{-19} C). E is the electric field, kV/mm. ε_0 is the vacuum permittivity (8.85×10^{-12} F/m), and ε_r is the relative permittivity of PP (2.2).

When E is set at 100 kV/mm, $\Delta\Phi(+V_{dc})$ (0.26 eV) is calculated by Eq. (2.3). This calculation demonstrates that the hole injection barrier (Al/PP) is significantly lower compared to the electron injection barrier (SC/PP). The electric field exerted on the system has a more pronounced impact on weakening the hole injection barrier compared to the electron injection barrier. Upon applying a positive voltage to the sample/electrode system, electrons from the cathode (SC/PP) are injected into the PP and migrate toward the anode. During migration, the deep traps introduced by the PP-g-AO can trap these electrons near the cathode and hinder their further injection. In a similar manner, a substantial quantity of hole charges is injected from the anode and subsequently trapped by deep traps near the anode. Consequently, grafting PP with antioxidants could introduce localized state traps and improve the crystalline morphology, enhance the ability to trap charge carriers, suppress space charge injection, and reduce the damage to molecular chains by high-energy electrons during collisional ionization and energy release, thus ultimately improving the breakdown performance of PP [81].

In conclusion, the grafted antioxidant method can enhance the oxidation induction time and the electrical properties. After 10, 30, and 60 minutes of hetero-polarity prestressing, PP-g-AO exhibits a superior effect in improving the hetero-polarity DC prestressing breakdown field of PP compared to PP/AO.

Particularly at 90 °C, PP-g-AO achieves a 5.3–6.7% increase after the same duration of DC-prestressing compared to PP.

2.5 Conclusion and Outlook

The improvement of the flexibility and impact resistance of PP cable insulation by elastomer blending is required without notable sacrifice of electrical and thermal properties. The balance of mechanical and electrical properties of PP blend insulation is constantly desirable. Although the electrical performance has decreased to some extent due to the blending of elastomers, inorganic nanofillers, and organic compound additives can further improve the electrical performance of PP-blended insulation, which provide an alternative solution for high-performance HVDC cables. In the future, further attention needs to be paid to the large-scale manufacturing and long-term stability of different modified polypropylene insulation.

References

1 Chen, G., Hao, M., Xu, Z. et al. (2015). Review of high voltage direct current cables. *CSEE Journal of Power and Energy Systems* 1 (2): 9–21. https://doi.org/10.17775/CSEEJPES.2015.00015.

2 Alassi, A., Bañales, S., Ellabban, O. et al. (2019). Technology review, market trends, and future outlook. *Renewable and Sustainable Energy Reviews* 112: 530–554. https://doi.org/10.1016/j.rser.2019.04.062.

3 Zhou, Y., Peng, S., Hu, J., and He, J. (2017). Polymeric insulation materials for HVDC cables: Development, challenges and future perspective. *IEEE Transactions on Dielectrics and Electrical Insulation* 24 (3): 1308–1318. https://doi.org/10.1109/TDEI.2017.006205.

4 Yue, L., Liu, L., Zhang, W. et al. (2020). Preparation and electrical properties of 4-acetoxystyrene grafted polypropylene for HVDC cable insulation. *Journal of Materials Science: Materials in Electronics* 31 (5): 3890–3898. https://doi.org/10.1007/s10854-020-02934-6.

5 Zha, J.W., Wu, Y.H., Wang, S.J. et al. (2016). Improvement of space charge suppression of polypropylene for potential application in HVDC cables. *IEEE Transactions on Dielectrics and Electrical Insulation* 23 (4): 2337–2343. https://doi.org/10.1109/TDEI.2016.7556511.

6 Calebrese, C., Hui, L., Schadler, L.S., and Nelson, J.K. (2011). A review on the importance of nanocomposite processing to enhance electrical insulation. *IEEE*

Transactions on Dielectrics and Electrical Insulation 18 (4): 938–945. https://doi.org/10.1109/TDEI.2011.5976079.

7 Zhang, C., Wang, T., Sun, W. et al. (2021). Grafting of antioxidant onto polyethylene to improve DC dielectric and thermal aging properties. *IEEE Transactions on Dielectrics and Electrical Insulation* 28 (2): 541–549. https://doi.org/10.1109/TDEI.2020.009261.

8 Kao, K.C. (2004). *Dielectric Phenomena in Solids*. Elsevier.

9 Du, B.X., Xu, H., and Li, J. (2017). Effects of mechanical stretching on space charge behaviors of PP/POE blend for HVDC cables. *IEEE Transactions on Dielectrics and Electrical Insulation* 24 (3): 1438–1445. https://doi.org/10.1109/TDEI.2017.006116.

10 Green, C.D., Vaughan, A.S., Stevens, G.C. et al. (2015). Thermoplastic cable insulation comprising a blend of isotactic polypropylene and a propylene-ethylene copolymer. *IEEE Transactions on Dielectrics and Electrical Insulation* 22 (2): 639–648. https://doi.org/10.1109/TDEI.2015.7076758.

11 Caramitu, A.R., Zaharescu, T., Mitrea, S. et al. (2015). Study regarding mechanical, thermal and structural properties of some elastomeric polyamide blends destined to cable insulating. *Proceedings of the International Symposium of Fundamentals Electrical Engineering (ISFEE)*, Bucharest, Romania (28-29 November 2014), 1–6. https://doi.org/10.1109/ISFEE.2014.7050629.

12 Zha, J.-W., Wang, Y., Li, W.-K. et al. (2017). Electrical properties of polypropylene/styrene-ethylene-butylene-styrene block copolymer/MgO nanocomposites. *IEEE Transactions on Dielectrics and Electrical Insulation* 24 (3): 1457–1464. https://doi.org/10.1109/TDEI.2017.006162.

13 Zhou, Y., He, J., Hu, J. et al. (2015). Evaluation of polypropylene/polyolefin elastomer blends for potential recyclable HVDC cable insulation applications. *IEEE Transactions on Dielectrics and Electrical Insulation* 22 (2): 673–681. https://doi.org/10.1109/TDEI.2015.7076762.

14 Gao, Y., Li, J., Li, Y. et al. (2017). Effect of elastomer type on electrical and mechanical properties of polypropylene/elastomer blends. *Proceedings of International Symposium on Electrical Insulating Materials* (ISEIM), Toyohashi, Japan (11-15 September 2017), 2, 574–577. https://doi.org/10.23919/ISEIM.2017.8166554.

15 Gao, Y., Huang, S., Chen, L.Y. et al. (2018). Effect of elastomer content on interface discharge behavior between polypropylene and silicone rubber under AC voltage. *High Voltage Engineering* 44 (9): 2874–2880. https://doi.org/10.1109/ICEMPE.2019.8727241.

16 Vistamaxx 6202 Propylene-Based Elastomer, Exxon Mobil Corp., Irving, TX, USA, 2010.

17 Gao, Y., Li, J., Yuan, Y. et al. (2018). Trap distribution and dielectric breakdown of isotactic polypropylene/propylene based elastomer with improved flexibility for DC

cable insulation. *IEEE Access* 6: 58645–58661. https://doi.org/10.1109/ACCESS.2018.2874826.

18 Xia, M.-S., Yao, Z.-T., Ge, L.-Q. et al. (2015). A potential biofiller: the substitution effect of furfural modified clam shell for carbonate calcium in polypropylene. *Journal of Composite Materials* 49 (7): 807–816. https://doi.org/10.1177/0021998314525981.

19 Hoang, A.T., Serdyuk, Y.V., and Gubanski, S.M. (2015). Mechanisms of surface potential decay on enamel wire coatings. *IEEE Transactions on Dielectrics and Electrical Insulation* 22 (6): 3470–3480. https://doi.org/10.1109/TDEI.2015.005223.

20 Alam, S., Serdyuk, Y.V., and Gubanski, S.M. (2015). Potential decay on silicone rubber surfaces affected by bulk and surface conductivities. *IEEE Transactions on Dielectrics and Electrical Insulation* 22 (2): 970–978. https://doi.org/10.1109/TDEI.2015.7076798.

21 Molinie, P. (2012). A review of mechanisms and models accounting for surface potential decay. *IEEE Transactions on Plasma Science* 40 (2): 167–176. https://doi.org/10.1109/TPS.2011.2171372.

22 Simmons, J.G. and Tam, M.C. (1972). Theory of isothermal currents and the direct determination of trap parameters in semiconductors and insulators containing arbitrary trap distributions. *Physical Review B: Condensed Matter and Materials Physics* 7 (8): 3706–3713. https://doi.org/10.1103/PhysRevB.7.3706.

23 Min, D., Cho, M., Li, S., and Khan, A.R. (2012). Charge transport properties of insulators revealed by surface potential decay experiment and bipolar charge transport model with genetic algorithm. *IEEE Transactions on Dielectrics and Electrical Insulation* 19 (6): 2206–2215. https://doi.org/10.1109/TDEI.2012.6396982.

24 Li, J.Y., Zhou, F.S., Min, D.M. et al. (2015). The energy distribution of trapped charges in polymers based on isothermal surface potential decay model. *IEEE Transactions on Dielectrics and Electrical Insulation* 22 (3): 1723–1732. https://doi.org/10.1109/TDEI.2015.7116370.

25 Kumara, J.R.S.S., Serdyuk, Y.V., and Gubanski, S.M. (2016). Surface potential decay on LDPE and LDPE/Al_2O_3 nano-composites: measurements and modeling. *IEEE Transactions on Dielectrics and Electrical Insulation* 23 (6): 3466–3475. https://doi.org/10.1109/TDEI.2016.005663.

26 Min, D.M. and Li, S.T. (2014). Simulation on the influence of bipolar charge injection and trapping on surface potential decay of polyethylene. *IEEE Transactions on Dielectrics and Electrical Insulation* 21 (4): 1627–1636. https://doi.org/10.1109/TDEI.2014.004271.

27 Zhou, T.-C., Chen, G., Liao, R.-J., and Xu, Z. (2011). Charge trapping and detrapping in polymeric materials: trapping parameters. *Journal of Applied Physics* 110: 043724. https://doi.org/10.1063/1.3626468.

28 Shen, W.-W., Mu, H.-B., Zhang, G.-J. et al. (2013). Identification of electron and hole trap based on isothermal surface potential decay model. *Journal of Applied Physics* 113: 083706. https://doi.org/10.1063/1.4792491.

29 Li, S., Yin, G., Chen, G. et al. (2010). Short-term breakdown and long-term failure in nanodielectrics: a review. *IEEE Transactions on Dielectrics and Electrical Insulation* 17 (5): 1523–1535. https://doi.org/10.1109/TDEI.2010.5595554.

30 Li, S., Min, D., Wang, W., and Chen, G. (2016). Linking traps to dielectric breakdown through charge dynamics for polymer nanocomposites. *IEEE Transactions on Dielectrics and Electrical Insulation* 23 (5): 2777–2785. https://doi.org/10.1109/TDEI.2016.7736837.

31 Huang, X.Y., Fan, Y., Zhang, J., and Jiang, P. (2017). Polypropylene based thermoplastic polymers for potential recyclable HVDC cable insulation applications. *IEEE Transactions on Dielectrics and Electrical Insulation* 24 (3): 1446–1456. https://doi.org/10.1109/TDEI.2017.006230.

32 Diao, J., Huang, X., Jia, Q. et al. (2017). Thermoplastic isotactic polypropylene/ethylene-octene polyolefin copolymer nanocomposite for recyclable HVDC cable insulation. *IEEE Transactions on Dielectrics and Electrical Insulation* 24 (3): 1416–1429. https://doi.org/10.1109/TDEI.2017.006208.

33 Du, B.X., Xu, H., Li, J., and Li, Z. (2016). Space charge behaviors of PP/POE/ZnO nanocomposites for HVDC cables. *IEEE Transactions on Dielectrics and Electrical Insulation* 23 (5): 3165–3174. https://doi.org/10.1109/TDEI.2016.7736882.

34 Boukezzi, L. and Boubakeur, A. (2013). Prediction of mechanical properties of XLPE cable insulation under thermal aging: neural network approach. *IEEE Transactions on Dielectrics and Electrical Insulation* 20 (6): 2125–2134. https://doi.org/10.1109/TDEI.2013.6678861.

35 He, D., Wang, X., Liu, H. et al. (2018). Space charge behavior in XLPE cable insulation under ac stress and its relation to thermoelectrical aging. *IEEE Transactions on Dielectrics and Electrical Insulation* 25 (2): 541–550. https://doi.org/10.1109/TDEI.2017.006958.

36 Kai, Z., Lunzhi, L., Lisheng, Z. et al. (2015). The mechanical properties of recyclable cable insulation materials based on thermo-plastic polyolefin blends. *Proceedings of International Conference on Properties and Applications of Dielectric Materials (ICPADM)*, Sydney, NSW, Australia (19-22 July 2015), 532–535. https://doi.org/10.1109/ICPADM.2015.7295326.

37 Zubova, E.A., Balabaev, N.K., and Manevitch, L.I. (2007). Molecular mechanisms of the chain diffusion between crystalline and amorphous fractions in polyethylene. *Polymer* 48 (6): 1802–1813. https://doi.org/10.1016/j.polymer.2006.12.032.

38 Wang, W., Takada, T., Tanaka, Y., and Li, S. (2017). Trap-controlled charge decay and quantum chemical analysis of charge transfer and trapping in XLPE. *IEEE*

Transactions on Dielectrics and Electrical Insulation 24 (5): 3144–3153. https://doi.org/10.1109/TDEI.2017.006637.

39 Wu, K. and Cheng, C. (2017). Interface charges between insulating materials. *IEEE Transactions on Dielectrics and Electrical Insulation* 24 (4): 2633–2642. https://doi.org/10.1109/TDEI.2017.006442.

40 Vu, T.T.N., Teyssedre, G., Vissouvanadin, B. et al. (2015). Correlating conductivity and space charge measurements in multidielectrics under various electrical and thermal stresses. *IEEE Transactions on Dielectrics and Electrical Insulation* 22 (1): 117–127. https://doi.org/10.1109/TDEI.2014.004507.

41 Fabiani, D., Montanari, G.C., Cavallini, A., and Mazzanti, G. (2004). Relation between space charge accumulation and partial discharge activity in enameled wires under PWM-like voltage waveforms. *IEEE Transactions on Dielectrics and Electrical Insulation* 11 (3): 393–405. https://doi.org/10.1109/TDEI.2004.1306718.

42 Mazzanti, G. and Montanari, G.C. (2001). Elemental strain and trapped space charge in thermoelectrical aging of insulating materials life modeling. *IEEE Transactions on Dielectrics and Electrical Insulation* 8 (6): 966–971. https://doi.org/10.1109/94.971453.

43 Dissado, L.A., Mazzanti, G., and Montanari, G.C. (1997). The role of trapped space charge in the electrical aging of insulating materials. *IEEE Transactions on Dielectrics and Electrical Insulation* 4 (5): 496–506. https://doi.org/10.1109/94.625642.

44 Montanari, G.C. (2011). Bringing an insulation to failure: the role of space charge. *IEEE Transactions on Dielectrics and Electrical Insulation* 18 (2): 339–364. https://doi.org/10.1109/TDEI.2011.5739438.

45 Fu, M., Dissado, L.A., Chen, G., and Fothergill, J.C. (2008). Space charge formation and its modified electric field under applied voltage reversal and temperature gradient in XLPE cable. *IEEE Transactions on Dielectrics and Electrical Insulation* 15 (3): 851–860. https://doi.org/10.1109/TDEI.2008.4543123.

46 Tanaka, T., Bulinski, A., Castellon, J. et al. (2011). Dielectric properties of XLPE/SiO_2 nanocomposites based on CIGRE WG D1.24 cooperative test results. *IEEE Transactions on Dielectrics and Electrical Insulation* 18 (5): 1484–1517. https://doi.org/10.1109/TDEI.2011.6032819.

47 Maezawa, T., Taima, J., Hayase, Y. et al. (2007). Space charge formation in LDPE/MgO nano-composite under high electric field at high temperature. *IEEE Conference on Electrical Insulation and Dielectric Phenomena* 271–274. https://doi.org/10.1109/CEIDP.2007.4451550.

48 Dakka, M.A., Bulinski, A., and Bamji, S.S. (2010). Space charge evolution in polypropylene containing synthetic and natural organoclays. *IEEE Conference on Electrical Insulation and Dielectric Phenomena* 1–4. https://doi.org/10.1109/CEIDP.2010.5724020.

49 Montanari, G.C., Fabiani, D., and Palmieri, F. (2004). Modification of electrical properties and performance of EVA and PP insulation through nanostructure by organophilic silicates. *IEEE Transactions on Dielectrics and Electrical Insulation* 11 (5): 754–762. https://doi.org/10.1109/TDEI.2004.1349780.

50 Tanaka, T., Kozako, M., Fuse, N., and Ohki, Y. (2005). Proposal of a multi-core model for polymer nanocomposite dielectrics. *IEEE Transactions on Dielectrics and Electrical Insulation* 12 (4): 669–681. https://doi.org/10.1109/TDEI.2005.1511092.

51 Takada, T., Hayase, Y., Tanaka, Y., and Tatsuki, O. (2008). Space charge trapping in electrical potential well caused by permanent and induced dipoles for LDPE/Mgo nanocomposite. *IEEE Transactions on Dielectrics and Electrical Insulation* 15 (1): 152–160. https://doi.org/10.1109/T-DEI.2008.4446746.

52 Tanaka, T. (2005). Dielectric nanocomposites with insulating properties. *IEEE Transactions on Dielectrics and Electrical Insulation* 12 (5): 914–928. https://doi.org/10.1109/TDEI.2005.1522186.

53 Nelson, J.K. and Fothergill, J.C. (2004). Internal charge behaviour of nanocomposites. *Nanotechnology* 15 (5): 586–595. https://doi.org/10.1088/0957-4484/15/5/032.

54 Roy, M., Nelson, J.K., MacCrone, R.K., and Schadler, L.S. (2007). Candidate mechanisms controlling the electrical characteristics of silica/XLPE nanodielectrics. *Journal of Materials Science* 42 (11): 3789–3799. https://doi.org/10.1007/s10853-006-0413-0.

55 Lewis, T.J. (2002). Polyethylene under electrical stress. *IEEE Transactions on Dielectrics and Electrical Insulation* 9 (5): 717–729. https://doi.org/10.1109/TDEI.2002.1038659.

56 Murakami, Y., Nemoto, M., Okuzumi, S. et al. (2008). DC conduction and electrical breakdown of Mgo/LDPE nanocomposite. *IEEE Transactions on Dielectrics and Electrical Insulation* 15 (1): 33–39. https://doi.org/10.1109/T-DEI.2008.4446734.

57 Hampton, R.N. (2008). Some of the considerations for materials operating under high-voltage, direct-current stresses. *IEEE Electrical Insulation Magazine* 24 (1): 5–13. https://doi.org/10.1109/MEI.2008.4455498.

58 Hanley, T.L., Burford, R.P., Fleming, R.J., and Barber, K.W. (2003). A general review of polymeric insulation for use in HVDC cables. *IEEE Electrical Insulation Magazine* 19 (1): 13–24. https://doi.org/10.1109/MEI.2003.1178104.

59 Li, J., Xu, H., Du, B.X. et al. (2016). Space charge accumulation characteristics in ethylene-vinyl acetate copolymer filled with ZnO nanoparticles. *Annual Report Conference on Electrical Insulation and Dielectric Phenomena (CEIDP)* 180–183. https://doi.org/10.1109/ICD.2016.7547574.

60 Yang, Y., Hu, J., and He, J.L. (2016). Mesoporous nano-silica serves as the degradation inhibitor in polymer dielectrics. *Science Reports* 6: 28749. https://doi.org/10.1038/srep28749.

61 Stankovich, S., Dikin, D.A., Dommett, G.H. et al. (2006). Graphene-based composite materials. *Nature* 442: 282–286. https://doi.org/10.1038/nature04969.

62 Pokharel, P., Lee, S.H., and Dai, S. (2015). Thermal, mechanical, and electrical properties of graphene nanoplatelet/graphene oxide/polyurethane hybrid nanocomposite. *Journal of Nanoscience and Nanotechnology* 15: 211–214. https://doi.org/10.1166/jnn.2015.8353.

63 Li, Z.L., Du, B.X., Han, C.L., and Xu, H. (2017). Trap modulated charge carrier transport in polyethylene/graphene nanocomposites. *Science Reports* 7: 4015. https://doi.org/10.1038/s41598-017-04196-5.

64 Li, Q., Zhang, G., Liu, F. et al. (2014). Solution-processed ferroelectric terpolymer nanocomposites with high breakdown strength and energy density utilizing boron nitride nanosheets. *Energy & Environmental Science* 8 (3): 922–931. https://doi.org/10.1039/c4ee02962c.

65 Wu, L.Y., Wu, K., Liu, D. et al. (2018). Largely enhanced energy storage density of poly(vinylidene fluoride) nanocomposites based on surface hydroxylation of boron nitride nanosheets. *Journal of Materials Chemistry A* 6 (17): 7573–7584.

66 Luo, S.B., Yu, J., Yu, S. et al. (2018). Significantly enhanced electrostatic energy storage performance of flexible polymer composites by introducing highly insulating-ferroelectric microhybrids as fillers. *Advanced Energy Materials* 9 (5): 1803204. https://doi.org/10.1002/aenm.201803204.

67 Li, Z.L., Fan, M.S., Zhou, S.F., and Du, B.X. (2021). BNNS encapsulated TiO_2 nanofillers endow polypropylene cable insulation with enhanced dielectric performance. *IEEE Transactions on Dielectrics and Electrical Insulation* 28 (4): 1238–1246. https://doi.org/10.1109/TDEI.2021.009574.

68 Du, B.X., Li, Z.L., and Li, J. (2016). Effects of direct fluorination on space charge accumulation in HTV silicone rubber. *IEEE Transactions on Dielectrics and Electrical Insulation* 23 (4): 2353–2360. https://doi.org/10.1109/TDEI.2016.7556513.

69 Zhou, Y., Hu, J., Dang, B., and He, J. (2017). Effect of different nanoparticles on tuning electrical properties of polypropylene nanocomposites. *IEEE Transactions on Dielectrics and Electrical Insulation* 24 (3): 1380–1389. https://doi.org/10.1109/TDEI.2017.006183.

70 Wang, W.W., Min, D.M., and Li, S.T. (2016). Dynamic charge transport characteristics in polyimide surface and surface layer under low-energy electron radiation. *IEEE Transactions on Dielectrics and Electrical Insulation* 23 (4): 564–572. https://doi.org/10.1109/TDEI.2016.7556518.

71 Liu, B., Yang, M., Zhou, W.Y. et al. (2019). High energy density and discharge efficiency polypropylene nanocomposites for potential high-power capacitor. *Energy Storage Materials* 27: 443–452. https://doi.org/10.1016/j.ensm.2019.12.006.

72 Zhang, X., Shen, Y., Zhang, Q. et al. (2014). Ultrahigh energy density of polymer nanocomposites containing $BaTiO_3$@TiO_2 nanofibers by atomic-scale interface engineering. *Advanced Materials* 27 (5): 819–824. https://doi.org/10.1002/adma.201404101.

73 Ashcraft, A.C., Eichhorn, R.M., and Shaw, R.G. (1978). Laboratory studies of treeing in solid dielectrics and voltage stabilization of polyethylene. *Proceedings of ISEI*, (14-16 June 1976), Montreal, Quebec, Canada, 213–218. https://doi.org/10.1109/EIC.1976.7464203.

74 Jarvid, M., Johansson, A., Bjuggren, J.M. et al. (2014). Tailored side-chain architecture of Benzil voltage stabilizers for enhanced dielectric strength of cross-linked polyethylene. *Journal of Polymer Science Part B-Polymer Physics* 52 (16): 1047–1054. https://doi.org/10.1002/polb.23523.

75 Jarvid, M., Johansson, A., Kroon, R. et al. (2015). A new application area for fullerenes: voltage stabilizers for power cable insulation. *Advanced Materials* 27 (5): 897–902. https://doi.org/10.1002/adma.201404306.

76 Hou, Z.H., Du, B.X., Li, Z.L., and Li, J. (2020). Effects of radical scavenger on space charge accumulation of PP/ULDPE composites for HVDC cable insulation. *IEEE Transactions on Dielectrics and Electrical Insulation* 27 (3): 989–997. https://doi.org/10.1109/TDEI.2020.008608.

77 Peña, J.M., Allen, N.S., Edge, M. et al. (2001). The use of microwave and FTIR spectroscopy for the characterisation of carbon blacks modified with stabilisers. *Polymer Degradation and Stability* 74 (1): 1–24. https://doi.org/10.1109/TDEI.2020.008608.

78 Du, B., Su, J., Tian, M. et al. (2018). Understanding trap effects on electrical treeing phenomena in EPDM/POSS composites. *Scientific Reports* 8 (1): 8481. https://doi.org/10.1038/s41598-018-26773-y.

79 Su, J., Du, B., Han, T. et al. (2019). Multistep and multiscale electron trapping for high-efficiency modulation of electrical degradation in polymer dielectrics. *Journal of Physical Chemistry C* 123 (12): 7045–7053. https://doi.org/10.1021/acs.jpcc.9b00349.

80 Takada, T., Kikuchi, H., Miyake, H. et al. (2015). Determination of charge-trapping sites in saturated and aromatic polymers by quantum chemical calculation. *IEEE Transactions on Dielectrics and Electrical Insulation* 22 (2): 1240–1249. https://doi.org/10.1109/TDEI.2015.7076827.

81 Du, B.X., Hou, Z.H., Li, J., and Li, Z.L. (2018). Effect of graphene nanoplatelets on space charge and breakdown strength of PP/ULDPE blends for HVDC cable insulation. *IEEE Transactions on Dielectrics and Electrical Insulation* 25 (6): 2405–2412. https://doi.org/10.1109/TDEI.2018.007271.

82 Du, B.X., Hou, Z.H., Li, Z.L., and Li, J. (2019). Temperature dependent space charge and breakdown strength of PP/ULDPE/graphene nanocomposites for HVDC extruded cable insulation. *IEEE Transactions on Dielectrics and Electrical Insulation* 26 (3): 876–884. https://doi.org/10.1109/TDEI.2019.007858.

83 Reingruber, E. and Buchberger, W. (2010). Analysis of polyolefin stabilizers and their degradation products. *Journal of Separation Science* 33 (22): 3463–3475. https://doi.org/10.1002/jssc.201000493.

84 Lee, S.H., Park, J.K., Lee, C.R., and Suh, K.S. (1997). The effect of low-molecular-weight species on space charge and conduction in LDPE. *IEEE Transactions on Dielectrics and Electrical Insulation* 4 (4): 425–432. https://doi.org/10.1109/94.625359.

85 Liu, Z., Zhang, F., Yu, J. et al. (2018). Research on key technologies in ±1100 kV ultra-high voltage DC transmission. *High Voltage* 3 (4): 279–288. https://doi.org/10.1049/hve.2018.5023.

86 Bu, J., Huang, X.Y., Li, S.T., and Jiang, P.K. (2016). Significantly enhancing the thermal oxidative stability while remaining the excellent electrical insulating property of low density polyethylene by addition of antioxidant functionalized graphene oxide. *Carbon* 106: 218–227. https://doi.org/10.1016/j.carbon.2016.05.020.

87 Chen, J., Yang, M.S., and Zhang, S.M. (2011). Immobilization of antioxidant on nanosilica and the aging resistance behavior in polypropylene. *Composites Part A – Applied Science and Manufacturing* 42 (5): 471–477. https://doi.org/10.1016/j.compositesa.2011.01.006.

88 Gugumus, F. (1994). New trends in the stabilization of polyolefin fibers. *Polymer Degradation and Stability* 44 (3): 273–297. https://doi.org/10.1016/0141-3910(94)90089-2.

89 Gao, X.W., Meng, X., Wang, H. et al. (2008). Antioxidant behaviour of a nanosilica-immobilized antioxidant in polypropylene. *Polymer Degradation and Stability* 93 (8): 1467–1471. https://doi.org/10.1016/j.polymdegradstab. 2008.05.009.

90 Xu, A., Roland, S., and Colin, X. (2020). Physico-chemical characterization of the blooming of Irganox 1076 antioxidant onto the surface of a silane-crosslinked polyethylene. *Polymer Degradation and Stability* 171: https://doi.org/10.1016/j.polymdegradstab. 2019.109046.

91 Zhou, Y., Hu, J., Dang, B., and He, J. (2016). Mechanism of highly improved electrical properties in polypropylene by chemical modification of grafting maleic anhydride. *Journal of Physics D: Applied Physics* 49 (41): 415301. https://doi.org/10.1088/0022-3727/49/41/415301.

92 Yang, J.M., Zhang, R., Yang, X. et al. (2023). Optimization method for carbon black distribution of polypropylene-based semi-conductive screen materials and its effect on charge emission behaviour at screen/insulation interface. *High Voltage* 8 (2): 239–250. https://doi.org/10.1049/hve2.12244.

93 Ran, Z.Y., Du, B., Xiao, M., and Liu, H. (2022). Breakdown performance of polypropylene and its long-chain branched blending for metalized film capacitors. *IEEE Transactions on Dielectrics and Electrical Insulation* 29 (1): 145–152. https://doi.org/10.1109/TDEI.2022.3146428.

94 Liang, Y., Weng, L., and Zhang, W. (2021). Preparation and electrical properties of 4-allyloxy-2-hydroxybenzophenone grafted polypropylene for HVDC cables. *Journal of Electronic Materials* 50 (11): 6228–6236. https://doi.org/10.1007/s11664-021-09147-5.

95 Meng, F.-B., Chen, X., Dai, C. et al. (2021). Effect of thermal ageing on physico-chemical and electrical properties of EHVDC XLPE cable insulation. *IEEE Transactions on Dielectrics and Electrical Insulation* 28 (3): 1012–1019. https://doi.org/10.1109/TDEI.2021.009449.

96 Boxue, D., Sun, G., and Wang, H. (2024). Hindered phenolic compounds grafting on tailoring the DC electrical characteristics of polypropylene cable insulation. *High Voltage* 1–10. https://doi.org/10.1049/hve2.12450.

97 Toyoda, K., Nakano, Y., Hamada, I. et al. (2009). First-principles study of the pentacene/Cu(111) interface: adsorption states and vacuum level shifts. *Journal of Electron Spectroscopy and Related Phenomena* 174 (1–3): 78–84. https://doi.org/10.1016/j.elspec.2009.04.005.

3

Electrical Treeing Phenomenon

3.1 Introduction

An electrical tree is a type of electrical aging and degradation phenomenon of insulating materials; once it crosses the insulation, electrical breakdown occurs, thereby affecting the operation safety of polymeric cables [1]. The global development of DC power transmission has brought many concerns on cable insulation under DC and associated voltage waveforms. Although much less reported than AC trees, the characteristics of DC electrical trees in solid insulation materials, especially the initiation properties, have attracted the attention of researchers and manufactures. It was found that electrical trees would not be triggered until the DC reached much higher values than those obtained with AC, and the calculated field strength at the needle tip far exceeded the expected values of the intrinsic breakdown strength of polymers. These behaviors were explained as a result of homo charge injection into the region surrounding the needle tip, when given enough time to develop, which reduced the maximum field strength sufficiently to prevent treeing [2–5]. However, the electrical treeing phenomenon of HVDC cable insulation has been investigated not only DC ramp voltage but also DC voltage with polarity reversal, grounding after DC prestressing, impulse superimposed on DC voltage, and impulse after DC prestressing since the 1970s. DC voltage with polarity reversal occurs in line-commutated converter (LCC) HVDC systems, to change the direction of energy transfer. Compared to DC voltage with polarity reversal, impulse superimposed on DC voltage is a more common operating condition, including switching overvoltage and lightning overvoltage [6, 7]. Under the impulse voltage, the electrical tree is easy to initiate and can easily cause breakdown [8]. Therefore, this chapter focuses on the electrical treeing initiation and breakdown phenomenon of PP cable insulation under DC and impulse combined voltages.

Polypropylene Cable Insulation, First Edition. Boxue Du and Zhonglei Li.

Published 2025 by John Wiley & Sons, Inc.

The electrical tree initiation characteristics under impulse voltage after DC prestressing in cross-linked polyethylene (XLPE) were reported [9]. It was found that the initialed tree length was mainly determined by the impulse amplitude and that the DC prestressing had little effect. The breakdown characteristics under DC and impulse superposed onto DC in modified high-density polyethylene (HDPE) were reported [10]. The results revealed that the breakdown voltage of an impulse superposed onto DC was lower than that of DC alone. The electrical treeing properties under DC and AC composite voltages in epoxy resin samples were reported [11]. The results indicated that under the positive and negative DC voltages, the electrical treeing properties were different. The electrical treeing properties under DC and AC composite voltages in XLPE were reported [12]. The AC amplitude was found to have a great influence on the electrical tree growth characteristics. The electrical aging under positive and negative impulse voltages in epoxy resin was reported [13]. It was found that the lifetime decreased with the impulse amplitude increasing, and the tree length under positive voltage was longer than that under negative voltage. The electrical treeing process in silicone rubber (SiR) with different impulse frequencies was reported [14]. It was found that with higher impulse frequencies, both the tree length and the fractal dimension show a tendency to increase. From these results, it can be seen that the amplitude and the polarity of the DC and impulse voltage, as well as the impulse frequency, have certain impact on the growth of the electrical tree. However, the electrical treeing initiation and breakdown phenomenon of PP cable insulation under DC and impulse-combined voltages rarely reported, and nearly no published information can be found about the effect of ambient temperature on electrical treeing.

High-voltage cables require good mechanical properties to withstand various extreme installation environments, such as high drop terrains and the laying process of submarine cables through towing pipes [15]. The occurrence of large bending deformations in cables, due to reduced turning radius, can lead to the concentration of mechanical stress and the development of microcracks. Consequently, partial discharges are more likely to occur and form electrical tree channels [16]. However, the existing research primarily applies mechanical stresses in a single direction, either parallel or perpendicular to the samples, which fails to capture the nonuniform and nonlinear stress distribution present in actual cables [17]. Recent studies have attempted to simulate the nonuniform strain distribution in cables by applying stresses to circular slices, investigating electrical tree characteristics under tensile and compressive stresses separately [18]. Yet, under bending operation of the cable, the outer cable layer experiences stretching while the inner layer undergoes compression, involving the coexistence of tensile and compressive force [19]. However, the electrical tree characteristics under such conditions remain unclear.

3.2 Electrical Treeing Under Impulse Superimposed on DC Voltage

In this subsection, the effects of DC voltage, impulse voltage, and impulse frequency on the electrical treeing under the DC and impulse combined voltages were investigated. The DC amplitude is in the range of −25 to +25 kV, and the impulse amplitude is in the range of −35 to +35 kV. The impulse frequency ranges from 400 to 800 Hz. The rise time of the single impulse voltage is 50 μs, and the tail time is 60 μs. The DC voltage amplitude range was from −25 to +25 kV, and the impulse voltage amplitude range was from −35 to +35 kV. During the experiment, the DC voltage was applied first. After one minute, the impulse voltage was applied, and the DC and impulse-combined electric field were generated. The treeing process is finished when the breakdown occurs, or the treeing time reaches 10 hours. A digital microscope and computer were employed to record the growth of the electrical tree. Figure 3.1 shows the diagrammatic waveforms of DC and impulse combined voltages. Figure 3.1a1, a2, and a3 represents a +25 kV impulse combined with +5, +15, and +25 kV DC voltages, respectively. Figure 3.1b1, b2, and b3 represents a +25 kV impulse combined with −5, −15, and −25 kV DC voltages, respectively. The impulse frequency is 400 Hz; the rise time of the single impulse voltage is 50 μs, and the tail time is 60 μs.

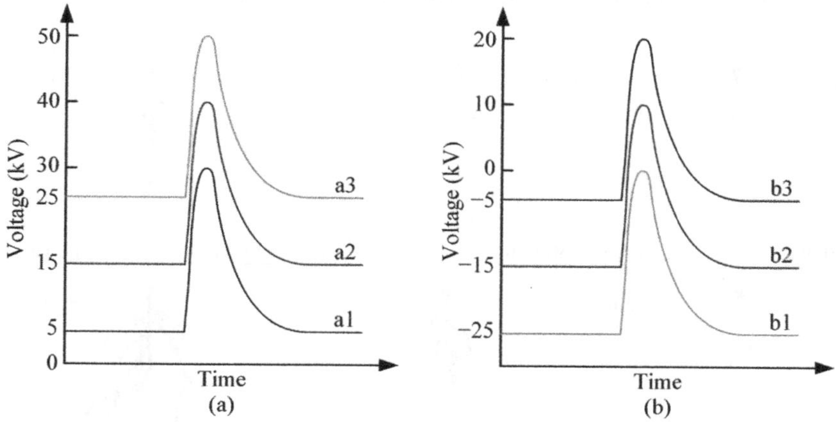

Figure 3.1 Diagram of the impulse superimposed on DC voltage. (a) Positive DC superimposed positive pulse. (b) Negative DC superimposed positive pulse. *Source:* Zhu et al. [66]/with permission from IEEE.

3.2.1 Effects of Impulse Amplitude

Figure 3.2 shows typical electrical tree morphologies of different DC amplitudes combined with +25 kV impulse voltages. The selected pictures are pre-breakdown electrical trees. Under the positive DC voltage, as shown in Figure 3.2a1, a2, and a3, the tree morphologies are similar with different amplitudes, implying that the positive DC voltage has little effect on the tree morphology. However, the tree morphology changes with different negative DC amplitudes. Under the negative DC amplitude of −5 and −15 kV, more pine branches are found compared with those under the positive DC voltages. Under DC voltage, space charges are formed around the needle tip or electrical tree branches tips. As the injection depth and range of electrons are greater than holes, a larger negative space charges layer will be formed around the electrical tree branches under a negative DC voltage [20]. There are more electron avalanche discharges in low-density areas, leading to the formation of more pine branches. Under the DC amplitude of −25 kV, the tree channels become significantly darker and wider, implying the different electrical treeing mechanisms.

Figure 3.3 shows the relationship between the tree length and the treeing time of different DC amplitudes and +25 kV impulse-combined voltages. As shown in Figure 3.3a, with the increase in the DC amplitude, the growth rate increases, implying that the positive DC voltage accelerates the treeing process. The charge transportation process under positive DC and impulse-combined voltages is shown in Figure 3.4a, corresponding to Figure 3.1a. During the application of the positive DC voltage, charges are formed around the needle or the electrical tree

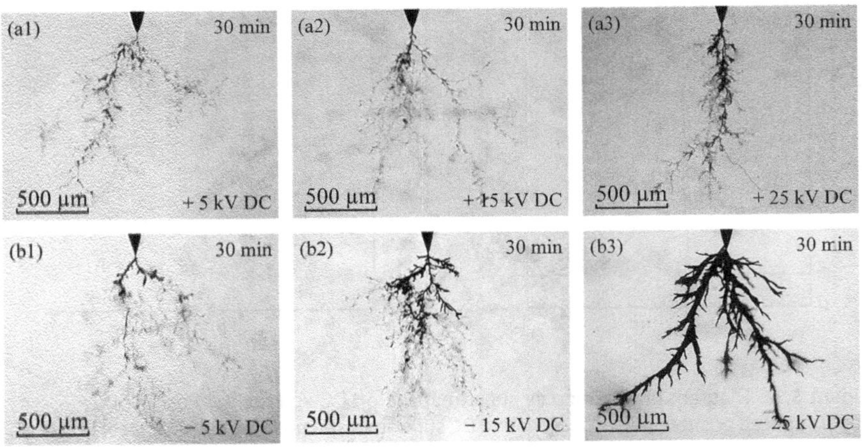

Figure 3.2 Typical electrical tree morphologies of different DC amplitudes combined with +25 kV impulse voltages. (a) Positive DC voltage. (b) Negative DC voltage. *Source:* Zhu et al. [66]/with permission from IEEE.

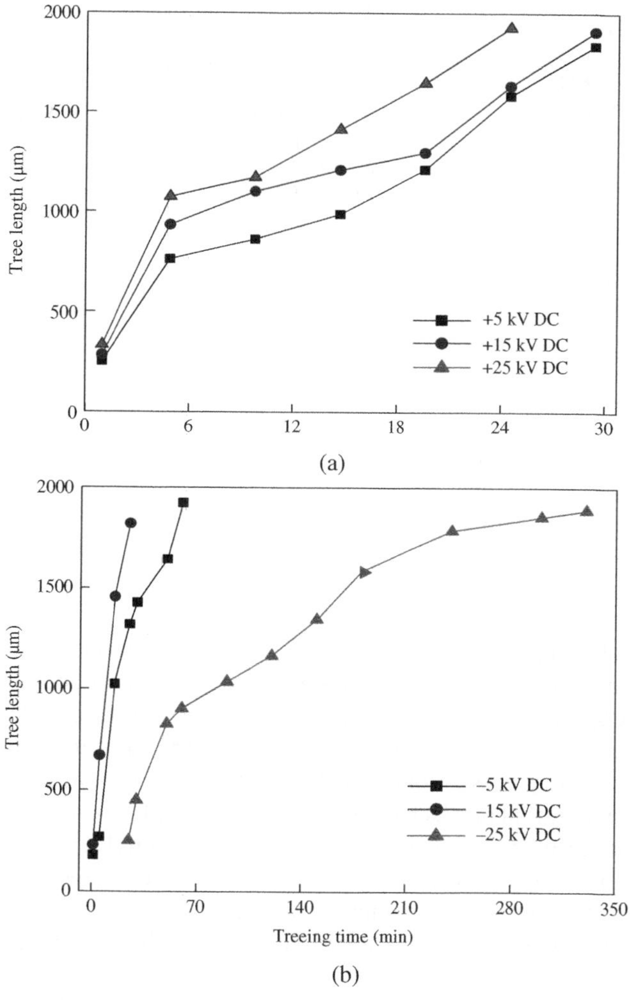

Figure 3.3 Relationship between the tree length and the treeing time of different DC amplitudes and +25 kV impulse-combined voltages. (a) Positive DC voltage. (b) Negative DC voltage. *Source:* Zhu et al. [66]/with permission from IEEE.

branches tips. When the impulse voltage is applied, a large number of charges are instantaneously injected; these charges stimulate the existed trapped charges to de-trap and move far away from the needle or the electrical tree branches tips and trap into other traps again. These injected charges themselves are trapped after one or more scattering around the needle tip or the electrical tree branches tips,

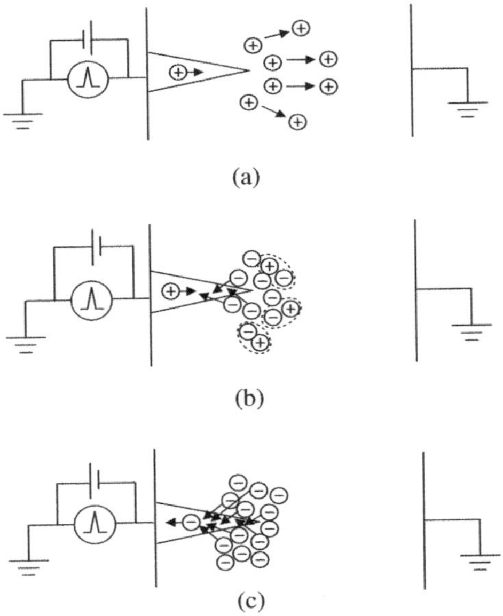

Figure 3.4 Charge transportation process under different DC amplitudes. (a) Positive DC and impulse-combined voltages. (b) −5 and −15 kV DC and impulse-combined voltages. (c) −25 kV DC and impulse-combined voltages. *Source:* Zhu et al. [66]/with permission from IEEE.

reducing the electric field strength and electron injection rate [9]. The states of the charges change from high-energy to low-energy state because of the trapping process. The excess energy is passed to other charges, which are converted into hot electrons via the Auger effect. Hot electrons break molecular chains to form low-density areas during this process. Charges collide with each other in low-density areas, releasing energy that breaks more molecular chains [21]. Avalanche discharges will also occur, resulting in the breaking of molecular chains [22]. With higher DC amplitude, more injected DC charges, and more moved charges are present. The larger effective values of the composite voltages also result in more serious electric field distortion and more discharges. Therefore, with the increase in DC amplitude, the growth rate increases. As shown in Figure 3.3b, with the DC voltage increasing from −5 to −15 kV DC, the growth rate increases. The charge transportation process of −5 and −15 kV is shown in Figure 3.4b, corresponding to Figure 3.1b1 and b2, respectively. During the application of the negative DC voltage, electrons are accumulated around the needle tip. After the positive impulse voltage is applied, some electrons are extracted, and when the voltage polarity changes from negative to positive, the positive charges are injected. Negative charges recombine with positive charges to produce energy that accelerates the growth of electrical trees [23]. With the voltage polarity changing from negative to positive, the electrons change from homo space charges to hetero space charges, leading to the electric field distortion that causes discharges and the breaking of molecular chains [24]. With the DC amplitude increasing from −5 to −15 kV, more homo space charges are injected, resulting in more serious discharges; therefore, the growth rate of −15 kV DC is larger than that of −5 kV DC [25]. However, with the DC voltage increasing from −15 to −25 kV DC, the growth rate decreases. The charge transportation process at −25 kV DC is different from that at −5 and −15 kV DC, as shown in Figure 3.4c, corresponding to the applied voltage of Figure 3.1b3. After the impulse voltage is applied, because of the electric potential reduction at the needle tip, electrons are extracted [26]. Such charge movement causes the breaking of molecular chains, leading to the electrical tree growth. The growth rate of −25 kV DC is smaller than that of −5 and −15 kV DC because of the lack of the polarity changing process.

Figure 3.5 shows the relationship between the accumulated damage and the treeing time of different DC amplitudes and +25 kV impulse-combined voltages. The accumulated damage is a means to evaluate the destruction degree caused by electrical tree aging of the tested sample. As shown in Figure 3.5a, with the same treeing time, the accumulated damage increases with higher positive DC amplitudes. As shown in Figure 3.5b, the accumulated damage of −15 kV DC is the greatest at the same treeing time; this observation is consistent with the growth rate shown in Figure 3.3b. However, the accumulated damage of pre-breakdown electrical trees of −25 kV DC at 330 minutes is significantly larger than that of −15 kV DC at 25 minutes, from which it can be inferred that once the breakdown occurs, the destruction areas caused by −25 kV DC is the greatest.

Figure 3.6 shows the breakdown time of different DC amplitudes and +25 kV impulse-combined voltages. With the increase in positive DC amplitude, the breakdown time decreases. However, with the increase in negative DC amplitude, the breakdown time decreases first and then increases. These breakdown results are consistent with the growth rate results in Figure 3.3. Figure 3.6 shows that the breakdown time of −15 kV DC is smaller than that of the positive DC voltage. However, the breakdown time of −5 kV DC is larger than that of the positive DC voltage. The effect of DC prestress on breakdown voltage was reported [27]. It was found that compared with the positive DC voltage, which has the same polarity as that of the applied impulse voltage, the breakdown strength is lower under the negative DC prestress, which has the opposite polarity with the applied impulse voltage. This result is attributed to the effect of space charges. With the voltage reversal, homo space charges become hetero space charges, thereby reducing the breakdown voltage. The electric field profile before and after polarity changes was reported [28]. It was found that the electric field changes from −50 to 130 kV/mm just after the polarity changes, indicating that the electric field distortion caused by polarity changes is serious. At voltages of −5 and −15 kV DC, there is a process of changing polarity. The electric field distortion in this process will lead to shorter breakdown time. However, the composite voltages of

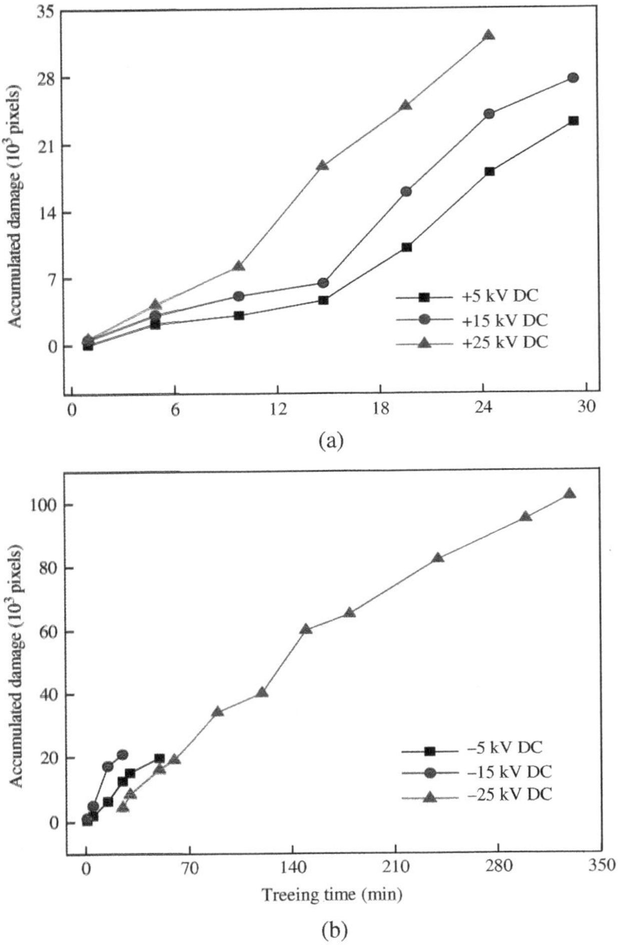

Figure 3.5 Relationship between the accumulated damage and the treeing time of different DC amplitudes and +25 kV impulse-combined voltages. (a) Positive DC voltage. (b) Negative DC voltage. *Source:* Zhu et al. [66]/with permission from IEEE.

−5 and −15 kV DC are both lower than that of the positive voltage, resulting in the longer breakdown time. The reason why the breakdown time of −5 kV DC is larger than that of the positive DC voltage and the breakdown time of −15 kV DC is smaller than that of the DC positive voltage depends on which factor is the dominant one. Under −5 kV DC, the DC voltage is small, resulting in a small quantity of injected electrons; therefore, the role of negative space charges is not obvious.

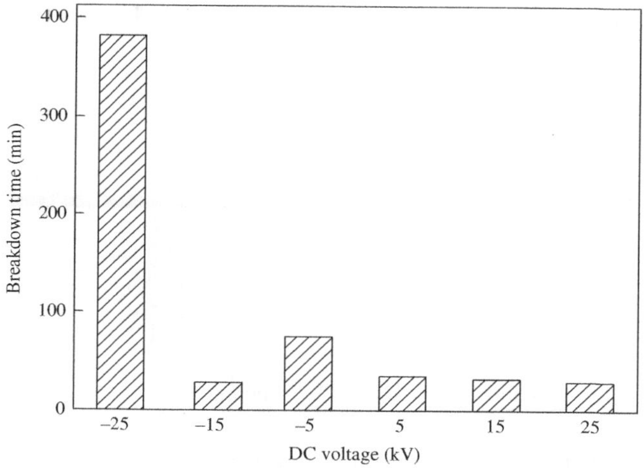

Figure 3.6 Breakdown time of different DC amplitudes and +25 kV impulse-combined voltages. *Source:* Zhu et al. [66]/with permission from IEEE.

In contrast, with the DC voltage increasing to −15 kV, the role of space charges occupies a dominant position; therefore, the breakdown time of −15 kV DC is smaller than that of the positive DC voltage.

3.2.2 Effects of Impulse Frequency

Figure 3.7 shows typical electrical tree morphologies of +5 and −15 kV DC and +25 kV impulse-combined voltages with different impulse frequencies. The treeing time is 10 minutes. As shown in Figure 3.7, the electrical tree channels become wider and darker with frequency increasing. The reason for this phenomenon will be explained later in the chapter.

Figure 3.8 shows the relationship between the tree length and the treeing time of +5 and −15 kV DC and +25 kV impulse combined voltages with different frequencies. Figure 3.9 shows the relationship between the accumulated damage and the treeing time of +5 and −15 kV DC and +25 kV impulse combined voltages with different frequencies. As shown in Figures 3.8 and 3.9, with the impulse frequency increasing, the growth rate and accumulated damage increase. When the electric field in the tube exceeds a critical value, avalanche discharge occurs. Discharges propagate along the tube. When the electric field in the tube is lower than the residual electric field, the propagation ends. The number of collision electrons generated by a starting electron in an avalanche is [17]

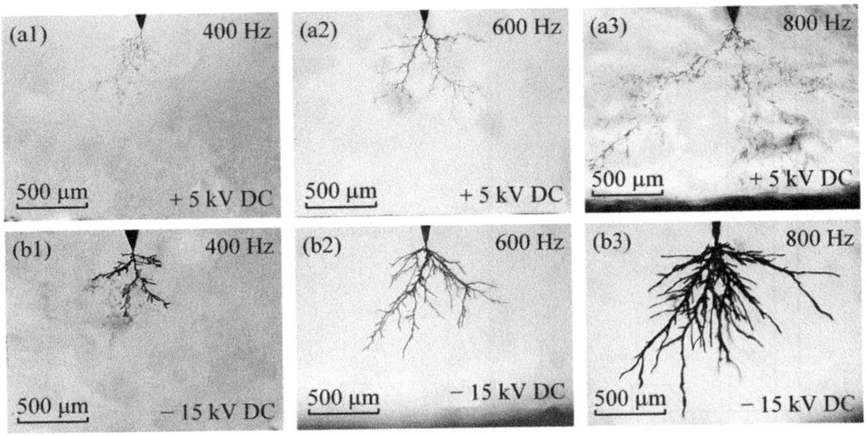

Figure 3.7 Typical electrical tree morphologies of (a) +5 and (b) −15 kV DC and +25 kV impulse-combined voltages with different impulse frequencies. *Source:* Zhu et al. [66]/with permission from IEEE.

$$N_A = \exp\left[\frac{L}{\lambda}\exp\left(-\frac{I}{\exp \lambda E}\right)\right] - 1 \tag{3.1}$$

where L is the path length occurring per electron, I is the ionization energy, and λ is the average free travel between two collisions. If the number of electrons in an avalanche in a certain area is n_0 and the number of discharges caused by each impulse voltage is b, then the number of ionizations N_t in time t is

$$N_t = ftbn_0N_A \tag{3.2}$$

With the impulse frequency increasing, N_t increases. Electrical tree channels propagate to the ground electrode, and a longer tree length is observed. N_t is proportional to the area destroyed by the avalanche discharges [17]. Therefore, the destruction caused by each avalanche zone within the same time increases, leading to wider branches and larger accumulated damage.

Figure 3.10 shows the breakdown time of −5 and −15 kV DC and +25 kV impulse-combined voltages with different frequencies. With the impulse frequency increasing, the breakdown time decreases significantly. The breakdown times of −15 kV DC with 600 and 800 Hz are shorter than those of +5 kV DC with 600 and 800 Hz, respectively, consistent with the results of 400 Hz in Subsection 3.2.1.

3.2.3 Effects of DC Voltage Amplitude

Figure 3.11 shows diagrammatic waveforms of DC and impulse-combined voltages. Figure 3.11a1, a2, and a3 represents −25 kV DC combined with −15, −25, and −35 kV impulse voltages, respectively. Figure 3.11b1, b2, and b3

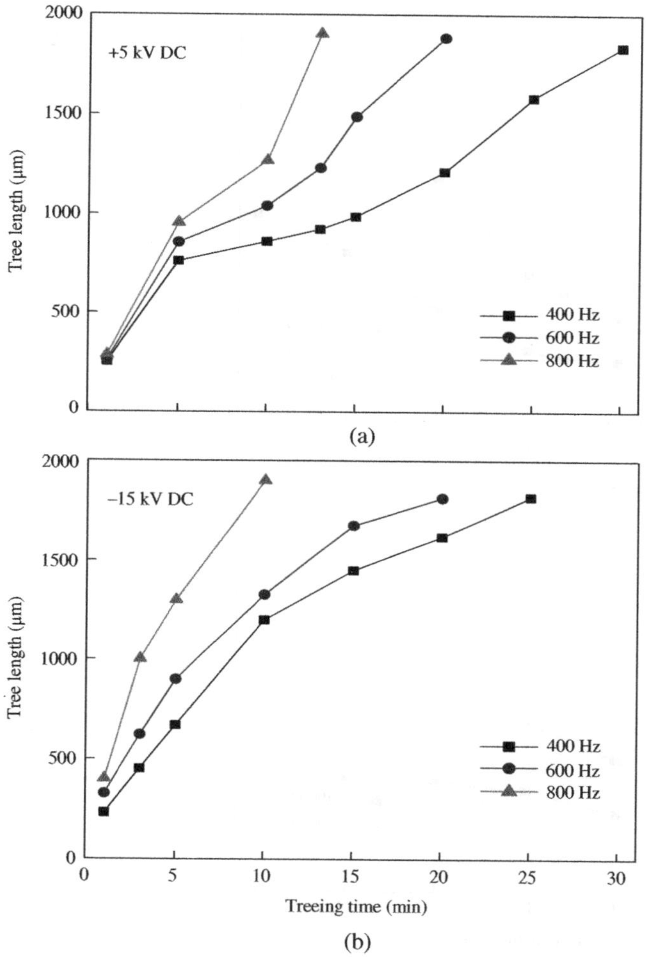

Figure 3.8 Relationship between the tree length and the treeing time of (a) +5 and (b) −15 kV DC and +25 kV impulse-combined voltages with different frequencies. *Source:* Zhu et al. [66]/with permission from IEEE.

represents −25 kV DC combined with +15, +25, and +35 kV impulse voltages, respectively. The impulse frequency is 400 Hz; the rise time of the single impulse voltage is 50 μs, and the tail time is 60 μs.

Figure 3.12 shows typical electrical tree morphologies of different impulse amplitudes and − 25 kV DC-combined voltages. The selected pictures represent pre-breakdown electrical trees. The impulse amplitude has effects on the tree morphology. With the voltage increasing, the depth of charge injection becomes deeper into the dielectric; as a result, the formed charge front is farther from the

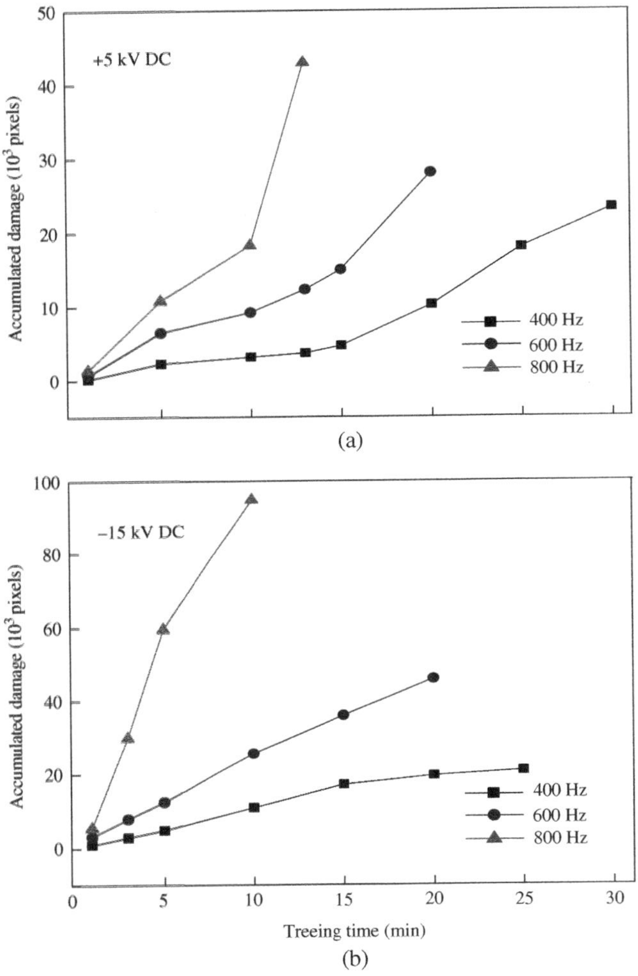

Figure 3.9 Relationship between the accumulated damage and the treeing time of (a) +5 and (b) −15 kV DC and +25 kV impulse-combined voltages with different frequencies. *Source:* Zhu et al. [66]/with permission from IEEE.

needle tip or main branches. An electrical tree can more easily bifurcate behind the charge front [18]. Therefore, with amplitude increasing, there are more side branches growing from the main branches.

Figure 3.13 shows relationship between the tree length and the treeing time of different impulse amplitudes and −25 kV DC-combined voltages. Figure 3.14 shows the relationship between the accumulated damage and the treeing time

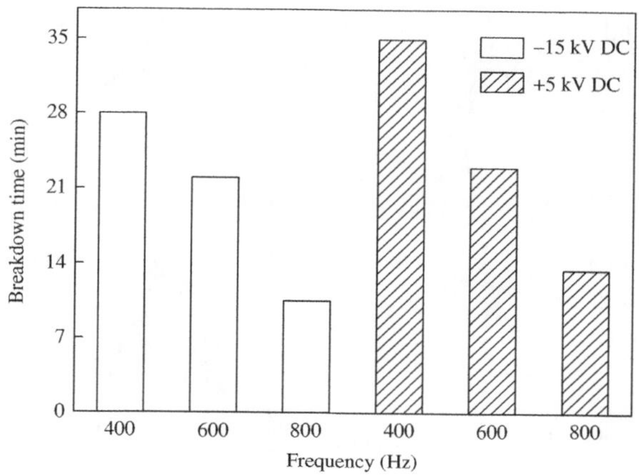

Figure 3.10 Breakdown time of −5 and −15 kV DC and +25 kV impulse-combined voltages with different frequencies. *Source:* Zhu et al. [66]/with permission from IEEE.

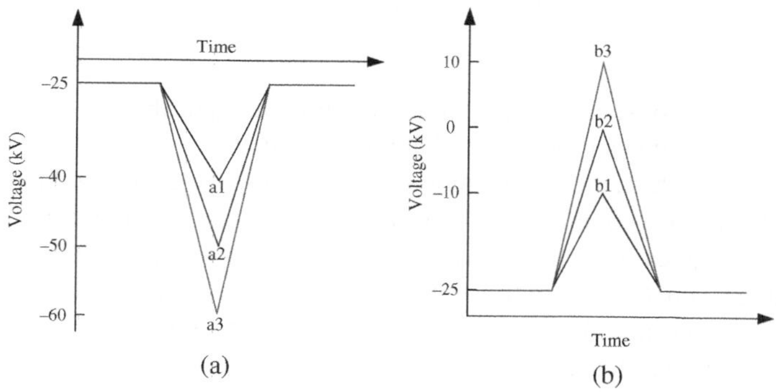

Figure 3.11 Diagram of the impulse superimposed on DC voltage in this subsection. (a) Negative DC superimposed negative pulse. (b) Negative DC superimposed positive pulse. *Source:* Zhu et al. [66]/with permission from IEEE.

of different impulse amplitudes and −25 kV DC combined voltages. As shown in Figures 3.13 and 3.14, with the impulse voltage increasing, the growth rate and the accumulated damage increase. After the impulse voltage is applied, charges are injected into the needle tip. Once a critical field F_c is exceeded, charge carriers change from low-energy states into high-mobility states and fast charge carrier transport occurs. When the self-field of the space charge distribution reduces the local electric field to $|F| \leq F_c$, fast charge carrier transport stops [29]. The number of the injected space charges is [30]

$$Q_s = \pi \varepsilon_r \varepsilon_0 F_c R^2 \left(\frac{V}{V_c} - 1 \right)^2 \tag{3.3}$$

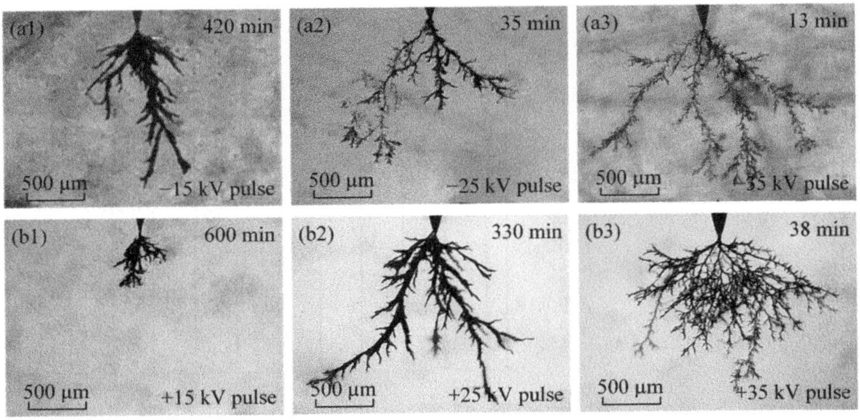

Figure 3.12 Typical electrical tree morphologies of different impulse amplitudes and −25 kV DC-combined voltages. (a) Superimposed negative pulse. (b) Superimposed positive pulse. *Source:* Zhu et al. [66]/with permission from IEEE.

where ε_r is the dielectric constant of the sample, R is the radius of the spherical electrode, and V_c is the voltage at which the field just equals F_c.

Figure 3.15 shows the schematic of the amount of injected space charges under the negative impulse-voltage combined with negative DC voltage. The amount of injected space charge increases at the rising edge, and it decreases at the falling edge [31]. According to Eq. (3.3), with the amplitude increasing, charge variation ΔQ generated by a single impulse increase ($\Delta Q_{+35} > \Delta Q_{+25} > \Delta Q_{+15}$, $\Delta Q_{-35} > \Delta Q_{-25} > \Delta Q_{-15}$). Therefore, the amounts of energy generated by charge trapping and detrapping increase. The number of hot electrons increases, and more polymer chains are broken. Therefore, with the amplitude increasing, the growth rate and accumulated damage increase.

Figure 3.16 shows the breakdown time of different impulse amplitudes and −25 kV DC-combined voltages. No breakdown occurs within 10 hours at +15 kV impulse. As shown in Figure 3.16, with the amplitude increasing, the breakdown time decreases significantly, implying that the influence of the impulse amplitudes particularly large on the breakdown time. The breakdown time of a negative impulse voltage is shorter than that of a positive impulse voltage. It can be concluded that the impulse voltage with the same polarity as that of the DC voltage is more likely to cause breakdown than the impulse voltage with the opposite polarity of that of the DC voltage. The mean effective values of the composite voltages of the negative impulse voltage are larger than the composite voltages of the positive impulse voltage. According to Eq. (3.1), with electric field E increasing,

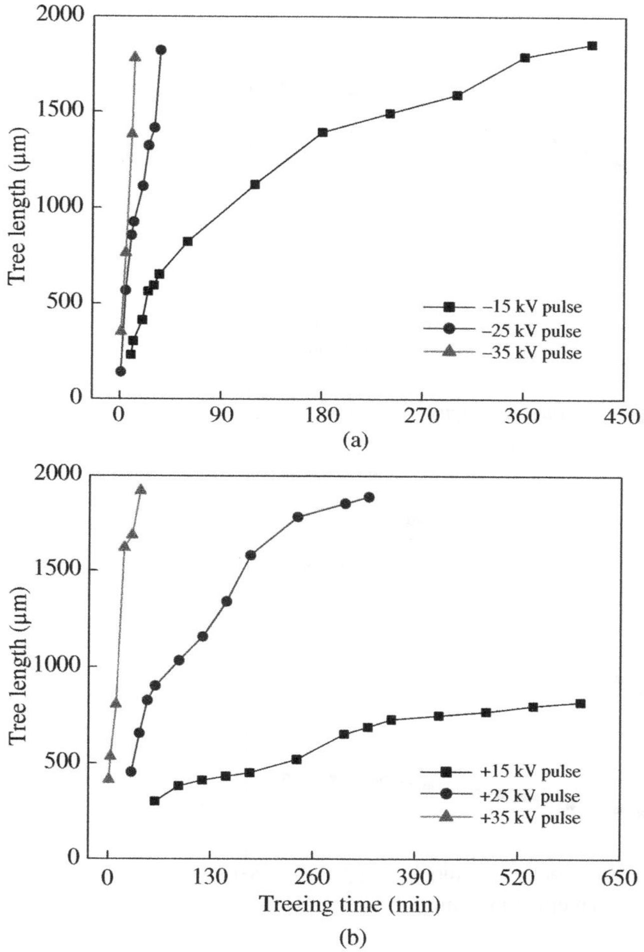

Figure 3.13 Relationship between the tree length and the treeing time of different impulse amplitudes and −25 kV DC-combined voltages. (a) Superimposed negative pulse. (b) Superimposed positive pulse. *Source:* Zhu et al. [66]/with permission from IEEE.

N_A increases. Therefore, the area destroyed by the avalanche discharges increases. Note that for the +35 kV impulse, a voltage polarity change occurs. Therefore, the electric field also increases significantly with the voltage changing from negative to positive [28]. However, the breakdown time of the −35 kV impulse is less than that of the +35 kV impulse. This finding may be because the effect of a larger effective electric field is greater than the effect of polarity changes.

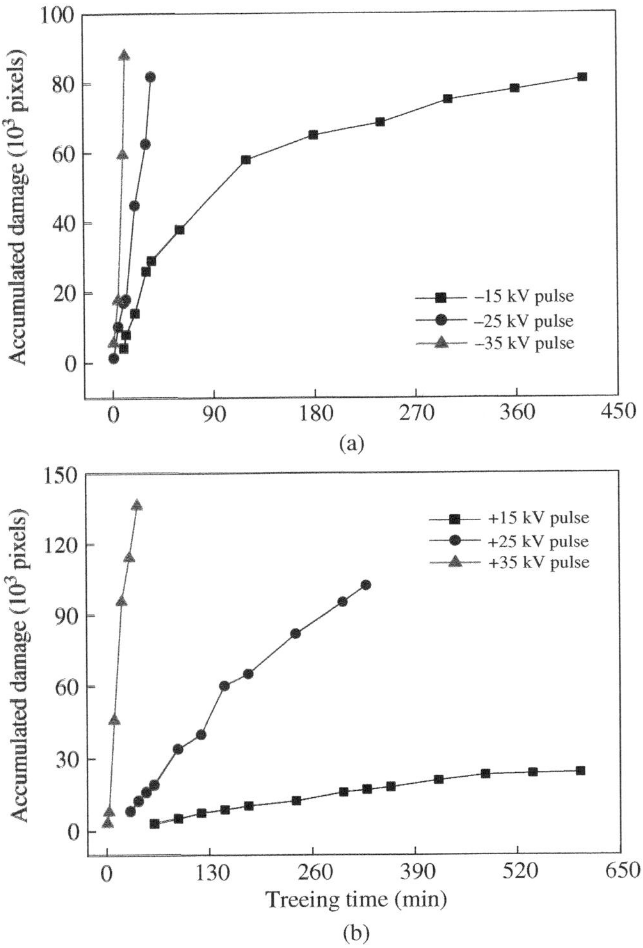

Figure 3.14 Relationship between the accumulated damage and the treeing time of different impulse amplitudes and −25 kV DC-combined voltages. (a) Superimposed negative pulse. (b) Superimposed positive pulse. *Source:* Zhu et al. [66]/with permission from IEEE.

In conclusion, the electrical tree in PP cable insulation is affected by the DC amplitudes, impulse amplitudes, and impulse frequencies. With the increase in the positive DC voltage, the treeing process is accelerated. However, with the increase in the negative DC amplitude, the breakdown time decreases first and then increases. This phenomenon is attributed to the different charge transportation processes under different DC amplitudes. With the impulse amplitude and frequency increasing, the growth rate and accumulated damage increase

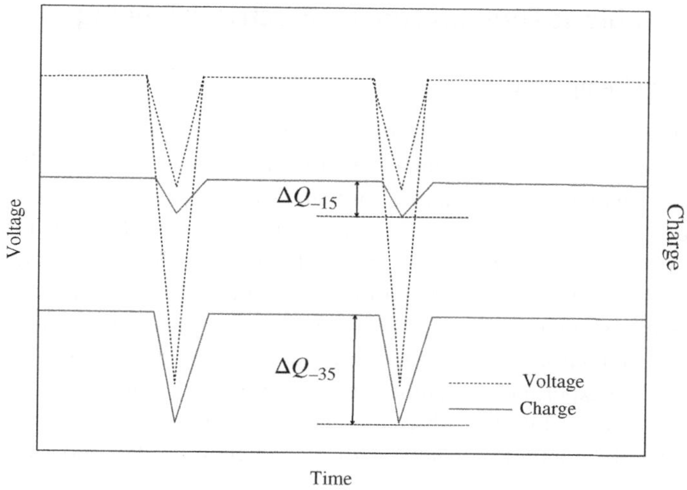

Figure 3.15 Schematic of the amount of injected space charges with the change of impulse voltage. *Source:* Zhu et al. [66]/with permission from IEEE.

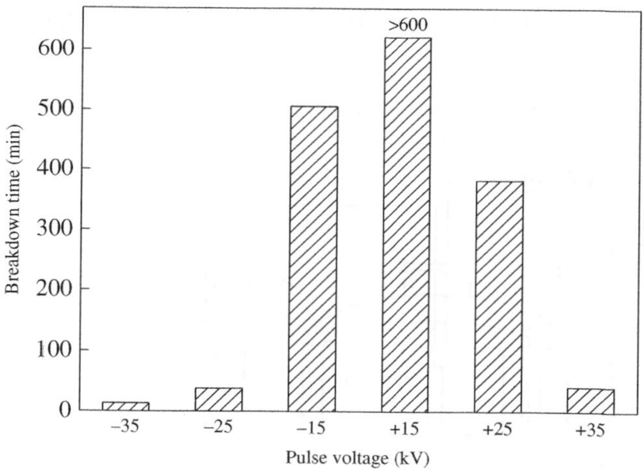

Figure 3.16 Breakdown time of different impulse amplitudes and −25 kV DC-combined voltages. *Source:* Zhu et al. [66]/with permission from IEEE.

significantly. The impulse amplitude and frequency affect significantly the breakdown time. The impulse voltage with the same polarity as that of the DC voltage is more likely to cause breakdown than the impulse voltage with opposite polarity of that of the DC voltage because of the larger effective values of the composite voltages.

3.3 Effect of Ambient Temperature on Electrical Treeing

3.3.1 Effect of Low Temperature

As mentioned earlier, PP insulation has a much higher glass transition temperature than XLPE insulation. Under low temperature, the mechanical characteristics of XLPE and PP will change. The mechanical property of XLPE becomes poorer with the decrease of temperature, and the hardness and elastic modulus of PP increase [32, 33]. This subsection focuses on the electrical treeing characteristics of PP and XLPE insulation under low-temperature environment. The schematic of the measuring system is shown in Figure 3.17. Liquid nitrogen was used to ensure inside temperature ranging from room temperature (30 °C) to liquid nitrogen temperature (−196 °C). The experimental temperature was set to 30, −30, −90, and −196 °C. The test voltage used in this subsection is the repetitive impulse voltage with the amplitudes of 12 and 15 kV, and with a frequency of 400, 800, and 1200 Hz.

Figure 3.18 shows the typical tree morphologies in the PP and the XLPE tested in different temperatures. The impulse amplitude and frequency is 12 kV and 400 Hz, respectively. It is obtained that the tree morphology changes with the temperature in the XLPE but is invariant with the temperature in the PP with the impulse

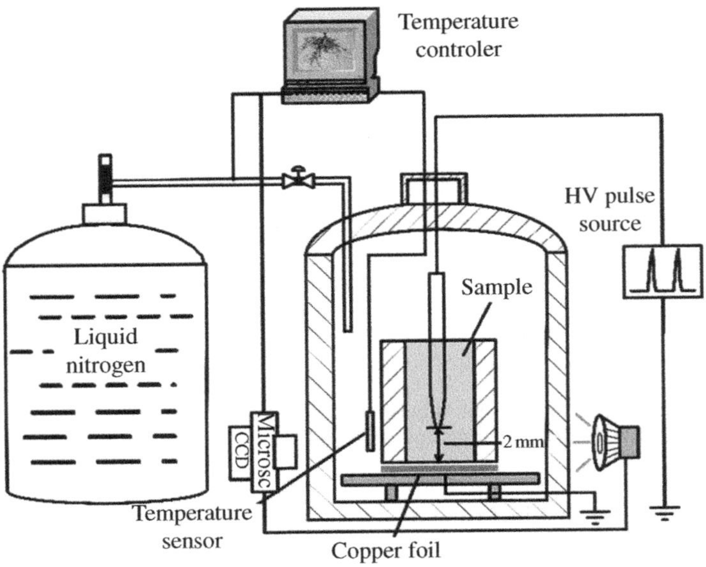

Figure 3.17 Schematic of the measuring system for electrical tree. *Source:* Du et al. [70]/ with permission from IEEE.

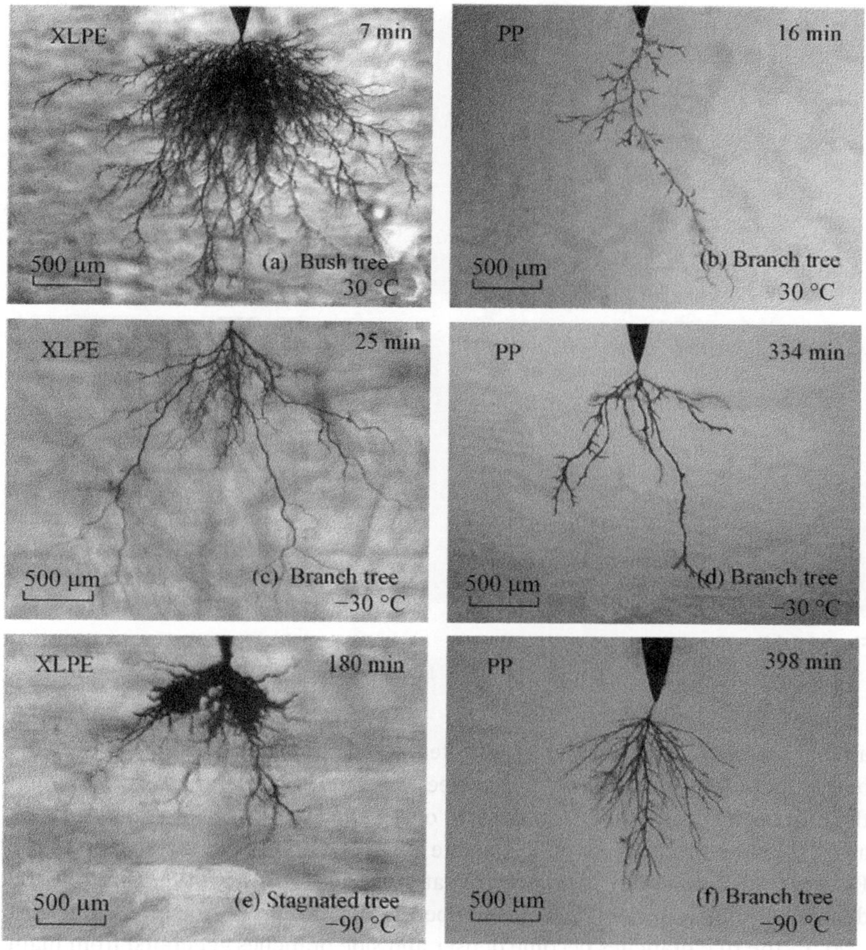

Figure 3.18 Typical profiles of electrical trees in the XLPE and the PP under different temperatures with the impulse voltage of 12 kV and 400 Hz. (a, c, e, g, i) XLPE insulation at difference temperature. (b, d, f, h, j) PP insulation at difference temperature. *Source:* Du et al. [70]/with permission from IEEE.

amplitude of 12 kV. In the XLPE, the bush tree morphology presents at 30 °C and the branch tree morphology presents at −30 and −196 °C. While it is at −90 °C, the stagnated and branch-pine tree are observed. The tree morphology is accounting for 50% of all samples respectively. At 30 °C, bush tree obtained is densely packed structures near the needle tip. At −30 and −196 °C, the branch trees are different from each other. At −30 °C, some channels are overlapped, while there are gaps

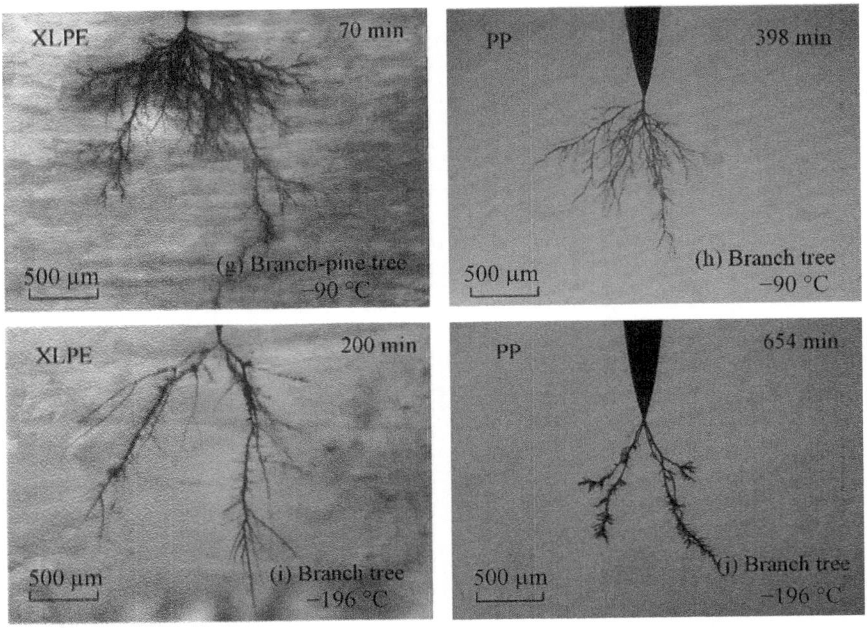

Figure 3.18 (Continued)

among the main branches when the temperature is −196 °C. At −90 °C, the branch-pine tree is double structure tree that one part is branch structure and the other is pine structure. Another tree observed at −90 °C is the stagnated tree, and the length of the tree would stagnate when it reaches a certain extent. In the PP, the tree morphology is branch tree at 30, −30, −90, and −196 °C. While the density of branch trees at different temperatures is different from each other. At −90 °C, the tree branches become denser and side branches generated from main branches are more and longer than these at −30 °C. At −196 °C, the number of main branches is smaller than these at −30 and − 90 °C, and gaps among main branches are larger. At the same temperature, tree morphology in the XLPE appears to be denser and a larger number of small branches are generated from main branches.

Figure 3.19 shows propagation characteristics of trees in the XLPE and the PP. The impulse voltage of Figure 3.19 is the same as that of Figure 3.18. The tree length is referring to the longest tree branch along the direction of the electric field. The fractal dimension can measure the branches complexity. Therefore, the box-counting method is employed, the value of fractal dimension varies between 1 and 2, which is consistent with the previous literatures [34, 35].

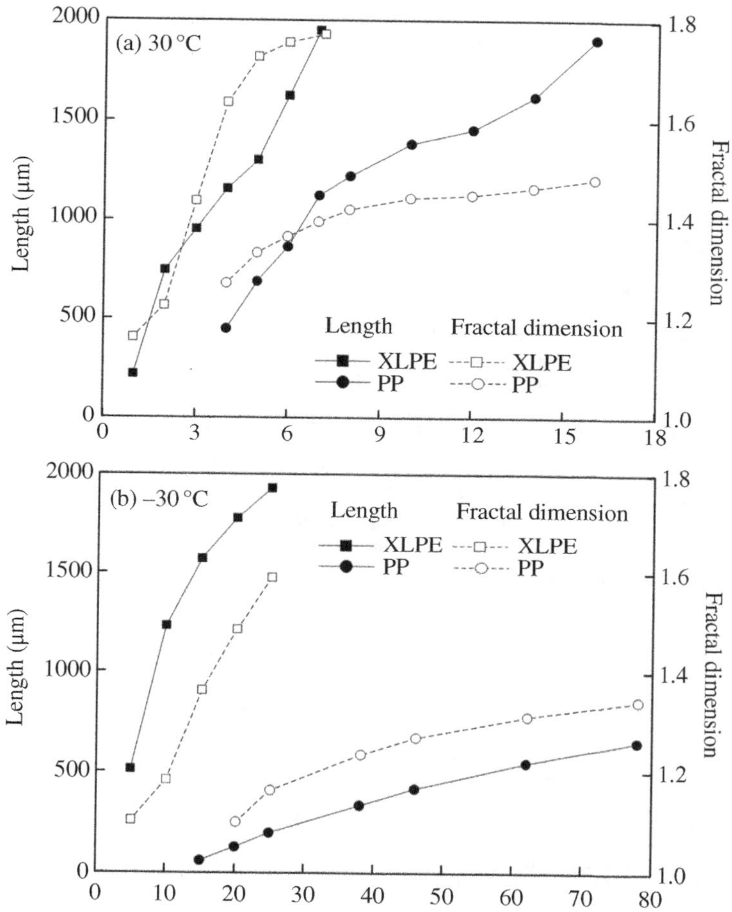

Figure 3.19 Typical propagation characteristics of electrical trees in the PP and the XLPE at (a) 30, (b) −30, (c) −90, and (d) −196 °C. *Source:* Du et al. [70]/with permission from IEEE.

At 30 °C, when the tree length reaches 2000 μm in the XLPE at seven minutes, the tree length of the PP is only 1120 μm. When the treeing time reaches 16 minutes, the tree length of the PP is 1907 μm. Just before the breakdown occurs in the XLPE at 30 °C, the fractal dimension is 1.767 in the XLPE, and the fractal dimension is only 1.392 in the PP at that time, and it increases slowly afterward. At −30 °C, when the tree breakdown occurs in the XLPE at 26 minutes, the tree length of the PP is only 198 μm, and it grows slowly afterward. When the treeing time reaches 78 minutes, the tree length of the PP is 650 μm. Just before the breakdown occurs in

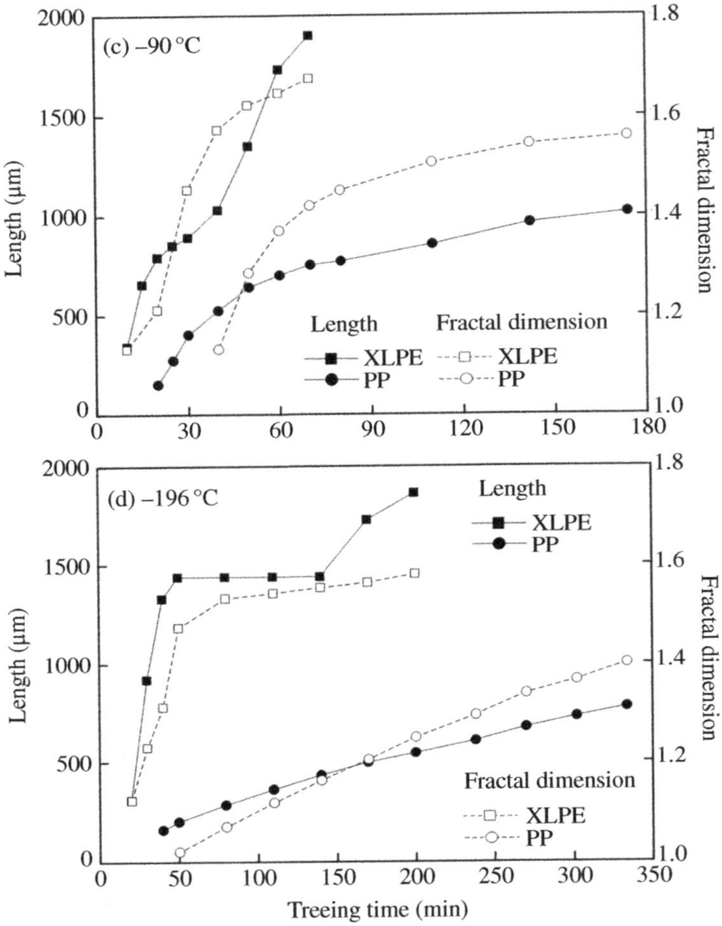

Figure 3.19 (Continued)

the XLPE at −30 °C, the fractal dimension is 1.59 in the XLPE, and the fractal dimension is only 1.16 in the PP at that moment. At −90 °C, there are two types of trees growing in the XLPE. The branch-pine tree in the XLPE has been chosen as the typical tree to compare with the branch tree in the PP. The breakdown of branch-pine tree occurs in the XLPE at 75 minutes, and the branch tree length of PP at −90 °C is only 760 μm at 75 minutes. When the treeing time reaches 174 minutes, the tree length of the PP at −90 °C is only 1016 μm. Just before the breakdown of branch-pine tree occurs in the XLPE at −90 °C, the fractal dimension is 1.67 in the XLPE, and the fractal dimension is only 1.42 in the PP at that time. At −196 °C, when the tree length reaches 2000 μm in the XLPE at 220 minutes, the tree length of the PP is only 570 μm, and it grows slowly afterward. When the treeing time reaches 334 minutes, the tree length of the PP is only 779 μm. Just before the breakdown occurs in the XLPE at −196 °C, the fractal dimension is 1.58 in the XLPE, and the fractal dimension is only 1.25 in the PP at that time. It can be seen from these figures that with the same treeing time, the tree length and fractal dimension in the PP are absolutely smaller than those in the XLPE and the electrical tree grows in a significant slower speed under the lower temperature. Due to the by-products produced during the crosslinking process, more defects would be produced in the XLPE. Charges would be trapped by these defects and accumulated in the bulk which leads to more partial discharges. Therefore, the molecular chains in XLPE are easier to be broken. That's why the tree length and fractal dimension in the XLPE are absolutely bigger than those in the PP with the same treeing time.

In order to investigate the difference between early tree growth in the PP and the XLPE under low temperature, the inception probability of electrical tree is calculated. It is the percentage of the initial tree that gets 30 μm generated for 10 minutes taking the proportion of 20 samples [36]. As shown in Figure 3.20, with temperature varying from −30 to −196 °C, the tree inception probability decreases obviously. Taking the tree inception probability in the PP as an example, it is only 5% and 65% at −196 and −90 °C, respectively, while the value can be up to 90% at −30 °C. It can be analyzed in Figure 3.20 that it is more difficult for the electrical

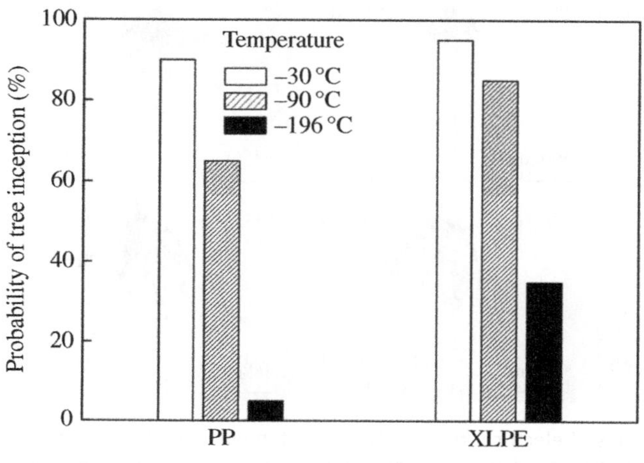

Figure 3.20 Electrical tree inception probability for 10 minutes in the PP and the XLPE under low temperature. *Source:* Du et al. [70]/with permission from IEEE.

trees to be generated from the needle tip at the lower temperature. It can be seen from Figure 3.20 that the probability is 95% in the XLPE at −30 °C, while 90% in PP. At −90 °C, it is 85% in the XLPE, while 65% in PP. At −196 °C, it is 35% in the XLPE, while 5% in the PP. It can be concluded that the inception probability of the electrical tree in the XLPE is higher than that in the PP which means that it is easier for a tree to be generated from the needle tip in the XLPE than the PP.

Electrical tree aging is a complex electrical, physical, and chemical process. Tree initiation and propagation depend on partial discharges activity, oxidization, and chemical decomposition in the micro tree channels [37]. With the change of ambient temperature and impulse frequency, the tree morphology may be completely different. Figure 3.21 shows the tree morphologies in the PP with different impulse frequencies under low temperature. The impulse amplitude is 15 kV and the frequencies are 400, 800, and 1200 Hz. The selected tree images are pre-breakdown to obtain the whole tree morphology images.

As shown in Figure 3.21, the tree morphology varies with different temperatures. Branch tree is observed at −30 and −196 °C and bush tree is observed at

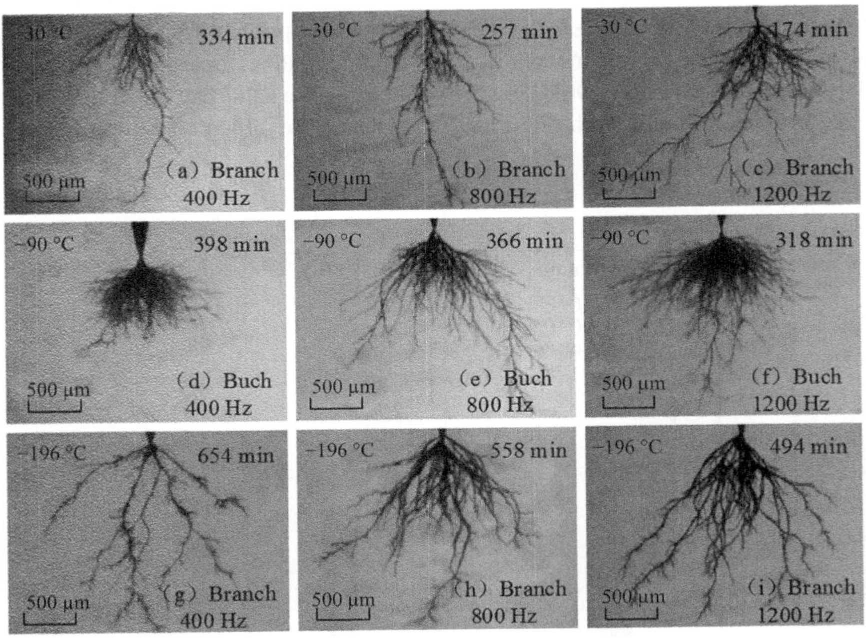

Figure 3.21 Typical profiles of electrical trees with different impulse frequencies under different temperatures with the impulse voltage of 15 kV. (a–c) At −30 °C with 400, 800, 1200 Hz voltage. (d–f) At −90 °C with 400, 800, 1200 Hz voltage. (g–i) At −196 °C with 400, 800, 1200 Hz voltage. *Source:* Du et al. [70]/with permission from IEEE.

−90 °C. At −196 °C, the branch tree is different from that at −30 °C. At −196 °C, the gaps between main branches are larger and more small branches are appeared among the main branches. Moreover, the tree channels become darker and thicker at the lower temperature. At −90 °C, bush tree is densely packed structure. At the same temperature, the tree morphology is invariant with frequency, while the density of trees changes. With higher frequency, the number of main branches and side branches gets larger and more tree channels are overlapping together resulting in denser trees.

When the impulse voltage is stressed, charges are injected into the insulation material from needle tip and the energy releasing from the process of trapping and detrapping of these charges would break molecule chains around the needle tip and the micro tree channels are formed. Besides, the partial discharges would release more charges into tree channels, releasing local heat and high pressures resulting in the further destruction to molecules chains [9]. Higher frequency promotes the partial discharges activity and the amount of space charge produced by partial discharges becomes larger. Moreover, more charges are injected in unit time, colliding the molecular chains. As a result, the number of main branches and side branches gets larger and more tree channels are overlapping together with higher frequency.

Figure 3.22 shows the propagation characteristics of trees in PP with different frequency. The impulse amplitude is 15 kV and the frequencies are 400, 800, and 1200 Hz. As shown in Figure 3.22, the electrical tree grows in an obviously slower speed under the lower temperature. Under the low temperature, the partial discharges are greatly weakened. In addition, the tree growth patterns are different with the variation of temperature and frequency. At −30 °C, the growing trends of tree length are independent of frequency. At first, the tree length increases with a high speed. And, then main branches grow slowly and new channels are generated from main branches and cross with each other. During the last period, the tree grows rapidly again. The tree conducting characteristics can be inferred from the tree channels color. The tree conductivity at −30, −90, and −196 °C is relatively high because of the black tree channels implying heavy carbon left in tree channels, which can act as an equivalent needle electrode [38]. According to the maximum electric field strength equation at the needle tip in the reference [2], the electric field would be higher as the distance between tip and ground electrode becomes shorter. With 800 and 1200 Hz, the growing trends of −90 °C are dependent on that of −30 °C. But, with a frequency of 400 Hz at −90 °C, the tree length increases rapidly at first and then nearly stops increasing, and at last increases again. The tree length nearly stops increasing because the bush area forms during this period. With higher frequency, the charges and partial discharges are enough to form the bush area and promote main branches to grow at the same time. That is why the tree length increases slowly during the second part with a frequency of 800

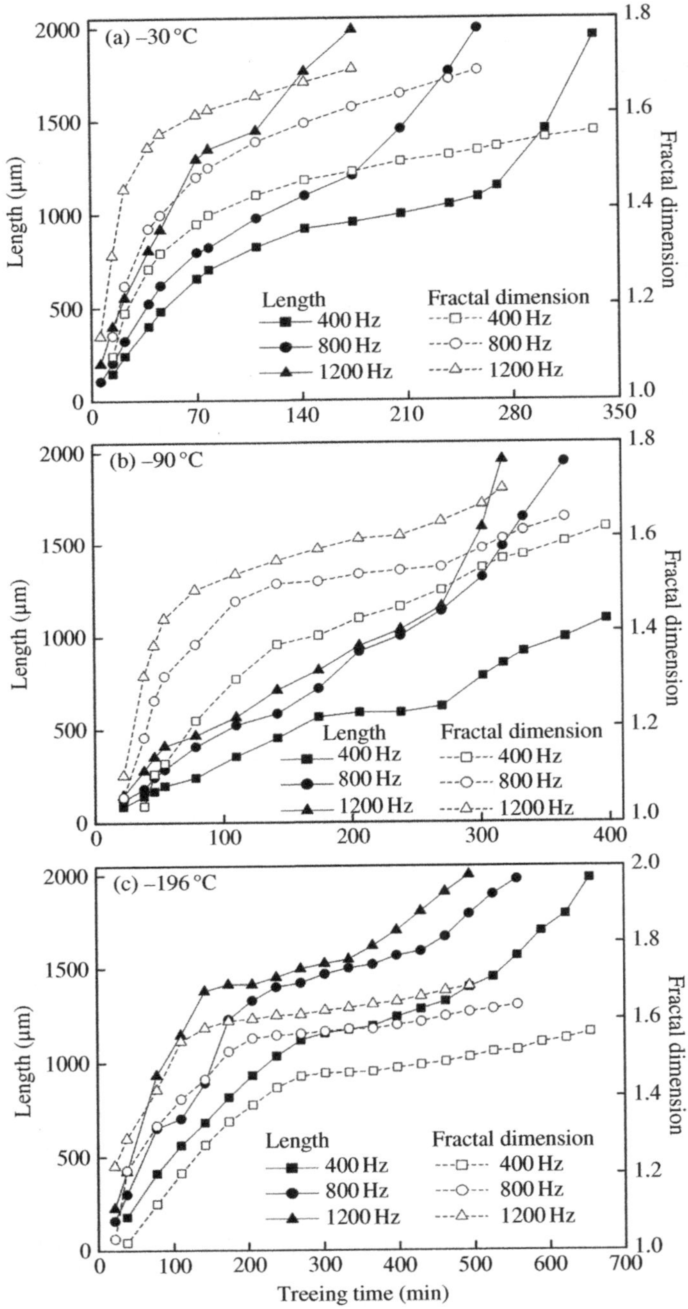

Figure 3.22 Typical propagation characteristics of electrical trees in the PP with the impulse frequency of 400, 800, and 1200 Hz at (a) −30, (b) −90, and (c) −196 °C. *Source:* Du et al. [70]/with permission from IEEE.

and 1200 Hz, but nearly stops increasing with a frequency of 400 Hz. The growing trends of −196 °C with different frequencies are dependent with that of −90 °C with 400 Hz. The partial discharges decrease with the decrease of temperature, and the heat and pressures during discharges are lower, and it is not adequate to promote the tree to grow [31]. When the local heat and high pressures accumulated during the partial discharges, the electric field would be higher as the distance between tree tip and ground electrode becomes shorter. The two factors promote trees to grow rapidly again at −196 °C. The conclusion of the treeing tendency is that the tree length increases in a high speed at first. And, then main branches grow slowly. During the last period, the tree grows rapidly again at −30, −90, and −196 °C.

As shown in Figure 3.22, the length increases at a higher speed with the frequency increasing at the same temperature. As the frequency increases from 400 to 1200 Hz, charges injected from the needle tip become more in unit time, and activity of partial discharges increases which accelerates the degradation of PP molecules. In this case, the electrical tree length tends to be longer with higher frequency.

As shown in Figure 3.22, the fractal dimension changing trends are similar at each condition. During the initial period, the fractal dimension increases rapidly and then continues to increase at a speed slower than the initial period in the later time. During the initial period, the strong partial discharges break molecule chains, and main branches are initialed to form a branch structure rapidly. During the propagation period, although the tree length increases slowly or nearly stops growing, more branches are generated leading to the slow increase of fractal dimension. It is noted that the fractal dimension at −196 °C increases much slower than that at −30 and −90 °C. Because the partial discharges decrease with temperature and the gap between main branches are larger at −196 °C, it is difficult for branches to be generated and cross with each other. During the last period, some slim channels grow rapidly to the ground electrode, which causes the fractal dimension increasing slowly. It can be seen from Figure 3.22 that the fractal dimension becomes larger with frequency increasing. As the frequency increases, numerous charges may collide with polymers, breaking the molecular chain. Thus, the new degradation zone forms. There are more side branches generated from main branches and these small branches may be overlapping with each other during their extension processes. This makes the tree branches packed and the complexity of electrical tree increases accordingly.

3.3.2 Effect of Operating Temperature

This subsection reports the treeing characteristics and breakdown phenomenon of polypropylene (PP) considering different temperatures (varying from 50 to 130 °C) under repetitive impulse voltage. Figure 3.23 shows typical morphologies of

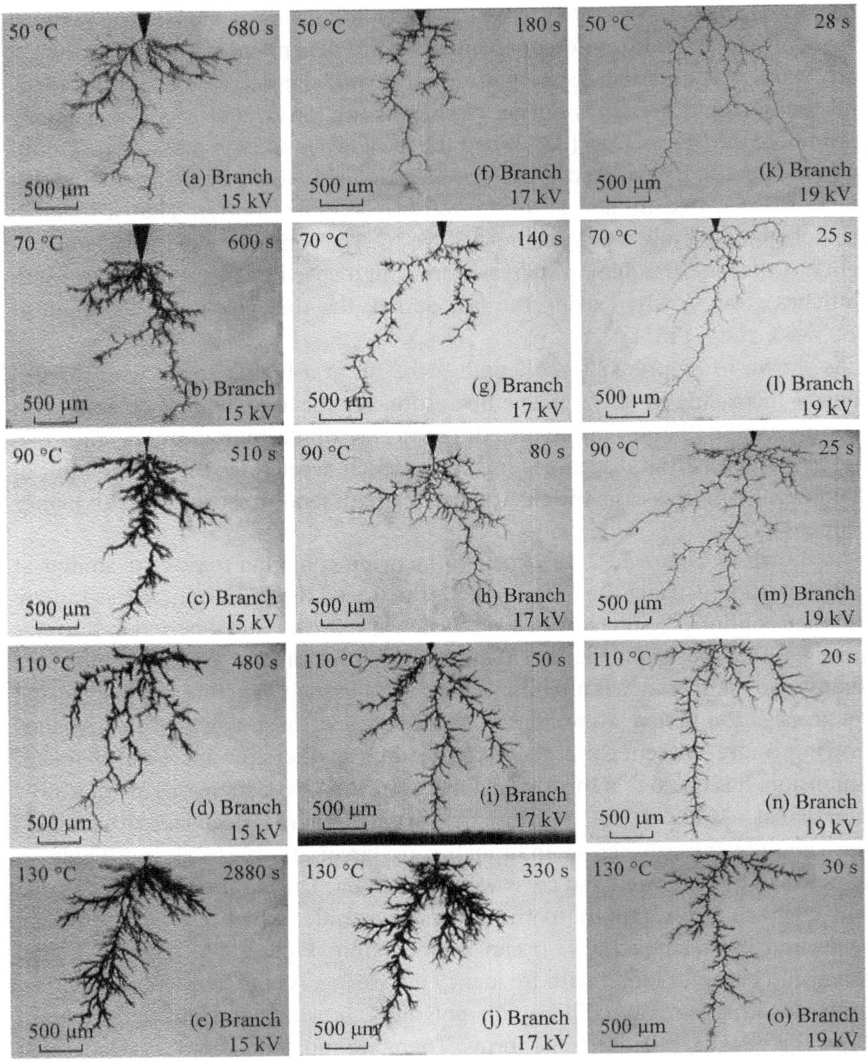

Figure 3.23 Typical morphologies of electrical trees under different conditions. (a–e) Under a voltage of 15 kV at difference temperature. (f–j) Under a voltage of 17 kV at difference temperature. (k–o) Under a voltage of 19 kV at difference temperature. *Source:* Du et al. [54]/with permission from IEEE.

electrical trees under different conditions. The selected tree images are pre-breakdown images to obtain the whole tree morphology images. The treeing time of the pictures in Figure 3.23 is different. The structure of electrical trees is branch tree that does not change with the temperature and impulse amplitude, but the density of electrical trees changes with the temperature. With higher temperature, the density of branches increases obviously and the branches become darker and larger. The results indicate that the temperature affects the tree morphology. While under the same temperature, with the amplitude increasing, more side branches extend toward all directions around the needle electrode.

Figure 3.24 shows the relationship between the inception probability and the temperature with different impulse amplitudes at five seconds. Calculating the inception probability when the applied time of the impulse voltage to the specimen is five seconds. With the temperature increasing from 50 to 130 °C, the inception probability varies from 5% to 45% in 15 kV. The inception probability varies from 30% to 75% in 17 kV. The inception probability varies from 55% to 95% in 19 kV. It indicates that the inception probability is dependent on the temperature.

After the impulse voltage is applied to the needle tip, the charges will be injected from the needle tip into the polymer. During the process of trapping, detrapping, and recombination of the charges, when the charges are transferred from the high-energy state to the low-energy state, excess energy is released by non-radiative effect and hot electrons are produced, which lead to the break of the molecular chain and produce the free radicals. The large molecule free radical initiates the chain reaction and the formation of new traps. The low-density areas are formed. In the areas, the partial discharges happen and partial high temperature and high

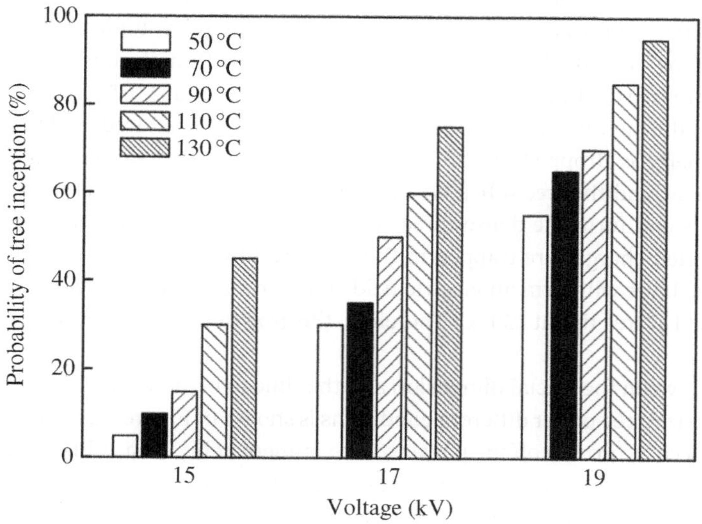

Figure 3.24 Relation between the inception probability and the temperature with different impulse amplitudes at five seconds. *Source:* Du et al. [54]/with permission from IEEE.

gas pressure are generated [36] which can produce thermal and mechanical damage to the samples. The damage accumulated to a certain extent leads to the initiation of electrical trees. With the temperature increasing, the thermal motion of the molecular chain is strengthened [23]. Therefore, partial discharges become stronger. More partial high temperature and high gas pressure are generated in the unit time. Therefore, the tree initiation probability increases with the temperature increasing from 50 to 130 °C. The tree inception probability increases with the amplitude increasing from 15 to 19 kV. It is noted that tree inception probability is 45% with 15 kV, while it changes to 75% and 95% with the amplitude of 17 and 19 kV at 130 °C. The injected charges will obtain higher energy at higher impulse amplitude, which generates severe damage around the needle tip.

Figure 3.25 shows the relation between the tree length and the treeing time under different temperatures. The treeing time of Figure 3.25 with different impulse amplitudes is in different scales. The impulse amplitude is, respectively, 15, 17, and 19 kV. With the temperature increasing from 50 to 110 °C, the growth rate increases. While the growth rate of 130 °C is lower than that of 110 °C.

The DSC curve is shown in Figure 3.26. An exothermal peak in the heat flow-temperature curves is observed during the cooling process, corresponding to the crystallization of the polymer chains, which is located at 122 °C. An endothermal peak in the heat flow-temperature curves is observed during the heating process, corresponding to the crystallites melting point of 148 °C. As is shown in Figure 3.26, with the temperature increasing from 110 to 130 °C, the crystal starts to melt. Therefore, the amorphous phase increases, contributing to the increase of free volume at 130 °C. It is easier for electrons to accelerate in free volume to form more "hot electrons" at 130 °C. These hot electrons collide with molecule chains and generate more tree channels, contributing to the denser area at 130 °C. Meanwhile, because of the melting of the crystal, the elastic modulus is reduced, and the mechanical property is dropped at 130 °C, the charges are easier to break the molecular chain around the tree tube, which contributes to generating the interconnect channels and form the denser area. With the formation of denser area, some of the injected charges are trapped in the matrix and well distributed [39]. The space charge layer will form an electric field shield, resulting in the decrease of the electric field at tree tips at 130 °C. Therefore, the growth rate of 130 °C is the slowest.

The relation between the fractal dimension and the time to breakdown (TTB) of breakdown electrical tree under different conditions is shown in Figure 3.27. Once the breakdown occurs, turning off the power source immediately. Then, taking the photo of the electrical trees with the camera and calculating the fractal dimension and accumulated damage. These results are the fractal dimension and accumulated damage of the breakdown electrical tree. The TTB tends to be shorter with the increase of impulse amplitude under the same temperature because high

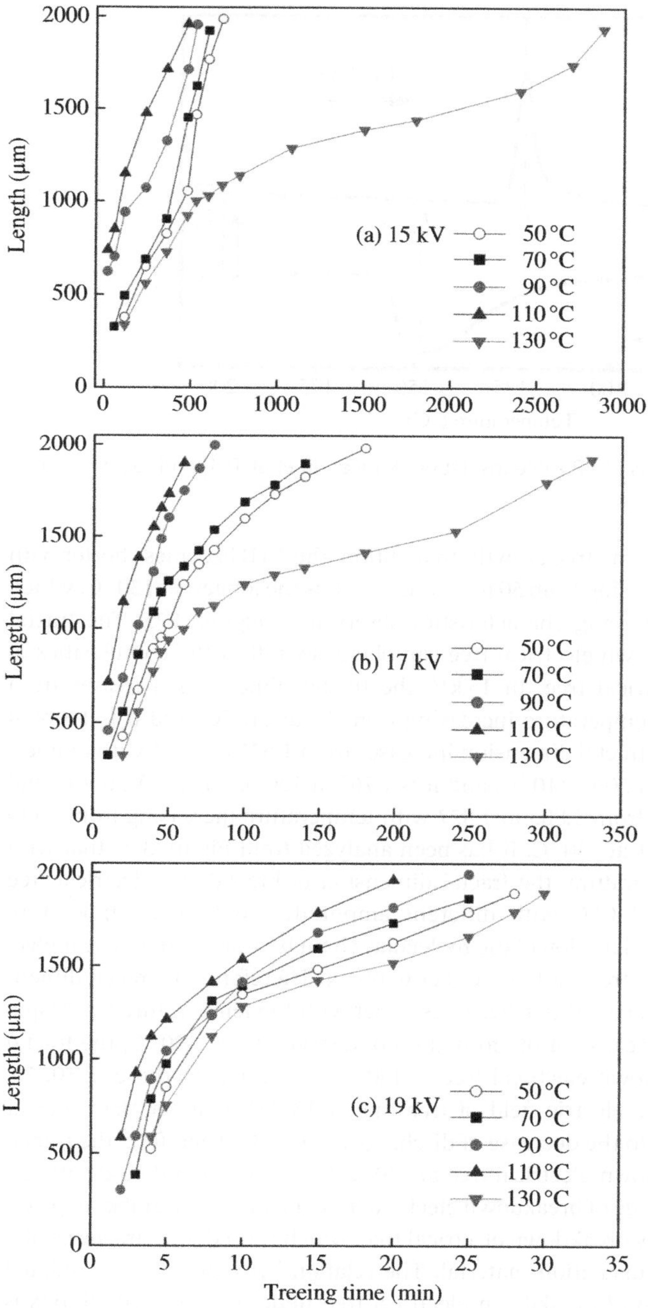

Figure 3.25 Relation between the tree length and the treeing time under different temperatures. *Source:* Du et al. [54]/with permission from IEEE.

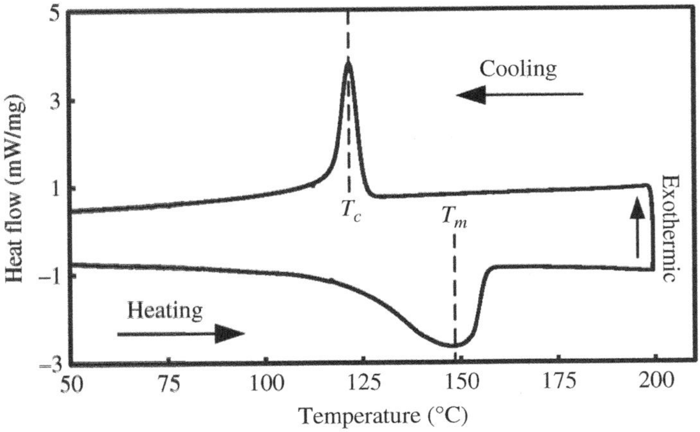

Figure 3.26 DSC curves of PP cable insulation. *Source:* Du et al. [54]/with permission from IEEE.

amplitude accelerates the tree growth. In addition, the TTB becomes shorter with the temperature increasing from 50 to 110 °C, and it is the longest at 130 °C, which are consistent with treeing characteristics observed in Figure 3.25. The fractal dimension of breakdown electrical tree can obviously reflect the complexities of the breakdown electrical tree. In 15 kV, the fractal dimension increases from 1.789 to 1.912 with temperature increasing from 50 to 110 °C, and it is 1.889 at 130 °C. In 17 kV, the fractal dimension increases from 1.675 to 1.951 with temperature increasing from 50 to 110 °C, and it is 1.767 at 130 °C. In 19 kV, the fractal dimension increases from 1.803 to 1.827 with temperature increasing from 50 to 110 °C, and it is 1.811 at 130 °C. It has been analyzed from Figure 3.27 that with the increase of temperature, the fractal dimension of breakdown electrical tree increases except at 130 °C with different amplitudes. With the temperature increasing, the thermal motion of the molecular chain becomes active. The molecular chain around the tree walls are easier to break. Therefore, the fractal dimension of breakdown electrical tree becomes larger with the temperature increasing from 50 to 110 °C. Because of the formation of denser area at 130 °C, the fractal dimension of breakdown electrical tree at 130 °C is larger than those at 50, 70, and 90 °C. While the electric field of tree tips at 130 °C is larger than that at 110 °C, contributing to the decrease of discharges at 130 °C. Therefore, the fractal dimension of breakdown electrical tree at 130 °C is smaller than that at 110 °C.

Accumulated damage of breakdown electrical tree refers to the number of pixels in the area covered by breakdown electrical tree, which is used to characterize the damage area of the insulation material. The relation between the accumulated damage and the TTB of breakdown electrical tree under different conditions is shown in Figure 3.28. In 15 kV, the accumulated damage decreases with

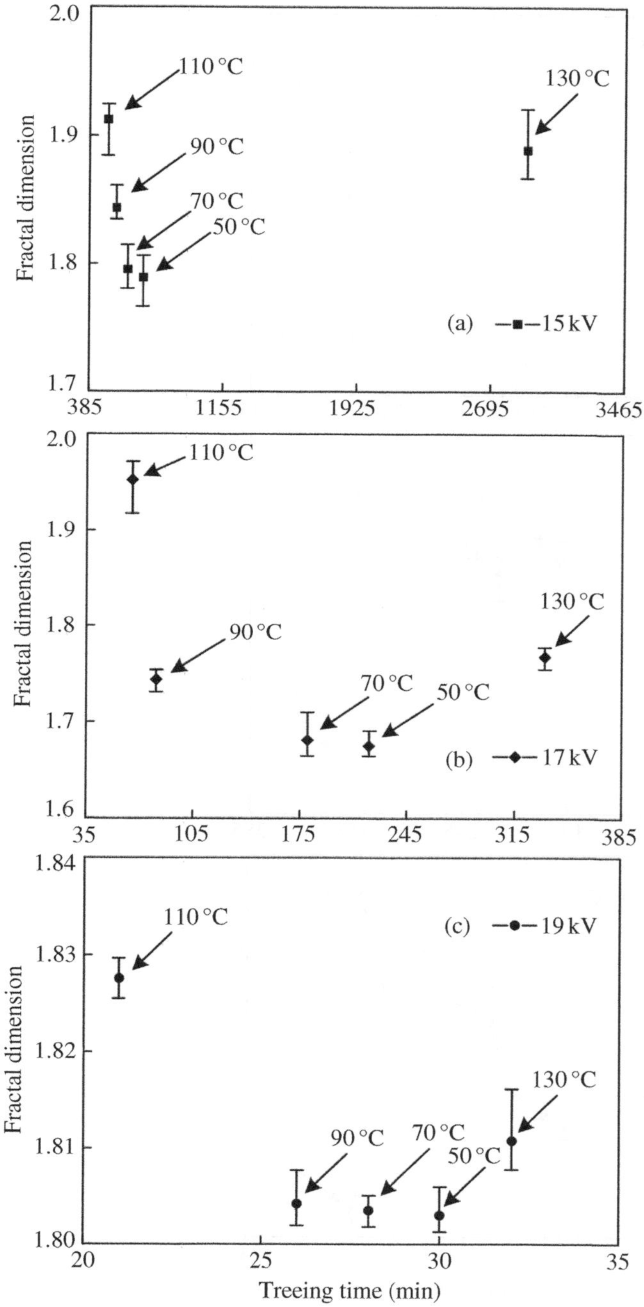

Figure 3.27 Relation between the fractal dimension and the TTB of breakdown electrical tree under different conditions. *Source:* Du et al. [54]/with permission from IEEE.

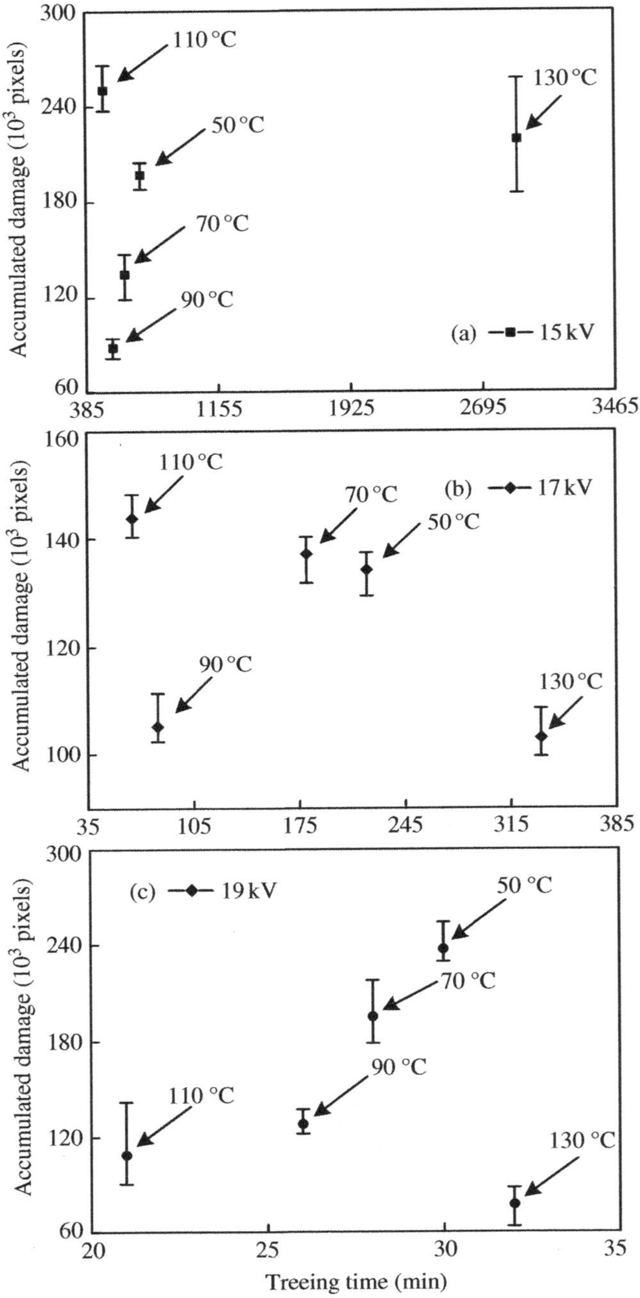

Figure 3.28 Relation between the accumulated damage and the TTB of breakdown electrical tree under different conditions. *Source:* Du et al. [54]/with permission from IEEE.

temperature increasing from 50 to 90 °C. It is smaller at 90 °C than 110 °C, and it is smaller at 130 °C than that at 110 °C. In 17 kV, the accumulated damage is larger at 70 °C than 90 °C. It is smaller at 90 °C than 110 °C, and it is smaller at 130 °C than 110 °C. In 19 kV, the accumulated damage at 50 °C is the highest, and it decreases with temperature increasing to 130 °C. It can be concluded that the accumulated damage of breakdown electrical tree is independent of high temperature.

The typical breakdown morphologies of electrical trees under different conditions are shown in Figure 3.29. The breakdown channel widths at the same temperature are different, but in general, the difference between them is small. Because the breakdown produces a lot of heat, PP around the needle tip is also carbonized to form the breakdown channels. As a result, the position of the needle tip is not displayed in the photo of the breakdown electrical trees. It can be observed that the widest breakdown channel is the breakdown channel at 110 °C. The breakdown channel at 70 °C is wider than that at 90 °C. The width of breakdown channel at 90 °C is almost the same as that at 130 °C. The results indicate that the breakdown channels width is independent of high temperature.

To further investigate the micro-structural difference between the original sample and the breakdown channels, they were, respectively, observed by SEM (Hitachi S4800). To investigate the elemental difference of the materials, EDAX analysis was also employed in the scanning area marked with red outlines. Figure 3.30a shows the surface morphology and EDAX result of the original sample. Figure 3.30b and c show the surface morphology and EDAX results of the typical breakdown channels. The magnitude of the applied voltage in Figure 3.30b is 17 kV and the temperature is 70 °C. The magnitude of the applied voltage in

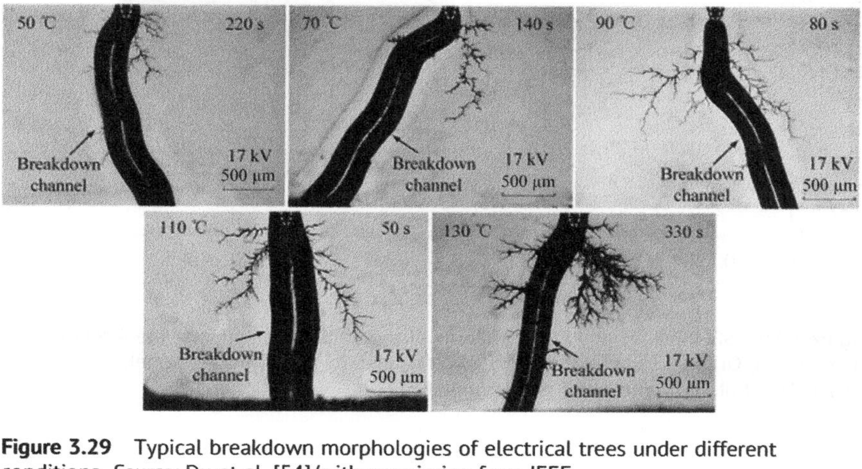

Figure 3.29 Typical breakdown morphologies of electrical trees under different conditions. *Source:* Du et al. [54]/with permission from IEEE.

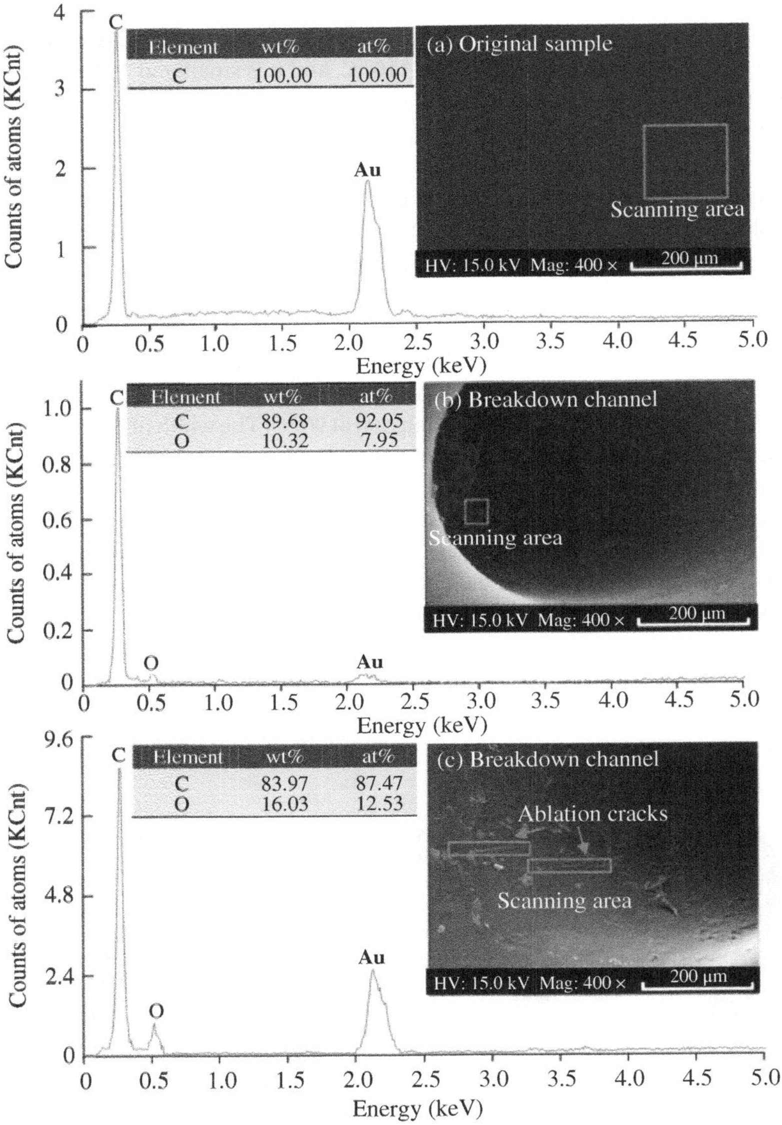

Figure 3.30 SEM images and EDAX analysis of the original sample and breakdown channels. (a) Original sample. (b) Breakdown channel. (c) Breakdown channel. *Source:* Du et al. [54]/with permission from IEEE.

Figure 3.30c is 17 kV and the temperature is 130 °C. The morphologies of particles in Figure 3.30b and c are different. As shown in Figure 3.30b, the morphology of observed particles is agglomerated, while in Figure 3.30c, the morphology of observed particles is dispersed and clear ablation cracks are found. It is noticed that because of the restriction of the EDAX analysis, the hydrogen atom of the samples could not be searched. As shown in Figure 3.30a–c, the carbon atom and the aurum atom are observed in the original sample and the breakdown channels. The aurum atom existing in the samples surface comes from the metallization process before SEM observation and is treated as impurity; therefore, the content of aurum atom is not being considered. Comparing with the analysis result of the original sample, the oxygen element is introduced into the breakdown channels. Oxygen elemenet accounts for 10.32 wt% and 7.95 at.% in Figure 3.30b and 16.03 wt% and 12.53 at.% in Figure 3.30c.

Figure 3.31a and b shows the ATR-IR spectrum of the original sample and the typical breakdown channel. The magnitude of the applied voltage in Figure 3.31 is 17 kV and the temperature is 70 °C. Absorption peaks of the common group are

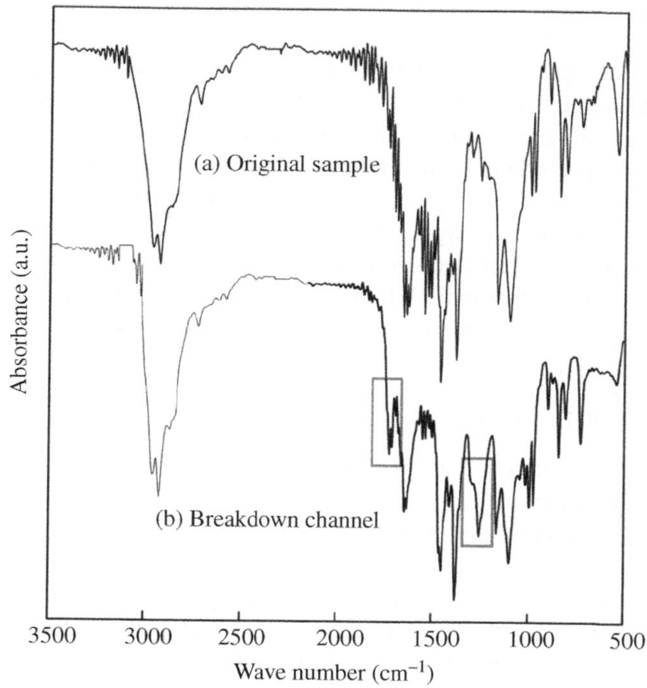

Figure 3.31 ATR-IR spectra of the (a) original sample and (b) breakdown channel. *Source:* Du et al. [54]/with permission from IEEE.

Table 3.1 Absorption peaks of common group. *Source:* Du et al. [54]/ with permission from IEEE.

Wave number (cm^{-1})	Group
3000–2850	C—H stretching vibration
3100–3010	C—H stretching vibration
1465–1340	C—H bending vibration
1000–675	C—H out of plane bending vibration
880–680	C—H out of plane bending vibration
1300–1000	C—O stretching vibration
1750–1700	C=O stretching vibration
1720–1706	C=O stretching vibration

shown in Table 3.1. Compared with the ATR-IR spectrum of the original sample, there are new absorption peaks at 1730 and 1253 cm^{-1} in the ATR-IR spectrum of the breakdown channel. These peaks are attributed to the stretching vibration of C=O groups and the stretching vibration of C—O groups [40–42]. These results show that the breakdown channel contains C=O and C—O groups, which do not exist in the original sample. In an oxygen-free environment, injected charges lead to the break of the molecular chain, resulting in the formation of small voids and electrical trees after long-time application of impulse voltage. In an oxygen-rich environment, oxygen plays an important role in the treeing process. It comes from air during the polymer production process and is deeply trapped in polymer matrix. When free radicals react with oxygen, autoxidation process begins. The free radical reacts with oxygen to produce hydroperoxide, and unstable hydroperoxide decomposes and reproduces the free radical. Repetition of this process breaks the polymer chain and forms the large voids and electrical tree channels [43].

$$\mathrm{RH} \rightarrow \mathrm{R}^* + \mathrm{H}^*; \mathrm{R}^* + \mathrm{O}_2 \rightarrow \mathrm{ROO}^*$$

$$\mathrm{ROO}^* + \mathrm{RH} \rightarrow \mathrm{ROOH} + \mathrm{R}^*; \mathrm{ROOH} \rightarrow \mathrm{RO}^* + \mathrm{OH}$$

Also, during this process, the C=O and C—O groups are formed, which are observed by the ATR-IR spectrum.

In conclusion, high temperature and impulse amplitude have great effects on the electrical tree morphologies, the inception probability, and tree growth process. The tree growth rate increases with temperature elevating from 50 to 110 °C. While it is smaller at 130 °C than at 110 °C. The tree growth rate and fractal dimension increase with the increase of impulse amplitude under the same high temperature. It is found that the breakdown channels width and accumulated damage of breakdown electrical tree are independent of high temperature. While, with the increase of temperature, the fractal dimension of breakdown electrical tree increases first, and then decreases. Agglomerated and dispersed particles are observed in the breakdown channels by SEM. It is also found that the oxygen element is introduced into the breakdown channel comparing with the original sample, which contains C=O and C—O groups by EDAX and ATR-IR.

3.4 Effect of Bending Deformation on Electrical Treeing

3.4.1 Effect of Bending Deformation

High-voltage cables require good mechanical properties to withstand various extreme installation environments, such as high drop terrains and the laying process of submarine cables through towing pipes [15, 44]. The occurrence of large bending deformations in cables, due to reduced turning radius, can lead to the concentration of mechanical stress and the development of microcracks. Consequently, partial discharges are more likely to occur and form electrical tree channels [16, 45]. However, the existing research primarily applies mechanical stresses in a single direction, either parallel or perpendicular to the samples, which fails to capture the nonuniform and nonlinear stress distribution present in actual cables [17, 46]. Recent studies have attempted to simulate the nonuniform strain distribution in cables by applying stresses to circular slices, investigating electrical tree characteristics under tensile and compressive stresses separately [18]. Yet, under bending operation of the cable, the outer cable layer experiences stretching while the inner layer undergoes compression, involving the coexistence of tensile and compressive force [19]. However, the electrical tree characteristics under such conditions remain unclear.

The poor mechanical properties of PP cable insulation, such as low toughness and high brittleness, constrain its utilization under bending conditions [47]. Thus, toughening modification of PP is pivotal, with current research focusing on introducing elastic phases into PP to enhance mechanical properties [48]. Studies have demonstrated that blending PP with isostructural PP (sPP) can enhance insulation toughness and reduce the growth rate of electrical trees [49]. In addition, research has indicated that impact PP copolymers (IPC) exhibit significantly enhanced mechanical strength and resistance to water trees compared to PP [50]. Research findings indicate that PP-based copolymers exhibit mechanical flexibility, limited space charge injection, and elevated breakdown strength [51]. Therefore, it is imperative to investigate whether the introduction of an elastic phase under the bending conditions of cables will continue to suppress the growth of electrical trees.

In this paper, the effect of bending deformation on the initiation, growth, and morphology of PP electrical trees is investigated. Additionally, PP blended with thermoplastic elastomer and PP copolymerized with ethylene are prepared. By employing finite element simulation and micro-morphological observation techniques, the effects of elastic phase introduction on the mechanical properties of PP and electrical tree growth under bending deformation are evaluated.

A platform is designed for electrical tree experiments conducted under bending conditions, as illustrated in Figure 3.32a. The thickness of electrical tree needle samples is 1 mm. According to the specifications outlined in the national standard GB/T 31489.1-2015, a voltage of −8 kV DC is applied, superimposed with a pulse voltage characterized by a frequency of 200 Hz and an amplitude of +17.6 kV [52]. The sample is installed on the mechanical bending device, and the width of the jaw can be adjusted by manipulating the nut. By establishing a rectangular coordinate system to delineate the bending curve, the curvature radii (R) of the needle electrode tail at coordinates (x, y) can be computed using Eq. (3.4):

$$R = \frac{1}{C} = \frac{\left(1 + \left(\frac{dy}{dx}\right)^2\right)^{\frac{3}{2}}}{\frac{d^2y}{dx^2}} \tag{3.4}$$

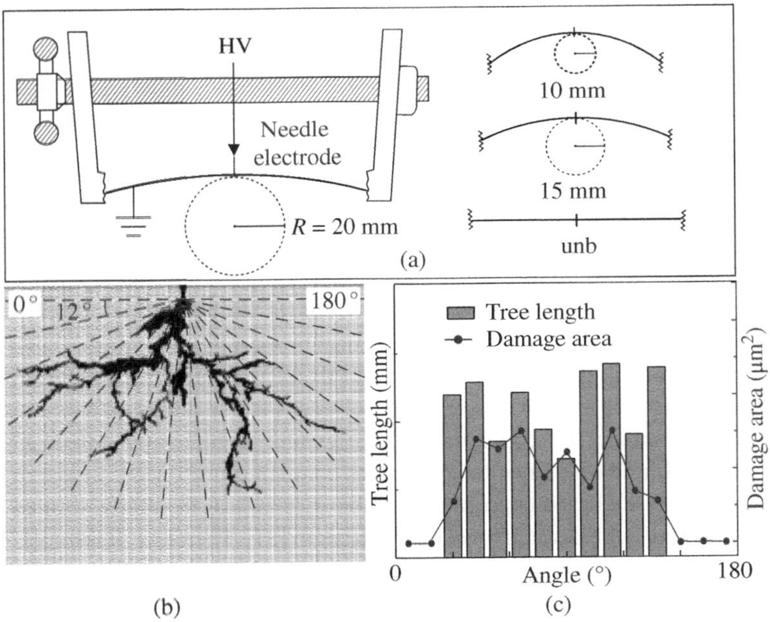

Figure 3.32 (a) Experimental setup of PP samples under bending deformation conditions. (b) Pixel recognition and segmented region. (c) Tree length and damage area. *Source:* Du et al. [71]/with permission from IEEE.

where C is the curvature. Based on the GB 50168-2018, the minimum bending radius of high-voltage single-core cables is 15–20 D [53]. So, the bending radius is set to 20, 15, and 10 mm for this experiment. Additionally, unbent samples (abbreviated as "unb") are selected for comparison. The experimental temperatures are set to 30, 70, and 90 °C to investigate the influence of temperature on the electrical tree phenomenon under bending conditions.

To compare the differences in the electrical tree growth morphologies under varying levels of bending, we utilized a specific characterization method outlined in Figure 3.32b. Initially, the original electrical tree image is processed using MATLAB, transforming it into a binarized graph. Subsequently, a polar coordinate system is established, where the pole is specified as the needle tip position and the direction of the polar axis is orientated horizontally to the left. A statistical range of 180° is defined in a counterclockwise direction, with an interval width set at 12°. By traversing the image pixels using a double loop, we counted the number of pixels with a grey value of 0 within each interval and measured the length of the polar path. This process allowed us to compute the distribution of the damage area and the length of the electrical tree across different regions, as depicted in Figure 3.32c.

Gaussian fit for the partitioned electrical tree length is performed. The fitting equation is as follows:

$$L_{\text{tree}} = L_{\text{max}} e^{-(x - x_c)^2/(2\sigma^2)} \tag{3.5}$$

where L_{max} is the amplitude, x_c is the peak center, and σ is the standard deviation. σ can reflect the dispersion of the data. The excess kurtosis (K) serves as a crucial metric for Gaussian curves, reflecting the steepness or smoothness of the data, calculated by the following equation:

$$K = \frac{1}{n}\sum_{i=1}^{n}\left[\left(\frac{L_i - \mu}{\sigma}\right)^4\right] - 3 \tag{3.6}$$

where μ is the mean. A K value of 0 indicates a normal distribution, with increasing K correlating to a steeper curve.

The mechanical tensile property is tested by the electronic tensile testing machine. A dumbbell-type sample, characterized by a uniform thickness of 4 mm, is deliberately chosen to ensure consistency in the assessment. Tensile speed controls at 20 mm/min. The breaking elongation (ε) is calculated as follows:

$$\varepsilon = \frac{L_1 - L_0}{L_0} \times 100\% \tag{3.7}$$

where L_1 is the length of PP at break, mm. L_0 is the effective length before stretching, mm. L_0 is 16 mm.

Impact strength is measured by cantilever beam impact testing machine (YH-YW-60). A strip sample with an A-notch on one side, penetrating to a depth of 0.2 mm, is prepared. The sample dimensions are controlled at 80 × 10 × 4 mm. Repeat the experiment five times and take the average value as the result.

The multiphysics field simulation software COMSOL multiphysics is used to build finite-element model of the electrical tree sample. In the solid mechanics module, horizontal inward symmetric loads are applied to the two side ends with the constraint of hinged support at both ends. Buckling analyses (incorporating geometrical nonlinearities) of materials are performed to obtain steady-state solutions for buckling deformations as well as stress and strain distributions under corresponding loads.

The electrical tree morphologies of PPH at different temperatures with different curvature radius are shown in Table 3.2. At room temperature, the unbent electrical tree shows a typical branch shape with discernible twig structures, as shown in Table 3.2(a1). Notably, the branches are gradually orientated toward the ground electrode as the radius decreases. At a curvature radius of 15 mm, the electrical tree grows straight along the direction of the electric field with minimal branching evident, which is defined as a straight-stick shape, as shown in Table 3.2a3. As the temperature increases to 70 °C and the radius is greater than 15 mm, a bush-branch form electrical tree can be observed, whose root is bush like and top is

Table 3.2 Electrical tree morphologies of PPH. *Source:* Du et al. [71]/with permission from IEEE.

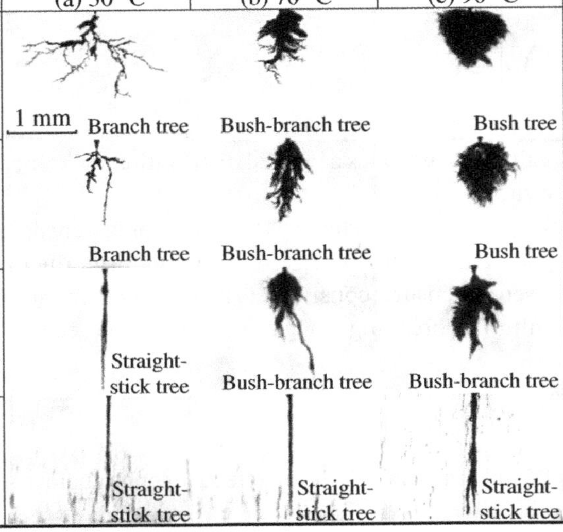

R (mm)	Electrical trees (a) 30 °C	Electrical trees (b) 70 °C	Electrical trees (c) 90 °C	Group
unb	Branch tree	Bush-branch tree	Bush tree	1
20	Branch tree	Bush-branch tree	Bush tree	2
15	Straight-stick tree	Bush-branch tree	Bush-branch tree	3
10	Straight-stick tree	Straight-stick tree	Straight-stick tree	4

branch like, as shown in Table 3.2b1–b3. Furthermore, the electrical tree trunks are densely packed in bush shape under a larger curvature radius at 90 °C.

The electrical trees finally develop to straight-stick morphology at different temperatures when the curvature radius is small enough and many linear microcracks can be observed on the surface of the samples, as shown in Table 3.2a4, b4, c4. It is due to the direction of the needle as the "ridge" of the bent sample, resulting in a concentration of mechanical stresses, which may exceed the yield strength of the material by applying a relatively small stress. As a result, vertically aligned defects are created and the electrical tree grows preferentially along the defects.

Figure 3.33a illustrates the inception time of electrical trees, and the error bands describe the 95% confidence intervals. It is defined that an electrical tree is considered initiated when a clearly observable branch emerges within the insulation sample (i.e. the length or width of the branch is ≥10 μm). The initiation time is

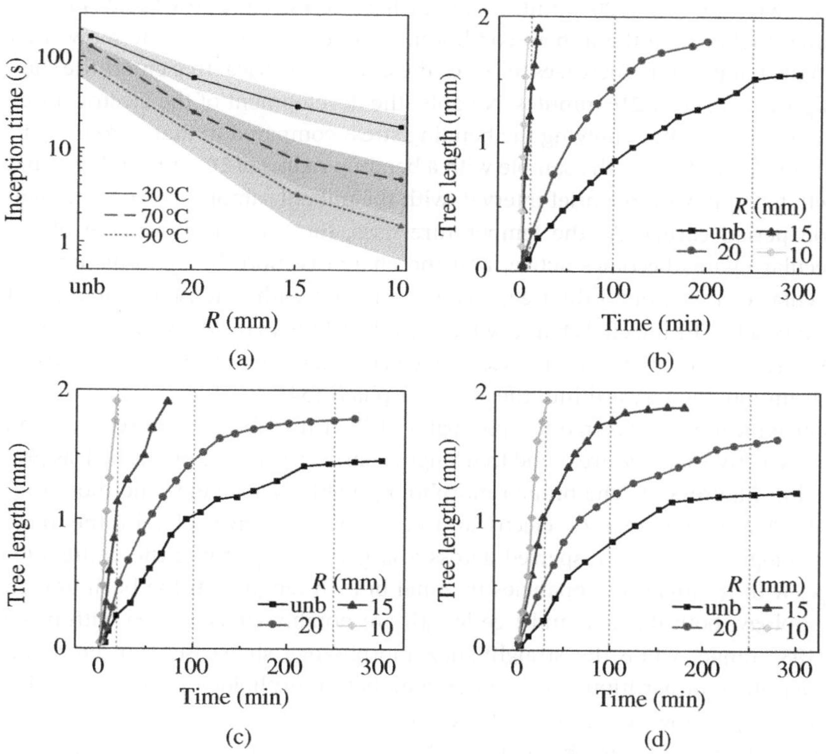

Figure 3.33 (a) Inception time of electrical trees. Tree length variation at different temperatures. (b) 30 °C. (c) 70 °C. (d) 90 °C. *Source:* Du et al. [71]/with permission from IEEE.

significantly reduced as the curvature radius decreases. The initiation time for the unbent electrical tree at room temperature is 168.4 seconds, whereas for a curvature radius of 10 mm, it diminishes to 17.75 seconds, indicating an 89.4% reduction. Elevated temperature is observed to accelerate the initiation of electrical trees, with the minimum initiation time recorded at 1.5 seconds for a curvature radius of 10 mm and a temperature of 90 °C. The needle electrode can form a tip defect, characterized by a stress concentration effect, rendering it susceptible to microcracks and rapid development under bending stresses, thus accelerating the initiation of electrical trees.

The trend of electrical tree length with respect to growth time is depicted in Figure 3.33b–d. The length from the tip of the needle to the tip of the branch is taken as the growth length of the electrical tree. At room temperature, the unbent sample exhibits a relatively uniform growth rate of approximately 6.02 μm/min over a duration of 250 seconds, as illustrated in Figure 3.33b. When the bending curvature radius is reduced to 15 and 20 mm, the samples progress to the critical breakdown state within 20 minutes, showcasing electrical tree growth rates 16 and 42 times higher than the unbent condition, respectively. Figure 3.33c shows that when the temperature increases to 70 °C, the unbent electrical tree enters the stagnant growth stage at 216 minutes. Notably, the development of the electrical tree also slows down after applying the bending stress compared to that at room temperature. At 90 °C, only the sample with a bending radius of 10 mm exhibits rapid growth to 2 mm within a brief interval, with the unbent sample entering the stagnation period earlier. As the temperature rises, the thermal movement of the molecular chains becomes active, and the charge is more likely to disrupt the molecular chains around the tree, creating a region with a dense distribution of fine channels. The injected charge will evenly distribute in the dense area, creating an electrical field shield that reduces the electric field and inhibits the electrical tree from moving forward into the stagnant phase [54].

Four typical electrical tree morphologies of PPH in Table 3.2 are selected and their respective damage areas and tree length characteristics are quantified, as presented in Figure 3.34. The branch morphology exhibits a smaller and more uniformly distributed area of electrical tree damage. Conversely, in the bush morphology, damage is dispersed across a larger area spanning the entire 180° range, with minimal discrepancies in zonal branch lengths. In the bush-branch morphology, both damage and tree length are notably concentrated within the 72°–108° range, while the straight-stick morphology showcases concentration within a singular partition at the center, with both morphologies exhibiting a distinctive cusp in the tree length distribution.

The Gaussian fit for the typical electrical tree length corresponding to Figure 3.35 is combined and the characteristic parameters σ and K are calculated, as shown in Figure 3.35. The straight-stick electrical tree is of a skinny-tall type

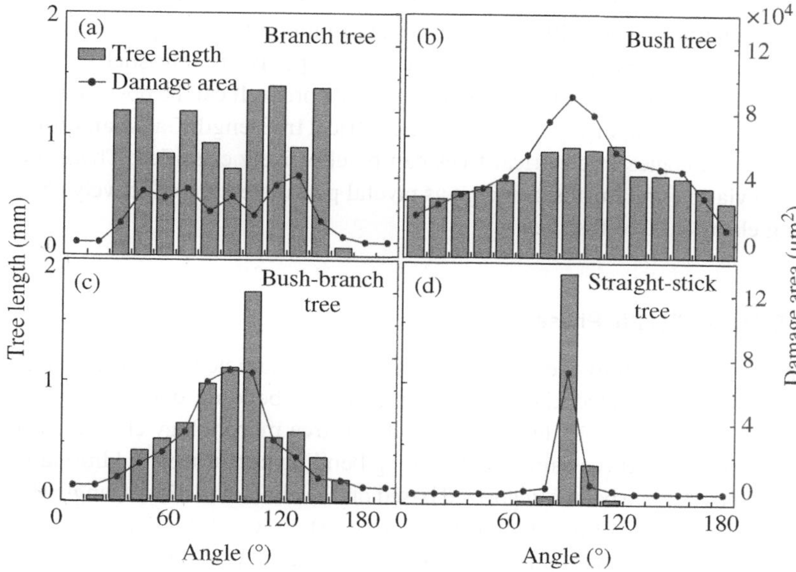

Figure 3.34 Morphological characterization of typical electrical trees corresponding to Table 3.2. (a) Table 3.2(a1). (b) Table 3.2(c1). (c) Table 3.2(b3). (d) Table 3.2(b4). *Source:* Du et al. [71]/with permission from IEEE.

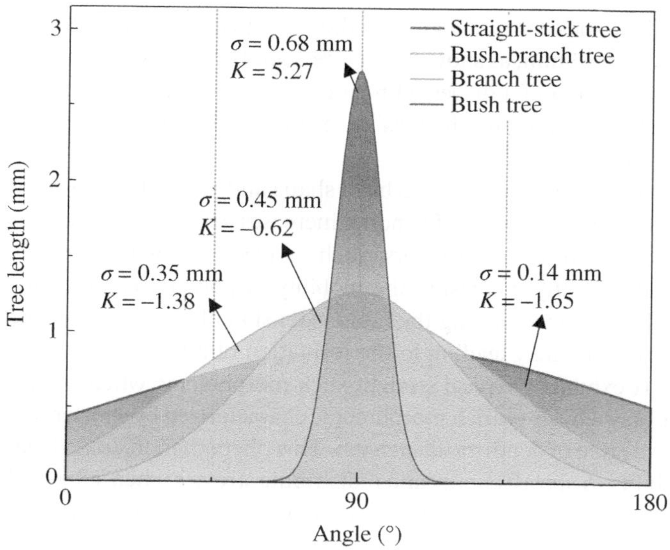

Figure 3.35 Gaussian fit for tree length classified by typical morphologies in Figure 3.34 and characteristic parameters. *Source:* Du et al. [71]/with permission from IEEE.

with maximum peak and branches concentrate around 90°. The bush tree distribution is of a stumpy type, with a more even distribution of tree length across the interval. The order of σ corresponding to the four shapes is: straight-stick < bush-branch < branch < bush, and the K is in the opposite order. It can be observed that by calculating and comparing σ and K of the electrical tree length Gaussian fitting, different morphologies of electrical trees can be effectively classified. Therefore, standard deviation and kurtosis emerge as pivotal parameters for effectively characterizing electrical tree features.

3.4.2 Effect of Elastic Phase

Further investigation into the electrical tree growth characteristics of PP block copolymer and PP blended elastomer samples under bending deformation has been conducted. Figure 3.36 depicts the electrical tree morphology characterization diagrams for the three samples at varying bending curvatures. Additionally, the electrical trees have been categorized based on the characteristics of length distribution and the value of the standard deviation. At 30 °C, PPB quickly changed from bush-branch to straight-stick as the radius increased. Concurrently, the randomness in the growth of PP/TPE branches escalates, accompanied by a notable increase in both damage area and tree length compared to both PPH and PPB. The standard deviation values in Figure 3.36c1 and c2 are close to bush-branch, but the tree lengths are uniformly distributed over 24°–144°, lacking significant vertices, so the shape is judged to be branch morphology. Furthermore, the electrical tree morphology is a straight-stick morphology with the largest standard deviation at curvature radius less than 15 mm, as shown in Figure 3.36c4. This observation can be attributed to the formation of a dispersed-phase structure of TPE, which may lead to an increase in free volume, thus increasing the mean free range of electrons and leading to enhanced degradation.

Up to 90 °C, the unbent samples develop a bush shape and gradually transform into a bush-branch shape as the degree of bending increases. Higher temperatures can enhance the mobility of charge carriers, such as electrons, within the PP matrix, while also significantly increasing the mobility of polymer chains. This phenomenon is conducive to facilitating the rapid extension of the electrical tree in multiple directions, ultimately leading to the formation of a bush shape. At a radius of 10 mm, PPH exhibits a typical straight-stick morphology, whereas PPB and PP/TPE only display a bush-branch morphology, characterized by shorter tree lengths and a higher degree of dendritic dispersion. This phenomenon arises from the elastic phase's ability to absorb energy and redistribute stresses when subjected to significant bending stresses. This mechanism reduces the likelihood of rapid crack propagation within the material, thereby suppressing deterioration

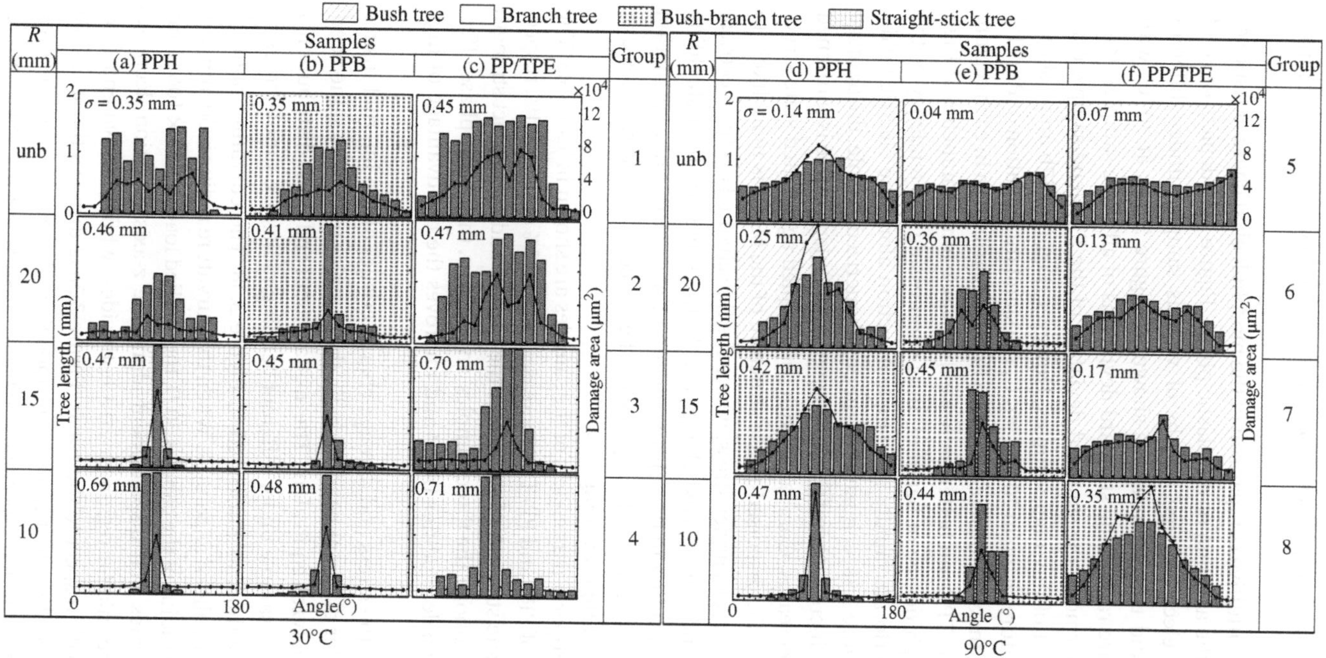

Figure 3.36 Morphological characterization of electrical trees and standard deviation σ of Gaussian fit for tree length. (a), (b), and (c) correspond to PPH, PPB, and PP/TPE at 30 °C. (d), (e), and (f) correspond to PPH, PPB, and PP/TPE at 90 °C. *Source:* Du et al. [71]/with permission from IEEE.

pathways along straight lines and enhancing the high-temperature electrical tree characteristics of PP.

To facilitate a more intuitive comparison of the morphological characteristics of electrical trees after the addition of the elastic phase, Figure 3.37 depicts the σ of the Gaussian fit for electrical tree lengths of various samples under different degrees of bending and temperatures. As shown in Figure 3.37a, at 30 °C, with a decrease in curvature radius, the standard deviations of all three samples gradually increase. Furthermore, the standard deviation of the PP blend sample consistently surpasses that of the other samples. However, Figure 3.37b and c shows that with the temperature rising to 70 and 90 °C, the standard deviation of PP/TPE decreases to a minimum with the most uniform branch distribution in each region. Compared to PPB, PP/TPE can better increase the dispersion of branches at high temperatures and high degree of bending, thus suppressing the tendency of the electrical tree to deteriorate rapidly in a single direction along the electric field.

Combined with Figure 3.36, compared to the copolymerized sample, the damage area of the blended is increased, but the final length of the electric tree is shortened and the dispersion increases. Compared with the coupled aggregated state structure of the copolymer, the dispersed-phase structure of the blend is more likely to form interfacial microcracks under stress, where the fine channels of the electrical tree develop faster, and the injected charge can form an electric-field shielding layer in a short time, leading to the advancement of the electrical tree stagnation period [36, 55]. Therefore, PP/TPE displays the best electrical tree resistance at high temperatures and high bending.

The mechanical property parameters of the samples are shown in Table 3.3. The breaking elongation of PPB and PP/TPE increases to 796.15% and 736.51%, respectively, while the tensile strength of PP/TPE is the highest and the impact strength of PPB is the highest among the three samples. In addition, the elastic phase effectively reduces the tensile modulus of PP and improves the disadvantage of high stiffness of PP.

To further explore the stress and damage characteristics of the internal insulation of the cable under different bending radii, the samples under the same bending conditions are analyzed equivalent. The displacement and stress distribution of the PPH electrical tree sample under different bending radius are simulated using the finite element method, as illustrated in Figure 3.38. The curvature radius is based on the value calculated at point A. When the curvature radius is 50 mm, corresponding bending radius of the cable is 50 *D*, the applied load is $1.49 \times 106\ \mathrm{N/m^2}$. Figure 3.38a1 illustrates that displacement along the *z*-axis direction decreases nonlinearly from the middle region toward both sides, while the two end faces remain stationary with respect to the *z*-axis. Figure 3.38a2 displays the stress distribution, where according to the bending model of the beam in engineering mechanics, the upper surface experiences tensile stress, while the lower surface

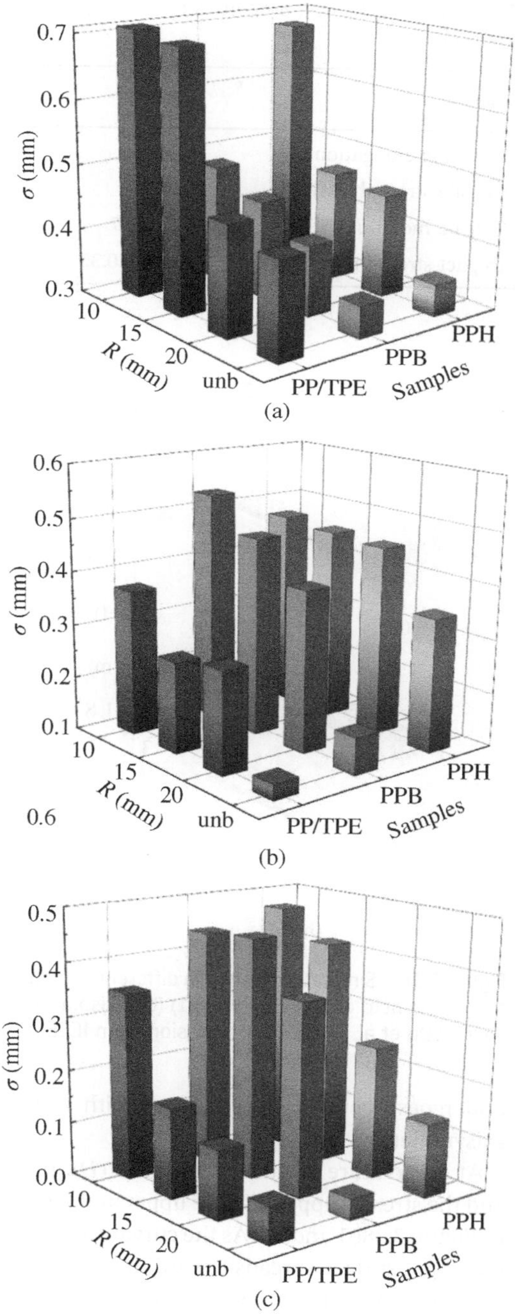

Figure 3.37 Standard deviation (σ) of Gaussian fit for the electrical tree lengths. (a) 30 °C. (b) 70 °C. (c) 90 °C. *Source:* Du et al. [71]/with permission from IEEE.

Table 3.3 Mechanical properties of PP samples. *Source:* Du et al. [71]/with permission from IEEE.

Sample	PPH	PPB	PP/TPE
Breaking elongation (%)	366.97	796.15	736.51
Tensile strength (MPa)	35.6	21.10	23.50
Tensile modulus (MPa)	1136.57	842.98	598.79
Impact strength (kJ/m^2)	3.08 ± 0.35	32.22 ± 0.14	28.88 ± 0.45

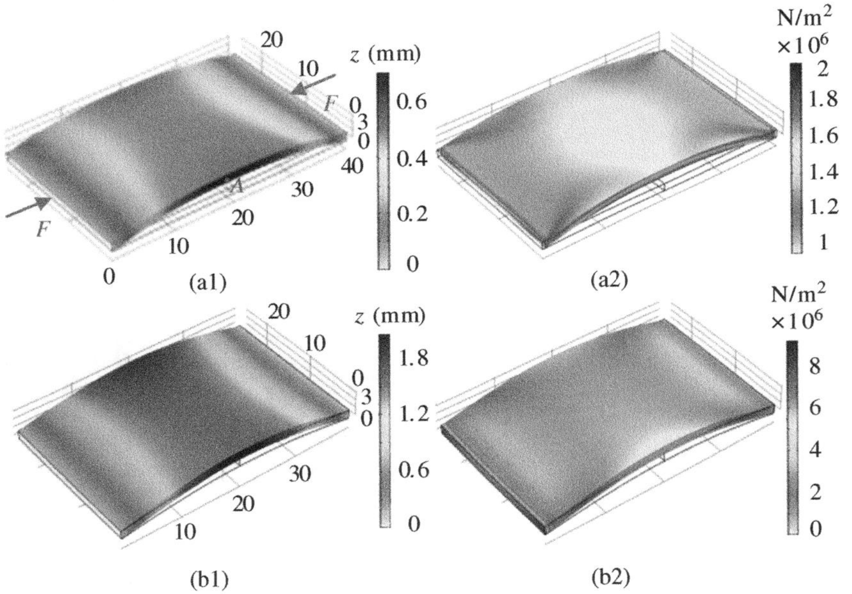

Figure 3.38 Simulation results in different curvature radius for PPH. (a1) (a2) Displacement and stress field at *R* of 50 mm. (b1) (b2) Displacement and stress field at *R* of 20 mm. *Source:* Du et al. [71]/with permission from IEEE.

undergoes compressive stress, with stress distribution being primarily axisymmetric.

At a curvature radius of 20 mm (20 *D*), the displacement at the sample center and the stresses applied to the upper and lower surfaces increase, as demonstrated in Figure 3.38b1 and b2. As the curvature radius decreases, stress concentration on the upper surface spreads toward the sides, while stress on the lower surface concentrates toward the center, reaching up to $8.45 \times 106\ N/m^2$. However, for a curvature radius of 15 mm (15 *D*), no steady-state solution can be calculated, possibly due to the maximum force of the sample exceeding the yield point of the PPH, which should be avoided as much as possible.

In contrast to compressive stresses, tensile stresses have a propensity to induce material defects, such as phase separation. Based on the length before and after bending at a point 1 mm at the center of the sample upper surface, the tensile strain near the needle electrode can be calculated. At a curvature radius of 10 mm (10 *D*), the tensile strain near the center of the upper surface is approximately 40%. To investigate the crystalline morphology of the material under 40% deformation, polarized optical microscopy (POM) is employed, as depicted

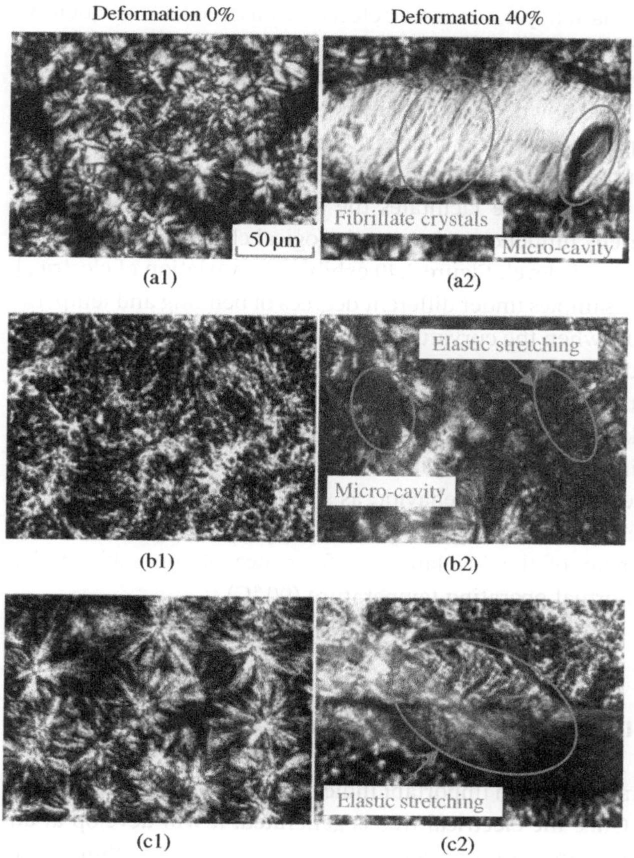

Figure 3.39 Effect of deformation on crystalline morphology. (a1) (a2) PPH. (b1) (b2) PPB. (c1) (c2) PP/TPE. *Source:* Du et al. [71]/with permission from IEEE.

in Figure 3.39. Figure 3.39a1 and b1 illustrates that following 10 minutes of isothermal crystallization and subjecting the sample to 40% deformation, some spherical crystals are observed to fibrillate, accompanied by the development of micro-cavity defects.

Upon introducing the elastic phase, both PPB and PP/TPE exhibit more densely connected spherical crystals. Under deformation, the elastic phase of the copolymer demonstrates an increased ability to withstand stress and deform prior to the crystalline region, as indicated in Figure 3.39b2. However, owing to the coupled structure of molecular chain entanglement, the degree of deformation is restricted under force application, rendering the molecular chain susceptible to fracture under high stresses, consequently leading to the micro-cavity. Figure 3.39c2 reveals that the TPE within the blend is the stress concentration point and undergoes a large elastic deformation, yet no severe defects are observed. Consequently, under substantial deformation, regions of PPH and PPB subjected to high tensile stresses may form severe physical defects, along which the electrical tree tends to develop rapidly until breakdown occurs. Conversely, PP/TPE demonstrates a reduced propensity to produce noticeable defects under equivalent deformation, thereby aiding in the suppression of electrical tree development.

The addition of the elastic phase facilitates the suppression of micro-defects under deformation, thereby influencing the morphological characteristics of electrical trees at the macroscopic level. Figure 3.40 exhibits the variation of electrical tree morphologies of the samples under different degrees of bending and temperatures. From Figure 3.40a, when the curvature radius is small enough, all the electrical trees of PPH will develop into the straight-stick shape, which is most likely to cause the development of the electrical trees and breakdown. Although PP/TPE show the straight-stick shape at low temperatures and high bending rates, it can rapidly transform into bush-branch with increasing temperatures, improving the drastic electrical tree degradation behavior, as shown in Figure 3.40c. Therefore, introducing the elastic phase, especially blending with elastomers, contributes to the improvement of the insulation performance of the cable under bending conditions at normal operating temperature (90 °C).

3.5 Methods for Suppressing Electrical Treeing

The electrical tree degradation is an important indicator for evaluating cable insulation properties [56]. Once the electrical tree is generated, it will develop at an extremely fast speed, resulting in insulation breakdown and power failure [57]. Electrical tree degradation restricts the development of cable insulation and limits the increase of transmission voltage level [58, 59]. At present, two ways to inhibit

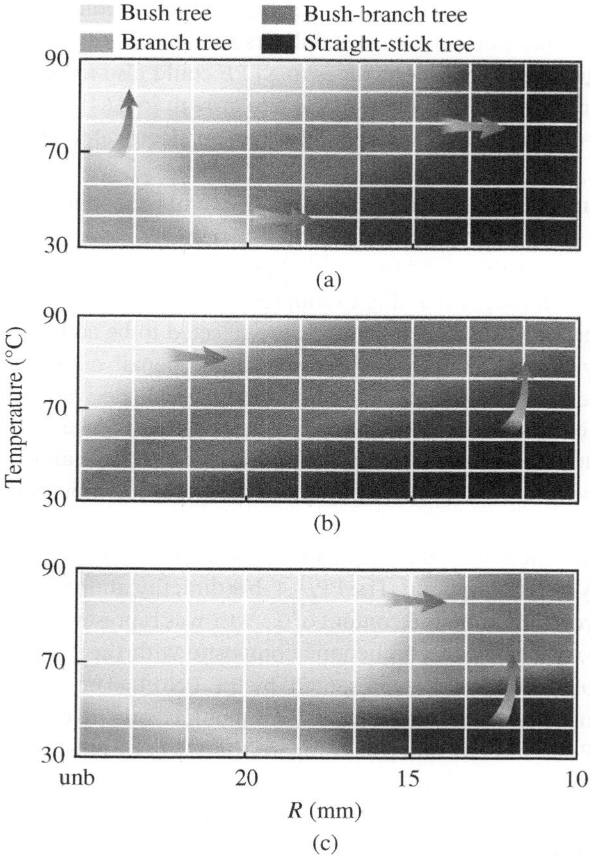

Figure 3.40 Variation of electrical tree morphologies with temperature and curvature radius. (a) PPH. (b) PPB. (c) PP/TPE. *Source:* Du et al. [71]/with permission from IEEE.

the growth of electrical tree are generally used, one is to add nanoparticles to the polymer matrix and another is to fill the polycyclic compounds [60, 61]. As an organic compound, the polycyclic compounds can be directly added into the polymer during the production process by mechanical blending, which has the advantages of low cost and simple production process. It was found that polycyclic compounds had an excellent effect on inhibiting electrical tree growth [59–62]. Simplex wire and cable reported a series of polycyclic compounds based on nitrotoluene structure that can significantly improve the electrical tree resistance in PE [61]. The inhibition efficiency of various polycyclic compounds with an added amount of 0.5 wt% on electrical trees were studied, and it was found that the highest efficiency of bismuth increased the electrical tree initiation voltage in LDPE by 1.6 times [62]. The addition of voltage stabilizers such as benzil, acetophenone derivatives, and benzophenone derivatives to XLPE could also increase the DC breakdown field strength and inhibit electrical tree growth [63, 64]. However, there are few studies on the effect of voltage stabilizers on electrical trees in PP insulation.

It was reported that adding benzil, acetophenone derivatives and benzophenone derivatives to XLPE could inhibit electrical tree growth under AC voltage. It was also found in our previous works that 4,4′-bis(dimethyl amino)benzil and benzophenone derivatives showed an excellent ability to suppress space charge in XLPE [65]. Therefore, three different polycyclic compounds are selected to be added to the PP matrix in this study. They are (a) 4,4′-bis(dimethylamino)benzil, a kind of benzil; (b) 2-hydroxy-2-phenylacetophenone, a kind of acetophenone derivatives; (c) 4-phenylbenzophenone, a kind of benzophenone derivatives. Figure 3.41 shows the chemical structural formula of them. An internal mixer and a thermoforming machine were used to prepare PP and PP/polycyclic compounds composites.

For the convenience of distinction, as shown in Table 3.4, PP was used to represent neat PP without polycyclic compound. The PP/4,4′-bis(dimethylamino)benzil composite with the polycyclic compound content of 0.1 wt% was represented by PP-AN0.1. The PP/2-hydroxy-2-phenylacetophenone composite with the polycyclic compound content of 0.1 wt% was represented by PP-BN0.1. The PP/4-phenylbenzophenone composite with the polycyclic compound content of 0.1 wt% was represented by PP-CN0.1. In order to compare the effect of polycyclic compound content on the growth characteristics of electrical tree, 0.2 and 0.3 wt% PP/2-hydroxy-2-phenylacetophenone composites were prepared, represented by PP-BN0.2 and PP-BN0.3, respectively.

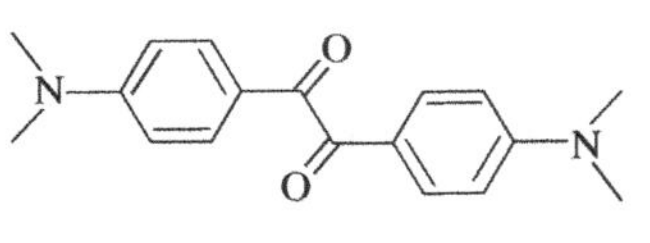

(a) 4,4′–Bis(dimethylamino)benzil

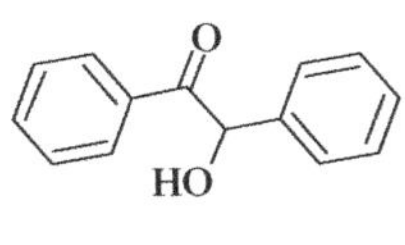

(b) 2–Hydroxy–2–phenylacetophenone

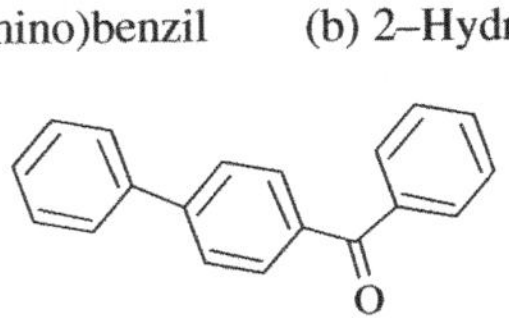

(c) 4–Phenylbenzophenone

Figure 3.41 Chemical structural formula of the chosen polycyclic compounds. (a) 4,4′–Bis (dimethylamino)benzil. (b) 2–Hydroxy–2–phenylacetophenone. (c) 4–Phenylbenzophenone. *Source:* Zhu et al. [72]/with permission from IEEE.

Table 3.4 Index for PP samples investigated in this work. *Source:* Zhu et al. [72]/with permission from IEEE.

Index	Polycyclic compounds kind	Content (wt%)
PP	/	0
PP-AN0.1	4,4′-Bis(dimethylamino)benzil	0.1
PP-BN0.1	2-Hydroxy-2-phenylacetophenone	0.1
PP-CN0.1	4-Phenylbenzophenone	0.1
PP-AN0.2	4,4′-Bis(dimethylamino)benzil	0.2
PP-BN0.2	2-Hydroxy-2-phenylacetophenone	0.2
PP-CN0.2	4-Phenylbenzophenone	0.2
PP-AN0.3	4,4′-Bis(dimethylamino)benzil	0.3

3.5.1 Effect of the Type of Voltage Stabilizer

Figure 3.42 shows the electrical tree morphologies in PP and PP-BN0.1 under different temperatures. The impulse voltage is +35 kV and the DC voltage is −25 kV. Figure 3.42a is tree morphologies in PP samples, and Figure 3.42b is tree morphologies in PP/polycyclic compounds composites to which the 0.1 wt% 2-hydroxy-2-phenylacetophenone polycyclic compound is added. The treeing time is two minutes. When the temperature is the same, the electrical tree length in PP-BN0.1 is shorter compared with that in PP. The electrical tree channel is thinner. As the temperature increases, the electrical tree channels become thicker, blacker, and the number of branches becomes more.

Figures 3.43 and 3.44 show the electrical tree growth characteristics with different kinds of polycyclic compounds under opposite polarity impulse-superimposed

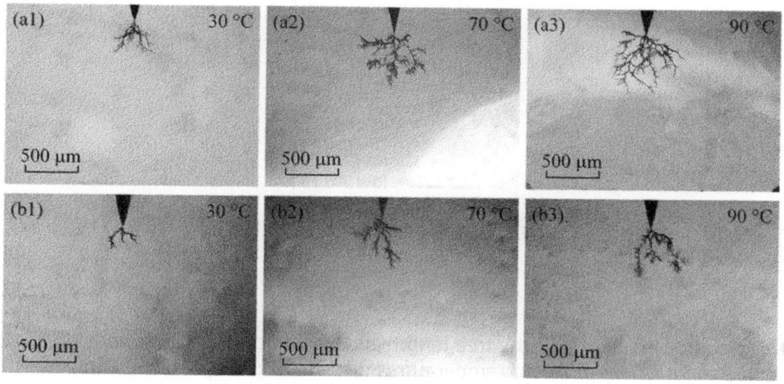

Figure 3.42 Tree morphologies in PP and PP-BN0.1 under different temperatures. (a) PP at difference temperature. (b) PP-BN0.1 at difference temperature. *Source:* Zhu et al. [72]/with permission from IEEE.

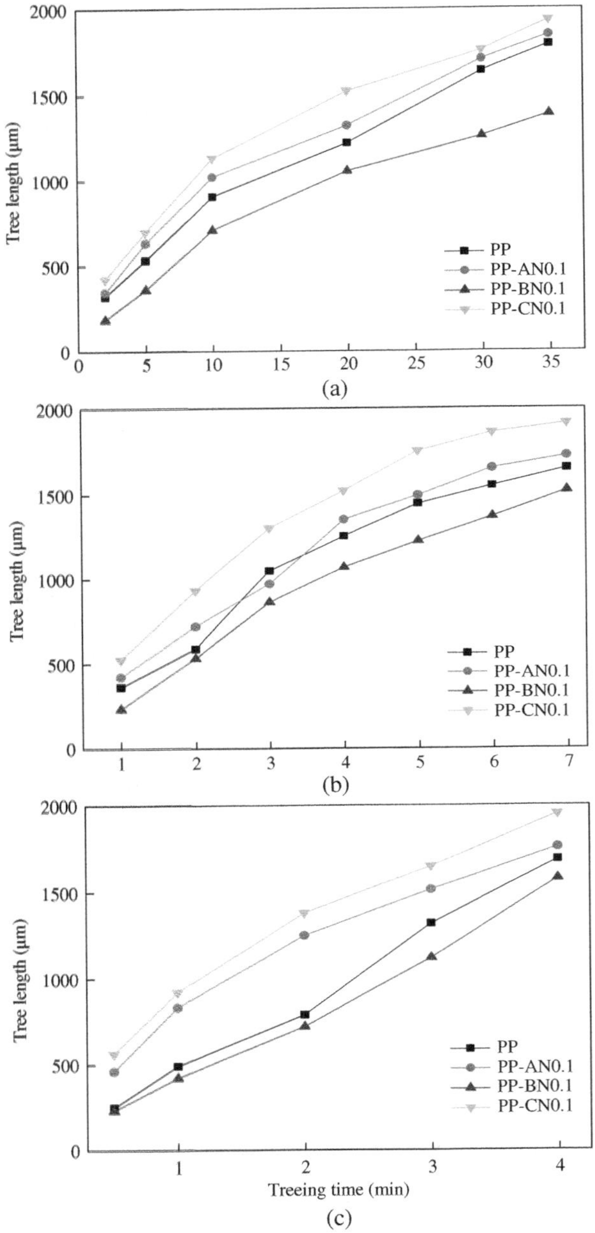

Figure 3.43 Relationship between the tree length and treeing time with different kinds of polycyclic compounds under different temperatures under +35 kV impulse superimposed −25 kV DC voltage. (a) 30 °C. (b) 70 °C. (c) 90 °C. *Source:* Zhu et al. [72]/with permission from IEEE.

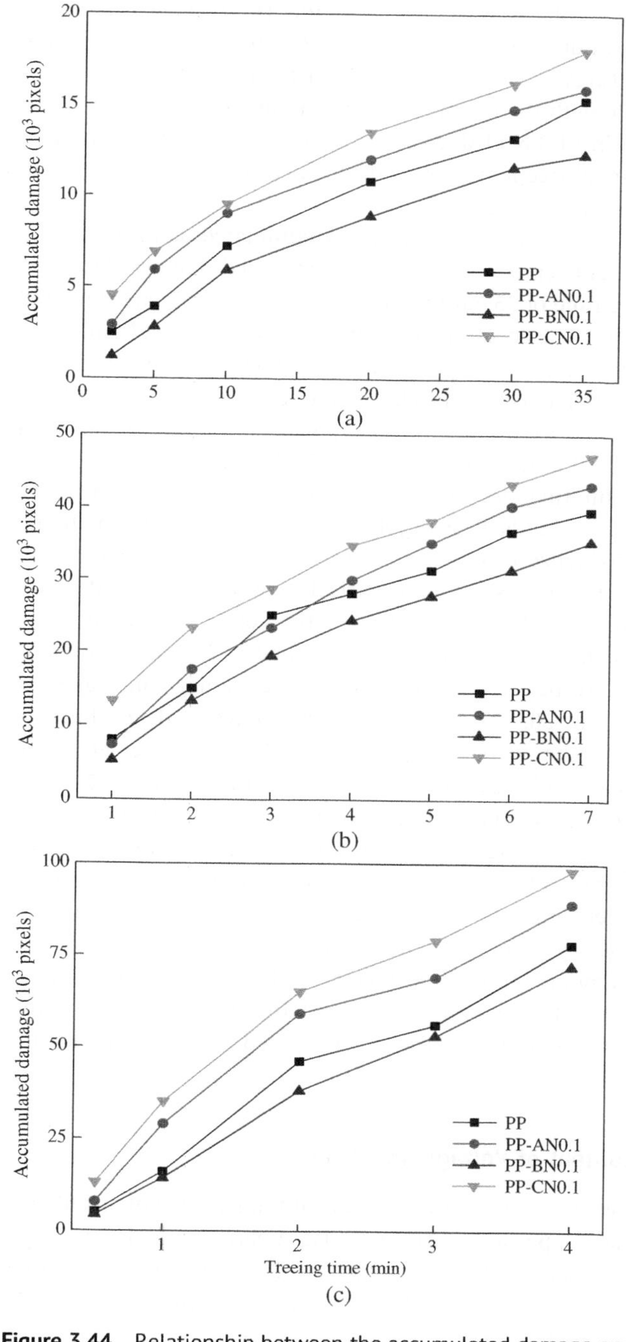

Figure 3.44 Relationship between the accumulated damage and treeing time with different kinds of polycyclic compounds under different temperatures under +35 kV impulse superimposed −25 kV DC voltage. (a) 30 °C. (b) 70 °C. (c) 90 °C. *Source:* Zhu et al. [72]/with permission from IEEE.

DC voltage. The impulse voltage is +35 kV and the DC voltage is −25 kV. Figure 3.43 shows the relationship between the tree length and the treeing time under different temperatures. The electrical tree length in PP-BN0.1 decreases by 19% compared with PP with the treeing time of 35 minutes at 30 °C. At 70 °C, the electrical tree length in PP-BN0.1 decreases by 7.8% with the treeing time of seven minutes. The electrical tree length in PP-BN0.1 decreases by 6.6% with the treeing time of four minutes at 90 °C.

Figure 3.44 shows the relationship between the accumulated damage and the treeing time under different temperatures. The accumulated damage decreases by 19.6% with the treeing time of 35 minutes at 30 °C. At 70 °C, the accumulated damage decreases by 10.9%. The accumulated damage decreases by 7.7% at 90 °C. It can be concluded that the 2-hydroxy-2-phenylacetophenone polycyclic compound has a certain inhibition effect on the electrical tree length and accumulated damage under opposite polarity impulse-superimposed DC voltage at different temperatures. It has the best inhibition effect at 30 °C. With the increase of temperature, the effect of inhibiting electrical tree degradation gradually decreases. However, when another two kinds of polycyclic compounds, 4,4′-bis(dimethylamino)benzil and 4-phenylbenzophenone, are added, the electrical tree length and accumulated damage increase. The electrical tree length in PP-AN0.1 increases by 3% compared with PP with the treeing time of 35 minutes at 30 °C, and it increases by 4.2% and 4.5% with the treeing time of 7 minutes at 70 °C and 4 minutes at 90 °C, respectively. The accumulated damage increases by 3.9%, 8.9%, and 14.1% at 30, 70, and 90 °C, respectively. The effect of 4-phenylbenzophenone is the worst. At 30 °C, the electrical tree length in PP-CN0.1 increases by 7.8% and the accumulated damage increases by 7.7% with the treeing time of 30 minutes. With the increase of temperature, the effect is even worse. At 70 °C, the electrical tree length in PP-CN0.1 increases by 15.8% and the accumulated damage increases by 18.9% when the treeing time is seven minutes. At 90 °C, the electrical tree length in PP-CN0.1 increases by 17.6% and the accumulated damage increases by 25.6% when the treeing time is four minutes.

It can be seen that among these three voltage stabilizer, the 2-hydroxy-2-phenylacetophenone polycyclic compound has a certain inhibition effect on the electrical tree degradation under the impulse-superimposed DC voltage under different temperatures.

3.5.2 Effect of the Content of Voltage Stabilizer

Figure 3.45 shows the electrical tree morphologies with different contents of polycyclic compounds under opposite polarity impulse-superimposed DC voltage. The impulse voltage is +35 kV and the DC voltage is −25 kV. Figure 3.45a is tree morphologies in PP samples, Figure 3.45b is tree morphologies in PP-BN0.1,

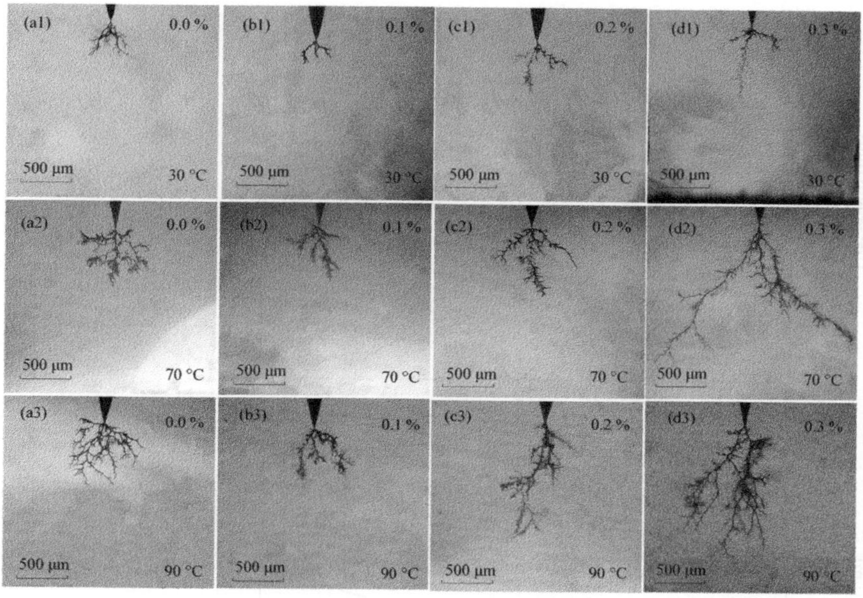

(a) PP (b) PP-BN0.1 (c) PP-BN0.2 (d) PP-BN0.3

Figure 3.45 Tree morphologies with different contents of polycyclic compounds under different temperatures. (a) PP (b) PP-BN0.1. (c) PP-BN0.2. (d) PP-BN0.3. *Source:* Zhu et al. [72]/with permission from IEEE.

Figure 3.45c is tree morphologies in PP-BN0.2, and Figure 3.45d is tree morphologies in PP-BN0.3. The treeing time is two minutes. The electrical tree morphologies are not affected by the content of polycyclic compound and temperature, which are branch trees under different conditions. However, the appearance is connected with the polycyclic compound content. As the content increases, the number of electrical tree channels decreases first and then increases. The length of the electrical tree first reduces and then increases.

Figures 3.46 and 3.47 show the electrical tree length and accumulated damage with different contents of polycyclic compounds under +35 kV impulse superimposed −25 kV DC voltage. It can be seen that the tree length and accumulated damage in PP-BN0.1 are the least compared with PP, PP-BN0.2, and PP-BN0.3, which means that the addition of 0.1 wt% 2-hydroxy-2-phenylacetophenone has the best inhibition effect on the electrical tree degradation. However, with the increase of the polycyclic compound addition filling ratio, polycyclic compound can no longer inhibit the growth of electrical tree but promote the growth of electrical tree. At 30 and 70 °C, PP-BN0.2 still has a certain inhibitory effect on the accumulated damage. With the treeing time of 30 minutes at 30 °C, the

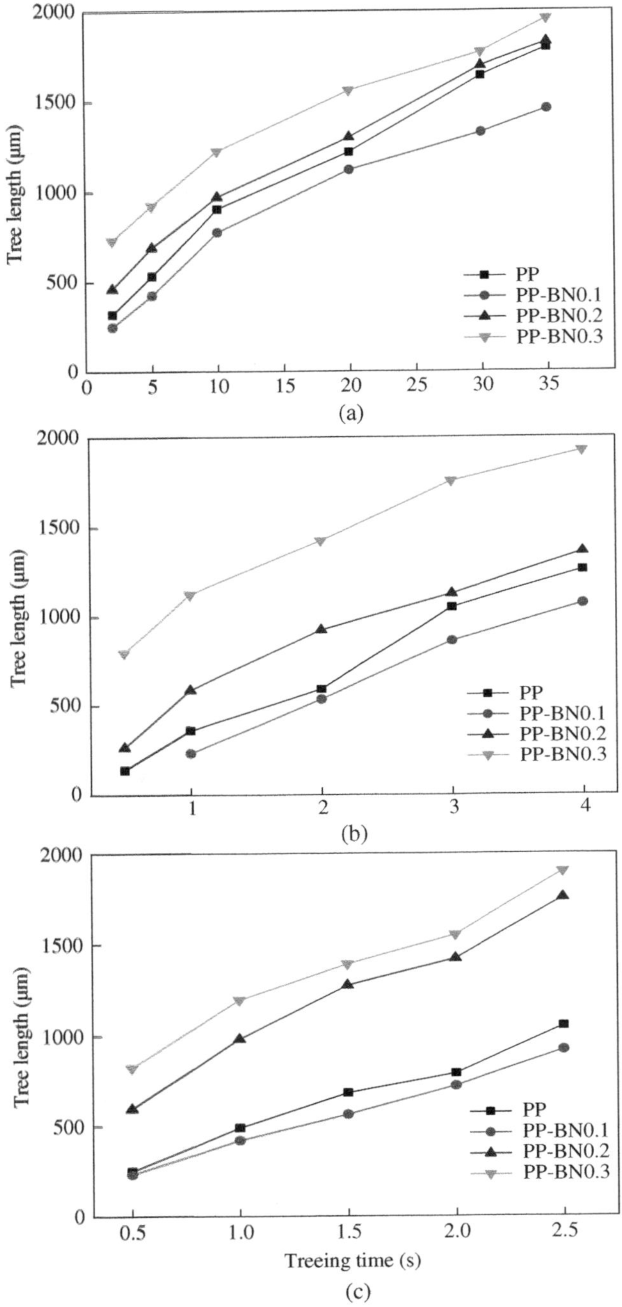

Figure 3.46 Relationship between the tree length and treeing time with different contents of polycyclic compounds under different temperatures under +35 kV impulse superimposed −25 kV DC voltage. (a) 30 °C. (b) 70 °C. (c) 90 °C. *Source:* Zhu et al. [72]/with permission from IEEE.

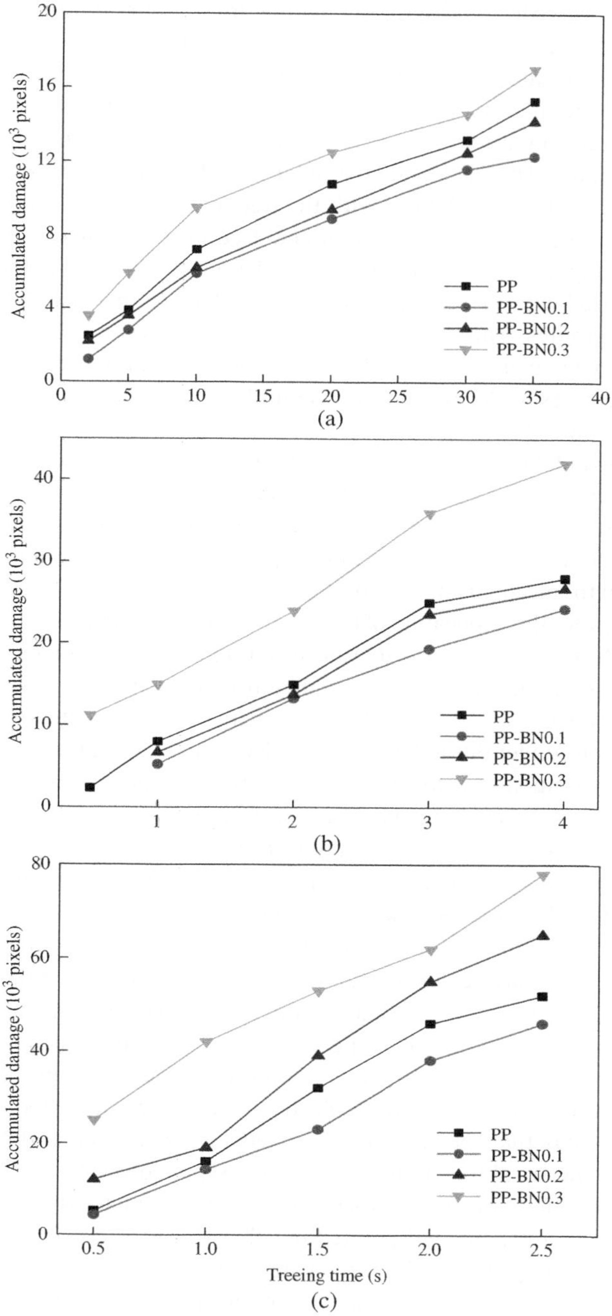

Figure 3.47 Relationship between the accumulated damage and treeing time with different contents of polycyclic compounds under different temperatures under +35 kV impulse superimposed −25 kV DC voltage. (a) 30 °C. (b) 70 °C. (c) 90 °C. *Source:* Zhu et al. [72]/with permission from IEEE.

accumulated damage decreases by 7.5%. With the treeing time of four minutes at 70 °C, the accumulated damage decreases by 7.1%. However, at 90 °C, the accumulated damage in PP-BN0.2 is larger than that in PP, which increases by 25.1%. PP-BN0.2 does not have the inhibitory effect on the electrical tree lengths, but increases the electrical tree lengths, which are larger in PP-BN0.2 than those in PP at different temperatures. The electrical tree lengths in PP-BN0.2 increase by 1.6%, 7.9%, and 66.5% at 30, 70, and 90 °C with the treeing time of 35, 4, and 2.5 minutes, respectively. The electrical tree length and accumulated damage in PP-BN0.3 are the largest, especially at high temperatures. The electrical tree length increases by 8.9% compared with PP, and the accumulated damage increases by 11.1% at 30 °C and 30 minutes. The electrical tree length increases by 52.9%, and the accumulated damage increases by 50.3% at 70 °C and four minutes. The electrical tree length increases by 80.2%, and the accumulated damage increases by 53.1% at 90 °C and 2.5 minutes.

The electrical tree degradation under impulse-superimposed DC voltage is closely associated with the thermal electron motion and trap distributions [66]. In our previous chapter, we discussed the charge motion process under impulse-superimposed DC voltage and the relationship between charge motion and electrical tree growth mechanism. When the applied voltage is +35 kV impulse and −25 kV DC, after the application of the impulse voltage, energy is produced by charges neutralization, exciting the thermal electrons to strike the molecular chain [67]. Since the energy required for excitation or ionization of most aromatic compounds is lower than that of PP, the aromatics will be excited or ionized by thermal electrons under the action of strong electric field, which makes it possible for polycyclic compound B to show the increased charge trap depth [65, 68]. During the process of injecting the charges, charges are easier to be trapped by the deep trap in PP-BN0.1, so the amount of the injected charges is reduced. At the same time, due to the effect of the polycyclic compound, the ability of samples to capture internal charges is enhanced with the increases of trap density and depth, resulting in a reduction in the amount of free charges. Therefore, the molecular chain will not easily be destroyed and the electrical tree resistance of PP-BN0.1 is improved. However, it was found from the experiments in this study that polycyclic compounds A and C could not inhibit the electrical tree degradation. In our previous research, it was found that the addition of 4-phenylbenzophenone could inhibit the XLPE electrical tree degradation [60], but it was found in this study that the addition of 4-phenylbenzophenone to PP accelerates the growth of electrical tree, indicating that the same polycyclic compound may have different effects on the electrical tree growth characteristics in different polymers. The literature [65, 69] pointed out that since the polycyclic compounds contained the group that could participate in the crosslinking reaction, they could be bonded to the XLPE molecular chain during the crosslinking process, thereby ensuring that the polycyclic compounds were not easily precipitated from the polymer matrix. However, there is no cross-linking reaction in the production process of PP. The polycyclic compounds are only added to PP by physical blending, and the compatibility of polycyclic compounds with PP molecules has a great influence on the electrical tree growth characteristics. Adding 4-phenylbenzophenone and 4,4′-bis(dimethylamino)benzil to PP accelerates the electrical tree growth rate, possibly due to the compatibility of these two polycyclic compounds with PP molecules is not high enough, and some additives migrated and precipitated, leading to the decreased charge trap depth and the degradation of the electrical tree resistance of the PP/polycyclic compounds composites. As the addition content of 2-hydroxy-2-phenylacetophenone increases from 0.1 to 0.3 wt%, the electrical tree growth rate increases. With the increase in the addition content of polycyclic compound, additives will be concentrated in one place, and it is easy to introduce physical defects, resulting in deep trap levels, reduced trap density, and increased carrier mobility, which in turn is equivalent to the introduction of impurities, resulting in reduced resistance to electrical tree.

3.6 Conclusion and Outlook

Although it is difficult to trigger the electrical tree under DC voltage in PP-based cable insulation, the characteristics of the electrical tree still require much attention, considering the operating conditions, such as impulse superimposed on DC voltage and DC voltage with polarity reversal. Especially, in the case of the superimpose of an opposite polarity impulse and DC voltage, the probability of electric tree initiation significantly increases, and the growth rate of electric tree increases, leading to insulation being highly susceptible to breakdown. The electrical tree degradation under impulse-superimposed DC voltage is closely associated with the thermal electron motion and trap distributions. Therefore, it is an effective method of introducing voltage stabilizers with localized state traps to suppress the initiation and growth of electrical tree branches. Although the method of nanocomposite insulation has not been introduced in this chapter, considering the potential aggregation issues that may occur during large-scale preparation, it is still an effective method to suppress the electrical treeing that has been widely proven. In addition to the factors affecting the voltage waveform and temperature of high-voltage cables discussed in this chapter, there are also many key factors that affect the initiation and development of electrical tree branches in PP cable insulation, such as tensile or bending mechanical stress. Preliminary research has found that mechanical stress can damage the crystalline structure of PP and form micro-void defects, which accelerate insulation degradation and breakdown. Electrical tree growth experiments are conducted under various bending curvature radii and temperatures. The results indicate that bending deformation promotes the initiation and accelerates the growth of electrical trees. Under extreme bending deformation, PPH tends to develop a straight-stick electrical tree morphology. Conversely, the presence of elastic phases introduced by copolymers and blends increases electrical tree dispersion at elevated temperatures and suppresses the tendency to develop along a straight line. Simulation and microscopic tensile experiments demonstrate that blended samples will not form significant physical defects under large deformation. The enhanced mechanical properties support the excellent electrical tree resistance of blended samples under bending deformation, making them promising for insulation in bending conditions.

References

1 Liu, Y. and Cao, X. (2015). Electrical tree growth characteristics in XLPE cable insulation under DC voltage conditions. *IEEE Transactions on Dielectrics and Electrical Insulation* 22 (6): 3676–3684. https://doi.org/10.1109/TDEI.2015.005222.

2 Noto, F., Yoshimura, N., and Ohta, T. (1977). Tree initiation in polyethylene by application of DC and impulse voltage. *IEEE Transactions on Electrical Insulation* 12 (1): 26–30. https://doi.org/10.1109/TEI.1977.298003.

3 Ieda, M. and Nawata, M. (1977). DC treeing breakdown associated with space charge formation in polyethylene. *IEEE Transactions on Electrical Insulation* 12 (1): 19–25. https://doi.org/10.1109/TEI.1977.298002.

4 Sekii, Y., Kawanami, H., Saito, M. et al. (2005). DC tree and grounded DC tree in XLPE. *IEEE Conference on Electrical Insulation and Dielectric Phenomena*, Nashville, TN, USA, 16 October 2005, 523–526. https://doi.org/10.1109/CEIDP.2005.1560735.

5 Selsjord, M. and Ildstad, E. (2006). Electrical treeing caused by rapid DC voltage grounding of XLPE cable insulation. *IEEE International Symposium on Electrical Insulation*, Toronto, ON, Canada, 11 June 2006, 502–505. https://doi.org/10.1109/ELINSL.2006.1665366.

6 Kumara, S., Hammarström, T., and Serdyuk, Y.V. (2021). Polarity effect on electric tree inception in HVDC cable insulation. *IEEE Transactions on Dielectrics and Electrical Insulation* 28 (5): 1819–1827. https://doi.org/10.1109/TDEI.2021.009682.

7 Han, T., Du, B., and Su, J. (2018). Electrical tree initiation and growth in silicone rubber under combined DC-pulse voltage. *Energies* 11 (4): 764. https://doi.org/10.3390/en11040764.

8 Du, B., Han, T., and Su, J. (2015). Electrical tree characteristics in silicone rubber under repetitive pulse voltage. *IEEE Transactions on Dielectrics and Electrical Insulation* 22 (2): 720–727. https://doi.org/10.1109/TDEI.2015.7076767.

9 Liu, Y. and Cao, X. (2013). Electrical tree initiation in XLPE cable insulation by application of DC and impulse voltage. *IEEE Transactions on Dielectrics and Electrical Insulation* 20 (5): 1691–1698. https://doi.org/10.1109/TDEI.2013.6633699.

10 Katsuta, G., Itaya, T., Nakatsuka, T. et al. (1997). DC and impulse treeing characteristics in insulating material for HVDC cable. *International Conference on Properties and Applications of Dielectric Materials*, Seoul, Korea (South), 25 May 1997, 422–425. https://doi.org/10.1109/ICPADM.1997.617626.

11 Iddrissu, I. and Rowland, S.M. (2015). The impact of DC bias on electrical tree growth characteristics in epoxy resin samples. *IEEE Conference on Electrical Insulation and Dielectric Phenomena*, Ann Arbor, MI, USA, 18 October 2015, 876–879. https://doi.org/10.1109/CEIDP.2015.7352082.

12 Liu, M., Liu, Y., Li, Y. et al. (2017). Growth and partial discharge characteristics of electrical tree in XLPE under AC-DC composite voltage. *IEEE Transactions on Dielectrics and Electrical Insulation* 24 (4): 2282–2290. https://doi.org/10.1109/TDEI.2017.006537.

13 Stone, G., van Heeswijkb, R., and Bartnikas, R. (1992). Electrical aging and electroluminescence in epoxy under repetitive voltage surges. *IEEE Transactions on Electrical Insulation* 27 (2): 233–244. https://doi.org/10.1109/14.135595.

14 Du, B., Su, J., and Han, T. (2017). Temperature-dependent electrical tree in silicone rubber under repetitive pulse voltage. *IEEE Transactions on Dielectrics and Electrical Insulation* 24 (4): 2291–2298. https://doi.org/10.1109/TDEI.2017.006461.

15 Lou, T.C., Zhang, S.F., and Huang, W. (2014). Theory and application of high voltage XLPE cable (vertical) laying. *Power and Energy* 35 (05): 643–646.

16 Chen, X., Xu, Y., Cao, X. et al. (2011). Effect of tree channel conductivity on electrical tree shape and breakdown in XLPE cable insulation samples. *IEEE Transactions on Dielectrics and Electrical Insulation* 18 (3): 847–860. https://doi.org/10.1109/TDEI.2011.5931074.

17 Du, B.X., Zhang, W., and Liang, H.C. (2023). Combined effects of electrical and mechanical stresses on insulation breakdown – Part I: tree growth of epoxy resin. *IEEE Transactions on Dielectrics and Electrical Insulation* 30 (5): 2362–2369. https://doi.org/10.1109/TDEI.2023.3267753.

18 Hu, F., Emersic, C., Smith, D. et al. (2023). The effect of mechanical strain on electrical tree development in XLPE. *IEEE Transactions on Dielectrics and Electrical Insulation* https://doi.org/10.1109/TDEI.2023.3337765.

19 Zhou, K., Li, K., He, Y., and Zhu, G. (2017). A new insight into the influence of mechanical orientation on water tree propagation in abnormal water tree shapes. *IEEE Transactions on Dielectrics and Electrical Insulation* 24 (6): 3878–3886. https://doi.org/10.1109/TDEI.2017.006436.

20 Kitani, I. and Arii, K. (1987). DC tree associated with space charge in PMMA. *IEEE Transactions on Electrical Insulation* 22 (3): 303–307. https://doi.org/10.1109/TEI.1987.298994.

21 Kao, K. (1984). New theory of electrical discharge and breakdown in low mobility condensed insulators. *Journal of Applied Physics* 55 (3): 752–755. https://doi.org/10.1063/1.333133.

22 Dissado, L.A. (2002). Understanding electrical trees in solids from experiment to theory. *IEEE Transactions on Dielectrics and Electrical Insulation* 9 (4): 483–497. https://doi.org/10.1109/TDEI.2002.1024425.

23 Tanaka, T. (1992). Charge transfer and tree initiation in polyethylene subjected to ac voltage stress. *IEEE Transactions on Electrical Insulation* 27 (3): 424–431. https://doi.org/10.1109/14.142702.

24 Huang, B., Hao, M., Hao, J. et al. (2016). Space charge characteristics in oil and oil-impregnated pressboard and electric field distortion after polarity reversal. *IEEE Transactions on Dielectrics and Electrical Insulation* 23 (2): 881–891. https://doi.org/10.1109/TDEI.2015.005413.

25 Kaneko, K. (1986). Effect of space charge on the breakdown strength under polarity reversal. *Electrical Engineering in Japan* 106 (3): 565–572. https://doi.org/10.1002/eej.4391060304.

26 Tanaka, T. (2001). Space charge injected via interfaces and tree initiation in polymers. *IEEE Transactions on Dielectrics and Electrical Insulation* 8 (5): 733–743. https://doi.org/10.1109/94.959692.

27 Tanaka, Y., Kodera, R., Kato, T. et al. (2015). Observation of space charge accumulation behavior in cross-linked polyethylene at voltage polarity reversal. *IEEE Conference on Electrical Insulation and Dielectric Phenomena*, Ann Arbor, MI, USA, 18 October 2015, 23–26. https://doi.org/10.1109/CEIDP.2015.7352115.

28 Wiesmann, H.J. and Zeller, H.R. (1986). A fractal model of dielectric breakdown and prebreakdown in solid dielectrics. *Journal of Applied Physics* 60 (5): 1770–1773. https://doi.org/10.1063/1.337219.

29 Hibma, T., Zeller, H.R., Pfluger, P., and Baumann, T. (1985). A model for space charge injection in dielectrics. *IEEE Conference on Electrical Insulation and Dielectric Phenomena*, Amherst, NY, USA, 20 October 1985, 259–265. https://doi.org/10.1109/CEIDP.1985.7728279.

30 Hibma, T. and Zeller, H.R. (1986). Direct measurement of space-charge injection from a needle electrode into dielectrics. *Journal of Applied Physics* 59 (5): 1614–1620. https://doi.org/10.1063/1.336473.

31 Du, B. and Zhu, L. (2015). Electrical tree characteristics of XLPE under repetitive pulse voltage in low temperature. *IEEE Transactions on Dielectrics and Electrical Insulation* 22 (4): 1801–1808. https://doi.org/10.1109/TDEI.2015.005183.

32 Chen, J., Bell, G., Dong, H. et al. (2010). A study of low temperature mechanical properties and creep behaviour of polypropylene using a new sub-ambient temperature nanoindentation test platform. *Journal of Applied Physics* 43 (42): 1–9. https://doi.org/10.1088/0022-3727/43/42/425404.

33 Lanca, M.C., Marat-Mendes, J.N., and Dissado, L.A. (2001). The fractal analysis of water trees - an estimate of the fractal dimension. *IEEE Transactions on Dielectrics and Electrical Insulation* 8 (5): 838–844. https://doi.org/10.1109/94.959714.

34 Kudo, K. (1998). Fractal analysis of electrical trees. *IEEE Transactions on Dielectrics and Electrical Insulation* 5 (5): 713–727. https://doi.org/10.1109/94.729694.

35 Du, B., Zhang, M., Han, T., and Su, J.G. (2015). Tree initiation characteristics of epoxy resin on LN_2 for superconducting magnet insulation. *IEEE Transactions on Dielectrics and Electrical Insulation* 22 (4): 1793–1800. https://doi.org/10.1109/TDEI.2014.004797.

36 Zheng, X. and Chen, G. (2008). Propagation mechanism of electrical tree in XLPE cable insulation by investigating a double electrical tree structure. *IEEE Transactions on Dielectrics and Electrical Insulation* 15 (3): 800–807. https://doi.org/10.1109/TDEI.2008.4543118.

37 Xie, A., Zheng, X., Li, S., and Chen, G. (2010). Investigations of electrical trees in the inner layer of XLPE cable insulation using computer-aided image recording monitoring. *IEEE Transactions on Dielectrics and Electrical Insulation* 17 (3): 685–693. https://doi.org/10.1109/TDEI.2010.5492239.

38 Vendik, I.B., Vendik, O.G., Afanasjev, V.P. et al. (2011). Correlation between electrical and mechanical properties of polymer composite. *IEEE Electronic Components and Technology Conference*, Lake Buena Vista, FL, USA, 31 May 2011, 1697–1702. https://doi.org/10.1109/ECTC.2011.5898740.

39 Miller, M.L., Gorur, R.S., Hendrickson, L.E., and Dyer, M.L. (1990). Water treeing characterization and tree retardant compound effects in XLPE power cables. *IEEE Conference on Electrical Insulation and Dielectric Phenomena*, Pocono Manor, PA, USA, 28 October 1990, 404–409. https://doi.org/10.1109/CEIDP.1990.201375.

40 Uchida, K., Asai, H., Shimizu, N., and Takahashi, T. (1991). Initiation mechanism of electrical tree and deteriorated region in polymers. *International Conference on the Properties and Applications of Dielectric Materials*, Tokyo, Japan, 8 July 1991, 240–243. https://doi.org/10.1109/ICPADM.1991.172018.

41 Shimizu, N., Uchida, K., and Rasikawan, S. (1992). Electrical tree and deteriorated region in polyethylene. *IEEE Transactions on Electrical Insulation* 27 (3): 513–518. https://doi.org/10.1109/14.142714.

42 Cui, Y., Wu, G.N., Wu, C., and Luo, Y. (2010). Possible mechanism of electrical treeing and breakdown for polyimide nanocomposite film used in inverter-fed motor. *IEEE Conference on Electrical Insulation and Dielectric Phenomena*, West Lafayette, IN, 1–4. https://doi.org/10.1109/CEIDP.2010.5724102.

43 Danikas, M.G. and Tanaka, T. (2009). Nanocomposites-a review of electrical treeing and breakdown. *IEEE Transactions on Electrical Insulation* 25 (4): 19–24. https://doi.org/10.1109/MEI.2009.5191413.

44 Chang, H.C. and Chen, B.F. (2019). Mechanical behavior of submarine cable under coupled tension torsion and compressive loads. *Ocean Engineering* 189: 106272.

45 Najafi, P.A. and Shahrtash, S.M. (2013). A comprehensive circuit model for partial discharge activities in cable insulation due to electrical treeing. *International Journal of Electrical Power & Energy Systems* 44 (1): 646–655.

46 Du, B.X., Zhang, Y., Kong, X.X. et al. (2022). Morphological feature analysis of electrical tree growth in epoxy resin under tensile and compressive stress. *IEEE Transactions on Dielectrics and Electrical Insulation* 29 (1): 343–346.

47 Hu, S., Zhang, W., Wang, W. et al. (2022). Comprehensive comparisons of grafting-modified different polypropylene as HVDC cable insulation material. *IEEE Transactions on Dielectrics and Electrical Insulation* 29 (5): 1865–1872.

48 Du, B.X., Xu, H., and Li, J. (2017). Effects of mechanical stretching on space charge behaviors of PP/POE blend for HVDC cables. *IEEE Transactions on Dielectrics and Electrical Insulation* 24 (3): 1438–1445.

49 Fan, M., Li, Z., Du, B., and Zhou, S. (2020). Effect of sPP content on dielectric relaxation and trap level characteristics of PP blend insulation. *2020 IEEE 3rd International Conference on Dielectrics (ICD)*, Valencia, Spain, 5 July 2020, 431–434.

50 Yang, K., Ren, Y., Jing, Z. et al. (2023). Enhancement in electrical properties of impact polypropylene copolymer for cable insulation induced by rapid cooling. *IEEE Transactions on Dielectrics and Electrical Insulation* 30 (1): 247–254.

51 Huang, X., Fan, Y., Zhang, J., and Jiang, P. (2017). Polypropylene based thermoplastic polymers for potential recyclable HVDC cable insulation applications. *IEEE Transactions on Dielectrics and Electrical Insulation* 24 (3): 1446–1456.

52 Y. H. Zhu, C. S. Wu, X. G. Xu et al. (2015). DC extruded cable systems for power transmission at a rated voltage up to and including 500kV – Part 1, GB/T 31489.1–2015. Standardization Administration of the People's Republic of China.

53 W. Gu, Y. Deng, R. K. Yang et al. (2018). Standard for construction and acceptance of cable line electric equipment installation engineering, GB 50168–2018. Standardization Administration of the People's Republic of China.

54 Du, B.X., Zhu, L.W., and Han, T. (2017). Effect of ambient temperature on electrical treeing and breakdown phenomenon of polypropylene with repetitive pulse voltage. *IEEE Transactions on Dielectrics and Electrical Insulation* 24 (4): 2216–2224. https://doi.org/10.1109/TDEI.2017.006229.

55 Zhu, L.W., Li, Z.L., and Hou, K. (2023). Effect of radical scavenger on electrical tree in cross-linked polyethylene with large harmonic superimposed DC voltage. *High Voltage* 8 (4): 739–748. https://doi.org/10.1049/hve2.12302.

56 Varlow, B.R. and Auckland, D.W. (1996). Mechanical aspects of electrical treeing in solid insulation. *IEEE Electrical Insulation Magazine* 12 (2): 21–26. https://doi.org/10.1109/57.486948.

57 Kosaki, M., Nagao, M., Shimizu, N., and Mizuno, Y. (1998). Solid insulation and its deterioration. *Cryogenics* 38 (11): 1095–1104. https://doi.org/10.1016/S0011-2275(98)00097-6.

58 Holto, J. and Ildstad, E. (2010). Electrical tree growth in extruded s-polypropylene. *IEEE International Conference on Solid Dielectrics*, Potsdam, Germany, 1–4. https://doi.org/10.1109/ICSD.2010.5568236.

59 Yamano, Y. (2014). Control of electrical tree at initiation stage in LDPE by mixed addition of Al_2O_3 nano-particle and azobenzoic compound. *IEEE Transactions on Dielectrics and Electrical Insulation* 21 (1): 209–216. https://doi.org/10.1109/TDEI.2013.003980.

60 Zhu, L., Du, B., Li, H., and Hou, K. (2019). Effect of polycyclic compounds fillers on electrical treeing characteristics in XLPE with DC-impulse voltage. *Energies* 12 (14): 1–15. https://doi.org/10.3390/en12142767.

61 Heidt, L.J. (1970). Solid dielectric polyolefin compositions containing various voltage stabilizers. US Patent 3, 522,183, July 1970.

62 Hunt, G.H. (1972). Cross-linked polyolefin compositions: US Patent 3, 632,680.

63 Jarvid, M., Johansson, A., Bjuggren, J.M. et al. (2014). Tailored side-chain architecture of Benzil voltage stabilizers for enhanced dielectric strength of cross-linked polyethylene. *Journal of Polymer Science Part B: Polymer Physics* 52 (16): 1047–1054. https://doi.org/10.1002/polb.23523.

64 Kisin, S., den Doelder, J., Eaton, R.F., and Caronia, P.J. (2009). Quantum mechanical criteria for choosing appropriate voltage stabilization additives for polyethylene. *Polymer Degradation and Stability* 94 (2): 171–175. https://doi.org/10.1016/j.polymdegradstab. 2008.11.009.

65 Du, B., Han, C., Li, Z., and Li, J. (2019). Improved DC conductivity and space charge characteristics of XLPE for HVDC cable application: effect of voltage stabilizers. *IEEE Access* 7: 66576–66583. https://doi.org/10.1109/ACCESS.2019.2917615.

66 Zhu, L., Du, B., Su, J. et al. (2019). Electrical treeing initiation and breakdown phenomenon in polypropylene under DC and pulse combined voltages. *IEEE Transactions on Dielectrics and Electrical Insulation* 26 (1): 202–210. https://doi.org/10.1109/TDEI.2018.007454.

67 Bamji, S.S., Bulinski, A.T., and Densley, R.J. (1991). Degradation mechanism at XLPE/semicon interface subjected to high electrical stress. *IEEE Transactions on Electrical Insulation* 26 (2): 278–284. https://doi.org/10.1109/14.78329.

68 Du, B., Han, C., Li, J., and Li, Z.L. (2019). Effect of voltage stabilizers on the space charge behavior of XLPE for HVDC cable application. *IEEE Transactions on Dielectrics and Electrical Insulation* 26 (1): 34–42. https://doi.org/10.1109/TDEI.2018.007390.

69 Deng, W., Ren, Y., Qu, K., and Zhao, H. (2022). Improved DC insulation performance of XLPE with graftable voltage stabilizer. *IEEE Transactions on Dielectrics and Electrical Insulation* 29 (5): 1857–1864. https://doi.org/10.1109/TDEI.2022.3198719.

70 Du, B. X., Zhu, L. W., & Han, T. (2016). Effect of low temperature on electrical treeing of polypropylene with repetitive pulse voltage. *IEEE Transactions on*

Dielectrics and Electrical Insulation, 23 (4), 1915–1923. https://doi.org/10.1109/tdei.2016.7556462.

71 Du, B., Sun, G., Wang, H., & Li, Z. (2024). Effect of Bending Deformation on Electrical Tree Properties of Polypropylene Copolymer and Blend Insulation for HVDC Cables. *IEEE Transactions on Dielectrics and Electrical Insulation*, 1–1. https://doi.org/10.1109/tdei.2024.3432092.

72 Zhu, L., Du, B., Li, Z. , Li, H., & Hou, K. (2020). Polycyclic Compounds Affecting Electrical Tree Growth in Polypropylene Under Ambient Temperature. *IEEE Access*, 8, 8886–8898. https://doi.org/10.1109/access.2020.2964036.

4

Insulation Thickness Optimization for HVDC Cables

4.1 Introduction

The larger transmission capacity and higher operating voltage levels bring additional issues and challenges to the structural design of HVDC cables, especially for the insulation thickness design of ±500 kV HVDC cables. Although a larger insulation thickness theoretically reduces the average electric field, it simultaneously changes the temperature of the insulation layer's both inner and outer sides, resulting in the increase of temperature gradient (ΔT). Suitable insulation thickness selection is essential to alleviate the cable's electric field distortion problem and improve its insulation reliability.

4.1.1 Development of Insulation Thickness of HVDC Cables

Currently, the most common voltage grade of HVDC extruded cables used in Europe is ± 320 kV, which are applied to the French-Spanish INELFE and the German DolWin3 projects to achieve the purpose of alternating current (AC) power grid interconnection and new energy grid connection [1]. The offshore wind power project in Rudong, China, uses flexible DC transmission technology to achieve the operational goal of ±400 kV HVDC-extruded cable [2]. Until now, the SuedLink cable projects located in the German corridor are being built with a target voltage level of ±500 kV, the highest in the world [3]. For achieving the sustainable development target of carbon neutrality, the future HVDC extruded cables' voltage level is developing toward ±640 kV or even ±800 kV with a transmission capacity of more than 8 GW [4, 5]. The larger transmission capacity and higher operating voltage levels also bring additional issues and challenges to the structural design of HVDC cables, especially for the insulation thickness design of ±500 kV HVDC cables. Europe and Japan use a relatively lower insulation

Polypropylene Cable Insulation, First Edition. Boxue Du and Zhonglei Li.

Published 2025 by John Wiley & Sons, Inc.

thickness ranging from 23 to 27 mm, while China has raised this parametric design value to approximately 30 mm [6].

4.1.2 Advantages of Insulation Thinning

Although a larger insulation thickness theoretically reduces the average electric field, it simultaneously changes the temperature of the insulation layer's both inner and outer sides, resulting in the increase of temperature gradient (ΔT). The generation of ΔT is mainly associated with the heat production and transfer caused by the cable conductor core loss during operation, and the uneven conductivity distribution in space caused by it can lead to severe distortion problems of the electric field occurring in the insulation layer's low-temperature side [7, 8]. The maximum distortion rate can reach 228% when ΔT comes to 40 °C [9]. Furthermore, there is a dynamic coupling process between charge behavior and electric field evolution during operation, which causes the electric field distortion problem as the primary reason for insulation degradation or even failure [10, 11]. To sum up, the different ΔT caused by the insulation thickness variation can lead to its internal charge and electric field distribution changes. Therefore, suitable insulation thickness selection is essential to alleviate the cable's electric field distortion problem induced by ΔT and improve its insulation reliability.

4.2 Electric Field Distribution Calculation Model for HVDC Cables

4.2.1 Classical Electromagnetic Theoretical Model

The electric field distribution in HVAC cables depends on the insulation geometry and relative permittivity of the insulation materials, which are insensitive to the electric field and operating temperature. According to Poisson Eq. (4.1) and electric field integral formula (4.2), the calculation formula of electric field distribution (4.3) of HVAC cable can be derived.

$$\nabla \cdot (\varepsilon \mathbf{E}) = \rho \tag{4.1}$$

$$U = \int E \mathrm{d}r \tag{4.2}$$

$$E_{\mathrm{AC}}(r) = \frac{U}{r \ln\left(\frac{r_{\mathrm{o}}}{r_{\mathrm{i}}}\right)} \tag{4.3}$$

where $\mathbf{E}$ is the electric field in V/m. ε is the dielectric constant in F/m. ρ is the electric charge density in C/m^3. E is the value of $\mathbf{E}$. U is the applied voltage in

V. r is the radius at the position of interest in m. $E_{AC}(r)$ is the AC electric field strength at radius r. r_i is the cable insulation inner radius in m. r_o is the cable insulation outer radius in m. It can be argued that the maximum field in an HVAC cable always occurs at the interface between insulation and inner-semiconductive layer, whereas the minimum field always occurs at the interface between the insulation and the outer-semiconductive layer.

For HVDC cables, the field distribution is dependent on the volume conductivity (or resistivity) of polymeric insulation, which varies markedly with the local electric field and temperature. At the initial pressurization time, the electric field distribution in the HVDC cable is similar to that in the HVAC cable, which satisfies the Laplace distribution. With the further extension of time, the electric field distribution changes from quasi-capacitance distribution to pure resistance distribution. The whole process can be described by Poisson Eq. (4.1), current continuity Eq. (4.4), and Ohm's law of current density (4.5):

$$\nabla \cdot \mathbf{j} = -\frac{\partial \rho}{\partial t} \tag{4.4}$$

$$\mathbf{j} = \sigma \mathbf{E} \tag{4.5}$$

where $\mathbf{j}$ is the current density in A/m^2. σ is the electrical conductivity in S/m. During the transition, the current and charge satisfy:

$$\nabla \cdot (\varepsilon E) = \nabla \cdot \left(\varepsilon \frac{\mathbf{J}}{\sigma}\right) = \mathbf{J} \cdot \nabla\left(\frac{\varepsilon}{\sigma}\right) + \frac{\varepsilon}{\sigma} \nabla \cdot \mathbf{J} = \mathbf{J} \cdot \nabla\left(\frac{\varepsilon}{\sigma}\right) + \frac{\varepsilon}{\sigma}\left(-\frac{\partial \rho}{\partial t}\right) \tag{4.6}$$

When the steady state is reached, the formula (4.5) changes as follows:

$$\nabla \cdot \mathbf{J} = 0 \tag{4.7}$$

Combine formulas (4.6) and (4.7) to obtain

$$\mathbf{J} \cdot \nabla\left(\frac{\varepsilon}{\sigma}\right) = \rho \tag{4.8}$$

As proved by relationship (4.8), the electromagnetic theory predicts that a dielectric sample where a steady DC current of density **J** is flowing, and where the ratio permittivity/conductivity varies with position, will acquire a charge density ρ [12].

Unfortunately, the insulation conductivity of solid dielectrics tends to be highly dependent on temperature and, to a lesser extent, on electrical field. Figure 4.1 presents the experimental conduction current versus electric field (I–E) at different temperatures for XLPE insulation [13]. The conduction current values, which were obtained after one hour of polarization, are proportional to the conductivity of XLPE. The conduction current proportional to the conductivity clearly exhibits a strong dependence on the temperature and electric field. The conductivity (σ) of

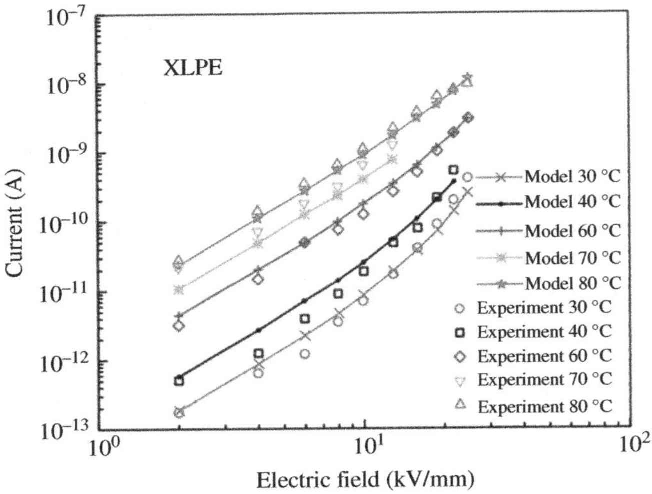

Figure 4.1 Experimental and fitted conduction current versus electric field at different temperatures for cross-linked polyethylene (XLPE) insulation. *Source:* Vu et al. [13]/with permission from IEEE.

polymeric insulation can be described as the following function of the electric field and temperature based on the electronic hopping conductivity model [14]:

$$\sigma(E,T) = A\exp\left(\frac{-\varphi\cdot q}{k_bT}\right)\frac{\sinh(B\cdot E)}{E} \tag{4.9}$$

where A and B are constants; φ is the thermal activation energy in eV; q is the elementary charge; T is the temperature in Kelvin; and E is the electric field in V/m. As shown in Figure 4.1, the conduction current of XLPE varies by two to three orders of magnitude with the temperature from 30 to 80 °C.

Boggs et al. proposed an analytical approximation for computing the field distribution in XLPE-insulated cables. The conductivities of two types of XLPE materials were defined as follows [15]:

$$\sigma_G(E,T) = 3.2781\times\exp\left(\frac{-0.56\cdot q}{k_bT}\right)\frac{\sinh(2.7756\times10^{-7}E)}{E} \tag{4.10}$$

$$\sigma_B(E,T) = 3.6728\times\exp\left(\frac{-0.98\cdot q}{k_bT}\right)\frac{\sinh(1.0860\times10^{-7}E)}{E} \tag{4.11}$$

The XLPE material G, which denotes "good" conductivity, has a lower thermal activation energy ($\varphi = 0.56$ eV) and a higher field dependence of conductivity. The material B, denoting "bad" conductivity, exhibits a higher thermal activation energy ($\varphi = 0.98$ eV) and a lower field dependence of conductivity.

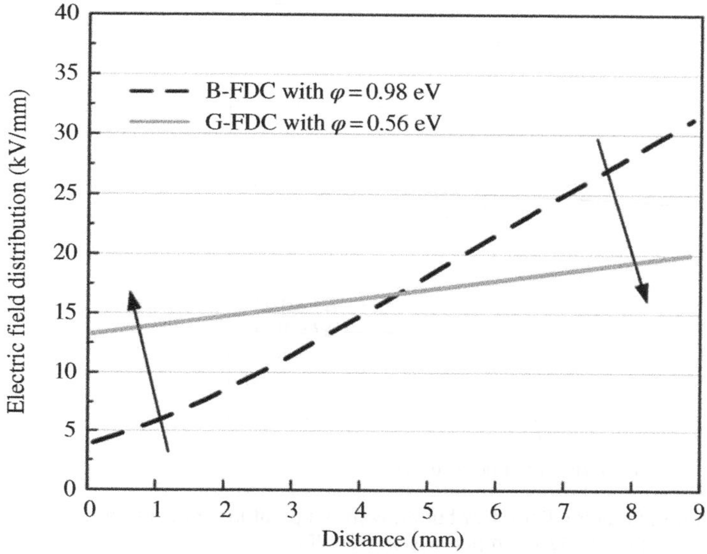

Figure 4.2 Electric field distribution in *G* and *B* materials considering the effect of the field-dependent conductivity (FDC) for a conductor and ground temperatures of 85 and 45 °C, respectively (the average field is 16.67 kV/mm). *Source:* Boggs et al. [15]/with permission from IEEE.

As the load current flows through the conductor in HVDC cables, the ohm heat formed in the conductor, as well as the ohmic heating in the insulation by the leakage current, generates a radial temperature gradient across the insulation material. As the temperature gradient increases, the conductivity of the insulation near the inner conductor shield increases sharply, resulting in a significant drop in the electric field at the inner side. As shown in Figure 4.2, for the XLPE insulation with an inner and outer shield temperature of 85 and 45 °C, the highest electric fields for the two types of insulation with field-dependent conductivity all appear at the outer insulation shield [15]. It is indicated that a lower thermal activation energy (0.56 eV) and a higher field dependence of conductivity could effectively uniform the electric field distribution across the insulation under the large thermal gradient. How to determine the optimal property of DC conductivity varying with the temperature and electric field is a key issue for the insulation design.

For the same insulating material, with the increase of temperature gradient, the electric field on the inner side of the insulation weakens and on the outer side of the insulation enhances. Finally, the phenomenon of "electric field inversion" appeared under the large temperature gradient. It is worth noting that when the temperature gradient is small ($\Delta T = 10\,°\text{C}$), the electric field distribution is uniform (Figure 4.3).

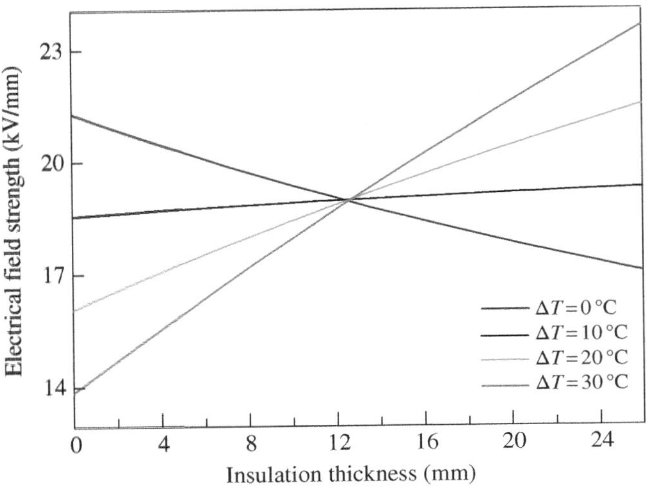

Figure 4.3 Variation of electric field distribution with temperature gradient based on Formula (4.9). *Source:* Li et al. [32]/with permission from IEEE.

Moreover, another method to uniform the electric field distribution is reducing the temperature gradient between the internal and the external insulation of the HVDC cable, by improving the thermal conductivity of polymeric insulation material [16, 17]. For polymeric insulation materials with low thermal conductivity, the introduction of particles with high thermal conductivity, including BN and AlN, has been proven as an effective but also a feasible method [17, 18]. But, with the enhancement of thermal conductivity, the electrical, mechanical, and other properties are also affected [19]. How to synergistically regulate electrical, thermal, and mechanical properties is an important issue for future studies.

4.2.2 Bipolar Electronic–Ionic Charge Transport Model

Classical electromagnetic theory model has been widely used to calculate the electric field distribution of HVDC cables. It should be emphasized that this macroscopic model ignores the transport operations such as injection, recombination, and impurity dissociation in the actual cable as in [20]. To overcome the limitations of the macroscopic models, a microscopic approach has been developed based on the so-called bipolar charge transport (BCT) model. Because of the limitations in the measuring range and accuracy of existing space charge measurement techniques, it is difficult to accurately measure space charge in ±500 kV HVDC cables. On the basis of the above problems, a modified BCT model considering the dissociation of impurity molecules is established in this chapter, the appropriate insulation thickness design parameters. In this chapter, based on the improved bipolar electronic–ionic charge transport (BEICT) model, the generation, transport, recombination, and extraction of electrons, holes, and positive and negative ions are considered comprehensively.

4.2.2.1 Charge Generation

Holes and electrons injected from the interface of the semi-conductive layer follow Schottky's theorem:

$$j_{\mathrm{h}}(r_{\mathrm{o}},t) = AT(r_{\mathrm{o}},t)^2 \exp\left(\frac{-w_{\mathrm{hi}}}{k_{\mathrm{B}}T(r_{\mathrm{o}},t)}\right) \exp\left(\frac{q\sqrt{\dfrac{eE(r_{\mathrm{o}},t)}{4\pi\varepsilon}}}{k_{\mathrm{B}}T(r_{\mathrm{o}},t)}\right) \tag{4.12}$$

$$j_{\mathrm{e}}(r_{\mathrm{d}},t) = AT(r_{\mathrm{d}},t)^2 \exp\left(\frac{-w_{\mathrm{ei}}}{k_{\mathrm{B}}T(r_{\mathrm{d}},t)}\right) \exp\left(\frac{q\sqrt{\dfrac{eE(r_{\mathrm{d}},t)}{4\pi\varepsilon}}}{k_{\mathrm{B}}T(r_{\mathrm{d}},t)}\right) \tag{4.13}$$

where A represents the constant of Richardson; j_{h} (r_{o}, t) and j_{e} (r_{d}, t), respectively, represent hole and electron current injection densities; t represents the polarization time; T represents the temperature; E represents the local electric field; q represents the elementary charge; k_{B} represents the Boltzmann constant; ε represents the polymers' permittivity; w_{hi} and w_{ei}, respectively, represent the hole and electron injection barriers.

While the impurity molecules have specific thermal vibration energy, they will overcome the dissociation barrier and be dissociated to form the ionic charge. The dissociation process satisfies the following equation:

$$D_{\mathrm{d}} = \nu \cdot \exp\left(\frac{-\Delta U_{\mathrm{dis}}}{k_{\mathrm{B}}T}\right) \tag{4.14}$$

where ν refers to the attempt for escape rate, and ΔU_{dis} refers to the dissociation potential of impurity molecules.

4.2.2.2 Charge Transport

In the process of polarization, the behavior of carriers on time and space scales can be described by

$$j_a(r,t) = \mu_a n_a(r,t)E(r,t) - D_{fa}\frac{\mathrm{d}n_a(r,t)}{\mathrm{d}r} \tag{4.15}$$

$$\nabla \cdot E(r,t) = \frac{n_{\mathrm{e}}(r,t) + n_{\mathrm{h}}(r,t) + n_{\mathrm{p}}(r,t) + n_{\mathrm{n}}(r,t)}{\varepsilon} \tag{4.16}$$

$$\frac{\partial n_a(r,t)}{\partial t} + \nabla \cdot j_a(r,t) = s_a(r,t) \tag{4.17}$$

where the subscript a corresponds to the types and states of charges. e, h, n, and p, respectively, represent electrons, holes, negative ions, and positive ions. μ and t, respectively, correspond to the two states of moving and trapped. j_a represent the density of current, μ_a represents the charges' mobility, n_a represents the charges' density, D_{fa} represents the diffusivity, s_a represents the source term, and ρ represents the density of net charge, which represents the charge density's change process.

The source terms of holes and electrons, respectively, correspond to

$$s_{\mathrm{ht}} = -S_{\mathrm{hte}\mu} \cdot n_{\mathrm{ht}} n_{\mathrm{e}\mu} - S_{\mathrm{htet}} \cdot n_{\mathrm{ht}} n_{\mathrm{et}} + n_{\mathrm{h}\mu} \cdot B_h \left(-\frac{n_{\mathrm{ht}}}{n_{\mathrm{oht}}} + 1 \right) - n_{\mathrm{ht}} \cdot D_{\mathrm{h}} \tag{4.18}$$

$$s_{\mathrm{h}\mu} = -S_{\mathrm{h}\mu\mathrm{et}} \cdot n_{\mathrm{h}\mu} n_{\mathrm{et}} - S_{\mathrm{h}\mu\mathrm{e}\mu} \cdot n_{\mathrm{h}\mu} n_{\mathrm{e}\mu} - n_{\mathrm{h}\mu} \cdot B_{\mathrm{h}} \left(-\frac{n_{\mathrm{ht}}}{n_{\mathrm{oht}}} + 1 \right) + n_{\mathrm{ht}} \cdot D_{\mathrm{h}} \tag{4.19}$$

$$s_{\mathrm{et}} = -S_{\mathrm{h}\mu\mathrm{et}} \cdot n_{\mathrm{h}\mu} n_{\mathrm{et}} - S_{\mathrm{htet}} \cdot n_{\mathrm{ht}} n_{\mathrm{et}} + n_{\mathrm{e}\mu} \cdot B_{\mathrm{e}} \left(-\frac{n_{\mathrm{et}}}{n_{\mathrm{oet}}} + 1 \right) - n_{\mathrm{et}} \cdot D_{\mathrm{e}} \tag{4.20}$$

$$s_{\mathrm{e}\mu} = -S_{\mathrm{hte}\mu} \cdot n_{\mathrm{ht}} n_{\mathrm{e}\mu} - S_{\mathrm{hue}\mu} \cdot n_{\mathrm{h}\mu} n_{\mathrm{e}\mu} - n_{\mathrm{e}\mu} \cdot B_{\mathrm{e}} \left(-\frac{n_{\mathrm{et}}}{n_{\mathrm{oet}}} + 1 \right) + n_{\mathrm{et}} \cdot D_{\mathrm{e}} \tag{4.21}$$

The source terms for positive and negative ions in the ionic charge can be written as

$$s_{\mathrm{pt}} = -S_{\mathrm{ptn}\mu} \cdot n_{\mathrm{pt}} n_{\mathrm{n}\mu} - S_{\mathrm{ptnt}} \cdot n_{\mathrm{pt}} n_{\mathrm{nt}} + B_{\mathrm{p}} \cdot n_{\mathrm{p}\mu} \left(1 - \frac{n_{pt}}{n_{opt}} \right) - D_{\mathrm{p}} \cdot n_{\mathrm{pt}} \tag{4.22}$$

$$s_{\mathrm{p}\mu} = -S_{\mathrm{p}\mu\mathrm{nt}} \cdot n_{\mathrm{p}\mu} n_{\mathrm{nt}} - S_{\mathrm{p}\mu\mathrm{n}\mu} \cdot n_{\mathrm{p}\mu} n_{\mathrm{n}\mu} - B_{\mathrm{p}} \cdot n_{\mathrm{p}\mu} \left(1 - \frac{n_{\mathrm{pt}}}{n_{\mathrm{opt}}} \right) + D_{\mathrm{p}} \cdot n_{\mathrm{pt}} + D_{\mathrm{e}} \cdot N_0 \tag{4.23}$$

$$s_{\mathrm{nt}} = -S_{\mathrm{p}\mu\mathrm{nt}} \cdot n_{\mathrm{p}\mu} n_{\mathrm{nt}} - S_{\mathrm{ptnt}} \cdot n_{\mathrm{pt}} n_{\mathrm{nt}} + B_{\mathrm{n}} \cdot n_{\mathrm{n}\mu} \left(1 - \frac{n_{\mathrm{nt}}}{n_{\mathrm{ont}}} \right) - D_{\mathrm{n}} \cdot n_{\mathrm{nt}} \tag{4.24}$$

$$s_{\mathrm{n}\mu} = -S_{\mathrm{ptn}\mu} \cdot n_{\mathrm{pt}} n_{\mathrm{n}\mu} - S_{\mathrm{p}\mu\mathrm{n}\mu} \cdot n_{\mathrm{p}\mu} n_{\mathrm{n}\mu} - B_{\mathrm{n}} \cdot n_{\mathrm{n}\mu} \left(1 - \frac{n_{\mathrm{nt}}}{n_{\mathrm{ont}}} \right) + D_{\mathrm{n}} \cdot n_{\mathrm{nt}} + D_{\mathrm{d}} \cdot N_0 \tag{4.25}$$

where S represents the charges' recombination coefficient. B_{h}, B_{e}, B_{p}, and B_{n}, respectively, represent holes, electrons, positive ions, and negative ions trapping coefficients. n_{oht}, n_{oet}, n_{opt}, as well as n_{ont} represent the densities of the trap. D_{e}, D_{h}, D_{n}, and D_{p} represent the coefficients of de-trapping. N_0 represents the residual impurity density.

In the polarization process, the residual impurity density also changes. The variation can be expressed as

$$\Delta N_0 = S_{\mathrm{ptn}\mu} \cdot n_{\mathrm{pt}} n_{\mathrm{n}\mu} + S_{\mathrm{p}\mu\mathrm{n}\mu} \cdot n_{\mathrm{p}\mu} n_{\mathrm{n}\mu} + S_{\mathrm{p}\mu\mathrm{nt}} \cdot n_{\mathrm{p}\mu} n_{\mathrm{nt}} + S_{\mathrm{ptnt}} \cdot n_{\mathrm{pt}} n_{\mathrm{nt}} - D_{\mathrm{d}} \cdot N_0 \tag{4.26}$$

The charge injected by the electrode migrates under the applied electric field, and the conduction of this type of charge is expressed by the apparent mobility which includes the charge trapping and de-trapping in the shallow trap. If the shallow traps of electrons and holes have a single energy level, the apparent mobility under shallow trap modulation can be

$$\mu_{e\mu} = \mu_{\mathrm{c}} \frac{N_{\mathrm{c}}}{N_{\mathrm{c}} + N_{\mathrm{se}}} \exp\left(-\frac{\Delta U_{\mathrm{stre}}}{kT}\right) \tag{4.27}$$

$$\mu_{h\mu} = \mu_{v} \frac{N_{v}}{N_{v} + N_{sh}} \exp\left(-\frac{\Delta U_{strh}}{kT}\right) \tag{4.28}$$

where μ_{c} and μ_{v} for carrier mobility in the conduction band and the valence band. N_{c} and N_{v} for effective density of states of the conduction band and the valence band. ΔU_{stre} and ΔU_{strh} for the depth of electron shallow trap and hole shallow trap.

In the process of polarization, ion carriers overcome the migration barrier and migrate to the opposite polarity electrode. The mobility of ions is defined as

$$\mu_a = \frac{\delta_a v}{6E(r,t)} \exp\left(-\frac{\Delta U_{ta}}{k_{\mathrm{B}}T}\right) \cdot \left(\exp\left(\frac{\Delta U_{na}}{k_{\mathrm{B}}T}\right) - \exp\left(-\frac{\Delta U_{na}}{k_{\mathrm{B}}T}\right)\right) \tag{4.29}$$

where ΔU_{na} represents the barrier height's change, ΔU_{ta} represents the barrier height of migration, and δ_a represents the mean hop distance. Among them, ΔU_{na} is defined as

$$\Delta U_{na} = 0.5 \cdot \delta_a \cdot e \cdot E(r,t) \tag{4.30}$$

The de-trapping coefficient can be written as

$$D_a = v \cdot \exp\left(-\frac{\Delta U_{\mathrm{dtra}}}{k_{\mathrm{B}}T}\right) \tag{4.31}$$

where the ΔU_{dtra} represents the deep trap depth for the carriers.

4.2.2.3 Charge Recombination

The various S of hole–electron pairs and positive–negative ion pairs need to be considered in the charge recombination process. These coefficients follow the charge mobility function and satisfy the Langevin type. Taking moving electrons and holes as an example, it can be written as:

$$S_{\mathrm{h}\mu\mathrm{et}} = \frac{\mu_{\mathrm{h}\mu}(r,t)}{\varepsilon} \tag{4.32}$$

$$S_{\mathrm{hte}\mu} = \frac{\mu_{\mathrm{e}\mu}(r,t)}{\varepsilon} \tag{4.33}$$

$$S_{\mathrm{e}\mu\mathrm{h}\mu} = \frac{\mu_{\mathrm{e}\mu}(r,t) + \mu_{\mathrm{h}\mu}(r,t)}{\varepsilon} \tag{4.34}$$

4.2.2.4 Charge Extraction

When the charges migrate to the opposite electrode, the ionic charges are considered to be completely blocked, so the extraction current is zero. While holes and electrons are considered to be blocked partially and satisfy the following formula:

$$j_{\mathrm{e}}(r_0,t) = E(r_0,t)C_{\mathrm{e}}\mu_{\mathrm{e}\mu}n_{\mathrm{e}\mu}(r_0,t) \tag{4.35}$$

$$j_{\mathrm{h}}(r_{\mathrm{d}},t) = E(r_{\mathrm{d}},t)C_{\mathrm{h}}\mu_{\mathrm{h}\mu}n_{\mathrm{h}\mu}(r_{\mathrm{d}},t) \tag{4.36}$$

where C_{e} and C_{h} represent the electron and hole coefficients of extraction, respectively.

The internal heat transfer mode of HVDC cable insulation is heat conduction

$$\rho_m c_m \frac{\partial T}{\partial t} = \nabla \cdot (\lambda \nabla T) + Q \tag{4.37}$$

where ρ_m is the insulation density; c_m is the specific heat capacity; λ is the thermal conductivity; Q is the heat source (W/m^3). The temperature of the outer semi-conductive layer is set at 40 °C, and the temperature of the inner semi-conductive layer is set at 40, 50, 60, and 70 °C. The temperature difference between the inner and outer sides of insulation is defined as the ΔT.

4.3 Space Charge and Electric Field Under DC Voltage

The classical electromagnetic theoretical model and the BEICT model are the same in the evolution trend of electric field, which shows that the results of the BEICT model are convincing [20]. The classical electromagnetic theoretical model (macroscopic model) mainly calculates the electric field by conductivity. It is difficult for a macroscopic model to accurately predict the distribution and transport of space charge. In the BEICT model (microscopic model), the dynamic behaviors of carriers in solid media are considered. Therefore, the follow-up discussion focuses on the calculation results of the BEICT model, and the effect of charge behavior on field distribution during polarity reversal is investigated. The different ΔT caused by the insulation thickness variation can lead to its internal charge and electric field distribution changes. Therefore, suitable insulation thickness selection is essential to alleviate the cable's electric field distortion problem induced by ΔT and improve its insulation reliability.

Figure 4.4 represents the dynamical process of (a) space charge behavior and (b) electric field evolution within the insulation layer during polarization under different ΔT, respectively. It can be seen that when $\Delta T = 0\,°\text{C}$, the initial electric field of the insulation layer's inner side is higher and gradually declines from inward to outward along the radial direction. At this time, with the low electric field acting, the electronic charges injection effect located at the semi-conductive layer's inner and outer interface is weak, and the injected holes and electrons amount is small and mainly distributed near the interface among the semi-conductive with the insulation layer. While the impurities inside the insulation material form massive positive and negative ionic charges through dissociation, which gradually accumulate on the insulation layer's inner and outer sides. Therefore, the ionic charge as heterocharges plays a dominant role, which enhances the field strength within the insulation's inner and outer sides to varying degrees.

As the conductor's temperature increases, ΔT within the insulation layer increases continuously. On the insulation layer's inner side with a higher

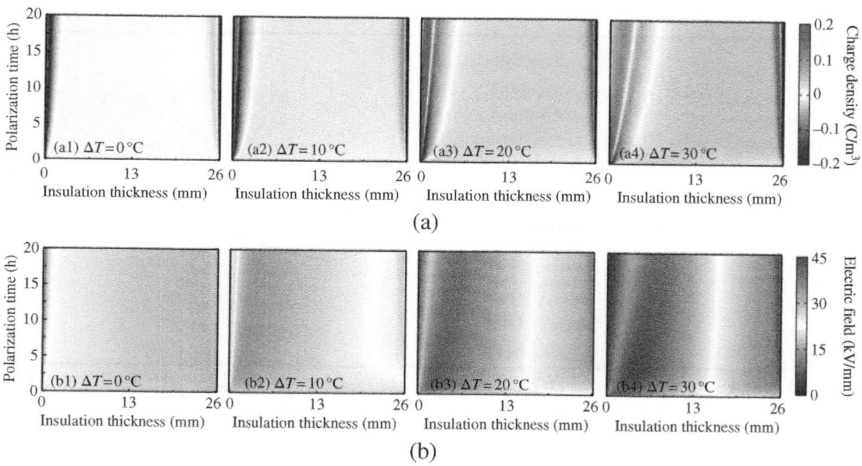

Figure 4.4 Effect of temperature gradient on (a) space charge and (b) electrical field distributions under +500 kV. *Source:* Li et al. [32]/with permission from IEEE.

temperature, the inner semi-conductive layer's higher temperature provides enough kinetic energy for the electronic charge to help it cross the interface barrier. Therefore, in the polarization process, many holes inject from the inner semi-conductive layer into the insulation layer and accumulate on its inner side in the form of homocharges. The effect is gradually more significant than that of the heterogeneous charges accumulated on the same side so that the initially enhanced field strength is suppressed.

Meanwhile, with the increase of ΔT, the accumulated heterocharges in the insulation layer gradually increase. This trend is most evident on the insulation layer's outer side with lower temperatures. The cause is that the "electric field reversal" effect induced by the ΔT strengthens the electric field strength in its vicinity so that the dissociation reaction of impurity molecules is enhanced. The generated heterocharges further enhance the field strength within the insulation layer's outer side.

By comparing the variation of the electric field within the insulation layer before and after the polarization process shown in Figure 4.5, it can be found that when $\Delta T = 0\,°C$, the electric field of the insulation layer on both the inner and outer sides is enhanced to different degrees with the effect of heterocharges. This enhancement effect is particularly evident on the insulation layer's inner side. After polarization for 20 hours, the maximal electric field value is 36.86 kV/mm, which increases by 68% compared with that before polarization.

However, when there is a ΔT in the cable, the electric field on the insulation layer's inner side is generally weakened by the injected holes, while the positive ionic charge further enhances the outer side electric field. The "electric field

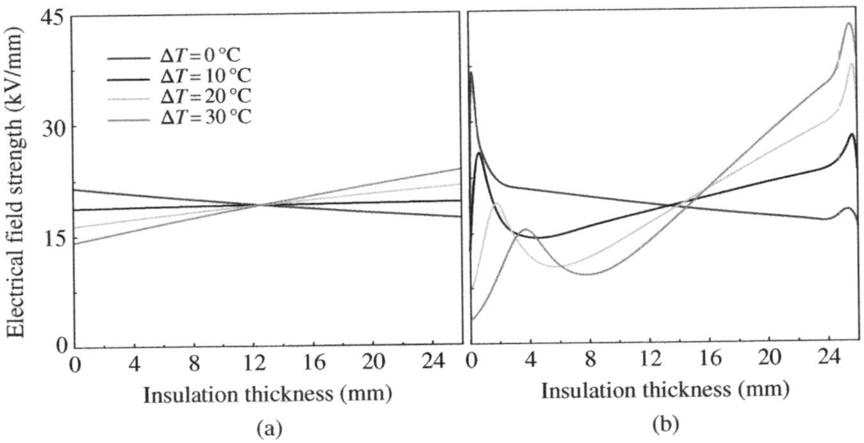

Figure 4.5 Electrical field distributions varying with temperature gradients. (a) Polarized for 0 h. (b) Polarized for 20 h. *Source:* Li et al. [32]/with permission from IEEE.

reversal" trend that low inner but high outer is also more significant with an increasing ΔT. When $\Delta T = 30\,°C$, the differential value between the insulation layer's inner and outer sides is only 9.6 kV/mm before polarization, but it reaches 39.35 kV/mm after polarization. It can be concluded that as the insulation layer's ΔT rises, the electric field distortion problem occurring within the radial direction of the cable becomes more and more serious.

For quantifying the influence of ΔT on the electric field evolution in cable, the maximal electric field strength (E_{max}) before and after polarization in the insulation layer under different ΔT is counted, and the data are shown in Figure 4.6. It can be found that after 20 hours polarized, the E_{max} within the insulation first decreases and then increases as ΔT increases. Among them, the value is the smallest at $\Delta T = 10\,°C$, only 27.85 kV/mm, and reaches the highest at 42.69 kV/mm when ΔT increases to 30 °C.

To analyze the influence mechanism of ΔT on the electric field evolution, the ΔT effect on both the value magnitude and the amplitude change of E_{max} before and after polarization is also compared in Figure 4.6.

By comparing the difference, it is found that when $\Delta T = 10\,°C$, the internal E_{max} of the cable is increased by about 42% compared with 19.54 kV/mm before polarization. And, the enhancement effect becomes more and more significant with the increase of ΔT. When $\Delta T = 30\,°C$, the lifting rate reaches 80%. The critical reason leading to this phenomenon is the space charge effect during polarization. Therefore, it can be concluded that temperature gradients can significantly affect the electric field evolution via influencing the space charge behavior within the cable.

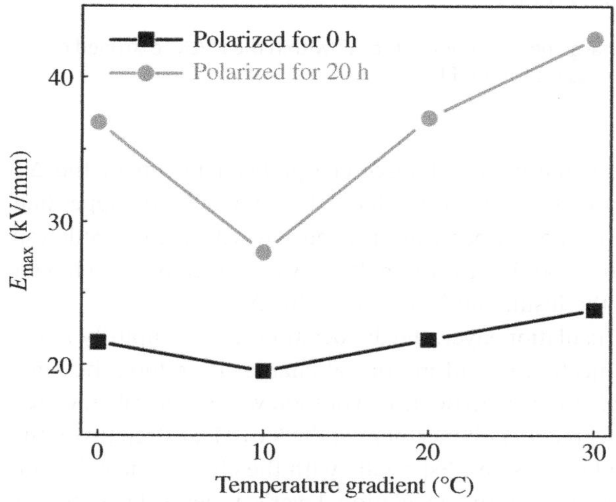

Figure 4.6 Relationship between the E_{max} and the temperature gradient. *Source:* Li et al. [32]/with permission from IEEE.

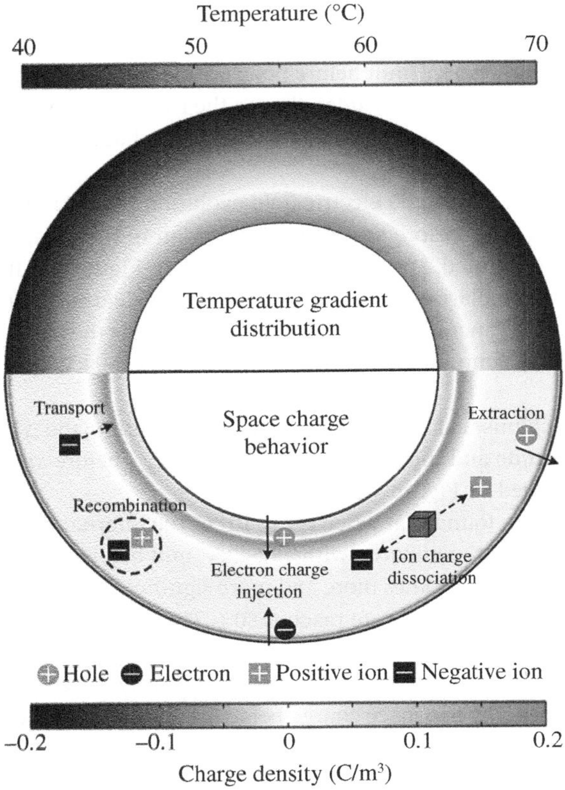

Figure 4.7 Model of space charge behavior under the ΔT within the HVDC-extruded cable. *Source:* Li et al. [32]/with permission from IEEE.

With the effect of the ΔT, the model of space charge behavior under the ΔT within the HVDC extruded cable is shown in Figure 4.7. In which, the upper half of the figure corresponds to the temperature distribution within the insulation layer in the case of $\Delta T = 30\,°C$. And, the lower half shows specifically the behavior of the four carriers inside the insulation layer under this ΔT.

When $\Delta T = 0\,°C$ in the insulation layer, the dissociation rate of impurity molecules is greater than the injection rate of interfacial charge. Therefore, the ionic charge plays the leading role in the electric field evolution within the cable, which is manifested in different degrees of enhancement of the local field strength on the insulation layer's inner and outer sides. Especially with the effect of negative ion charge, the local field strength of the insulation layer's inner side is greatly enhanced. The main reason leading to this result is related to the electric field distribution within the insulation. At this point, it gradually decreases from the inner to the outer side along the radial direction. Therefore, compared with the positive ions migrating to the insulation layer's outer side, the negative ions migrate to the insulation layer's inner side at a relatively faster rate under the influence of the strong local field and accumulate around in large amounts as heterocharges.

The ΔT between the insulation layer's inner and outer sides becomes larger with the conductor temperature rising. At this time, the space charge injection behavior, which appears at the inner semi-conductive layer interface with the higher temperature, is gradually enhanced. The injected holes accumulate on the insulation layer's inner side as homocharges, and the effect of it is increasingly more significant than that of the heterocharges on the same side, making the initially enhanced field strength in turn suppressed.

When $\Delta T = 30\,°\mathrm{C}$ within the cable, for the insulation layer's inner side with higher temperature, the influence of the difference in voltage polarity on the carrier behavior is mainly reflected in the charge generation process. Similarly, at the insulation layer's inner side with higher temperature, the injection process of interfacial charge is significantly stronger for positive voltage than negative voltage. And, this is mainly related to the injection barrier height of the interfacial charge. Since the holes' injection potential barrier is lower than electrons, the number of holes injected at the interface of the inner semiconducting layer during positive voltage action is higher than electrons injected under the effect of negative voltage.

While for the insulation layer's outer side with lower temperature, the effect of the difference in voltage polarity on the carrier behavior is mainly reflected in the charge migration process. Due to the higher trapping rate of positive ion charges, negative ion charges migrate relatively faster. Therefore, in the same polarization time, the negative ion charges that migrate and accumulate at the low-temperature side under the effect of negative voltage are more than the positive ion charges under the positive voltage. In contrast, the enhancement effect on the nearby electric field is also more significant.

4.4 Space Charge and Electric Field Under Polarity Reversal Voltage

In the LCC–HVDC transmission system, voltage polarity reversal is necessary to meet the bidirectional power flow demand. In the process of polarity reversal, the superposition of the induced electric field, generated by accumulated space charges, and the applied electric field may cause serious deformation of the electric field in the insulation [20]. The violent charge behavior and the strong electric field will lead to a sharp decline in the breakdown property of polymers and shorten the operating life [21, 22]. According to CIGRE TB-496, the reversal period of the polarity reversal test shall not exceed 120 seconds. The shorter the polarity reversal period, the higher enhancement of the electric field in the insulation [23]. However, applying relaxation time (voltage is 0 kV) can alleviate this problem [24]. Hence, it is crucial to investigate the impact of the polarity reversal period on the charge transport behaviors and electric field distribution. Due to the combined impact of the temperature gradient, space charge, and applied voltage, the electric field distribution in the insulation becomes challenging to predicate. In particular, the working condition of polarity reversal voltage superimposed with temperature gradient poses a serious threat to cable safety.

4.4.1 Effect of Temperature Gradients

As shown in Figure 4.8, the voltage during polarity reversal is set. To investigate the impact of the polarity reversal period, two types of polarity reversal periods have been established. For the type shown in Figure 4.8a, the variation of charge and the distribution of the electric field when the polarity reversal period is 300 ms, 10 seconds, and 120 seconds are discussed. To investigate the impact of the relaxation period, the polarity reversal period is set to 60 seconds, and the relaxation

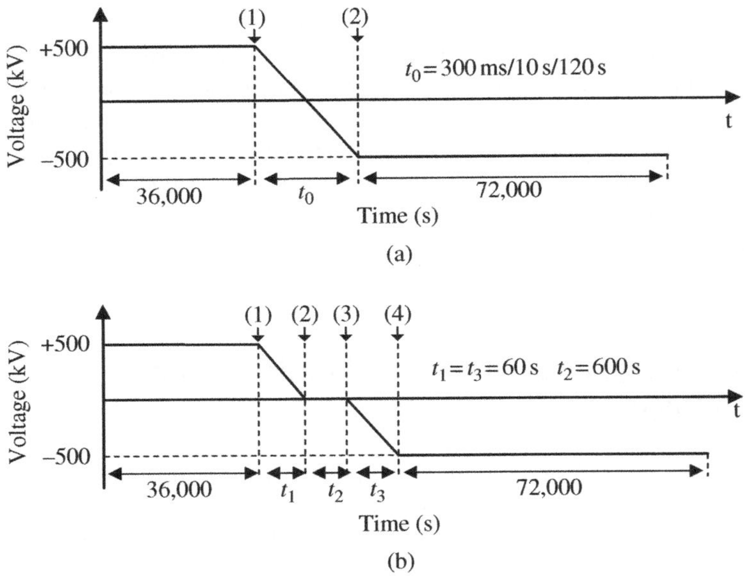

Figure 4.8 Schematic diagram of polarity reversal voltage (a) without relaxation period and (b) with relaxation period. *Source:* Li et al. [33]/with permission from IEEE.

period (voltage is 0 V) is set to 600 seconds, corresponding to the type shown in Figure 4.8b, which can be denoted as 120 seconds with a relaxation period of 600 seconds. Before starting the polarity reversal, +500 kV is applied to the cable core for 36,000 seconds, followed by −500 kV for 72,000 seconds after the polarity reversal. The polarity reversal period is defined as the PRP.

Figure 4.9 shows the distribution of space charge and electric field along the insulation radius under ΔT. The polarity reversal period is 120 seconds, and the dotted line indicates the beginning and the end moment of the polarity reversal. To show the distribution of space charge and the evolution of electric field during polarity reversal, the abscissa of Figure 4.9a and b is taken as the logarithmic coordinate axis.

As shown in Figure 4.9a1, when the ΔT is 0 °C and the DC voltage is applied for 36,000 sceonds, the space charge mainly accumulates within 0.3 mm near the inner and outer semi-conductive layers. And, there is almost no accumulated space charge in the middle of the insulation. As shown in Figure 4.9b1, the E_{max} occurs on the insulation layer's interior side, which is 27.1 kV/mm. During the 120 seconds polarity reversal process, the accumulated space charge in the insulation cannot dissipate in time, resulting in the electric field consisting of the capacitive field generated by the applied voltage and the induction field generated by the space charge. The E_{max} occurs at the interface between the insulation and the inner semi-conductive layer, reaching 39.8 kV/mm, which is 107.2% larger than the average electric field (19.2 kV/mm). When −500 kV voltage is applied, the E_{max} also occurs on the interior side of the insulation, with a value of 25.3 kV/mm. In addition, it is noteworthy that under different ΔT in Figure 4.9a, the distribution of space charge in the steady state before and after the reversal exhibits the "mirror effect."

When the ΔT increases to 10 °C, as shown in Figure 4.6a2, during +500 kV pre-loading, the accumulation range of space charge on the inner side of the insulation raises to 0.8 mm. There is a small amount of positive ion migrating from the high-temperature region to the low-temperature region in the bulk of the insulation. As a result, in Figure 4.9b2, the E_{max} on the inner side drops to 20 kV/mm, while that on the outer side increases to 22.9 kV/mm. The electric field is uniformly distributed along the insulation radius. In the process of polarity reversal, the E_{max} appears at the interface between the insulation and inner semi-conductive layer, and its value reaches 44.5 kV/mm.

In Figure 4.9, as the ΔT is further increased under DC voltage, in the vicinity of the inner semi-conductive layer, the accumulation range of the space charge expands while the space charge density decreases. There is minimal variation observed in the accumulation range of space charge on the outer side. Moreover, the temperature rise accelerates the migration of positive ion toward the insulation layer's outer side. When the ΔT is up to 30 °C, the E_{max} on the interior side of the

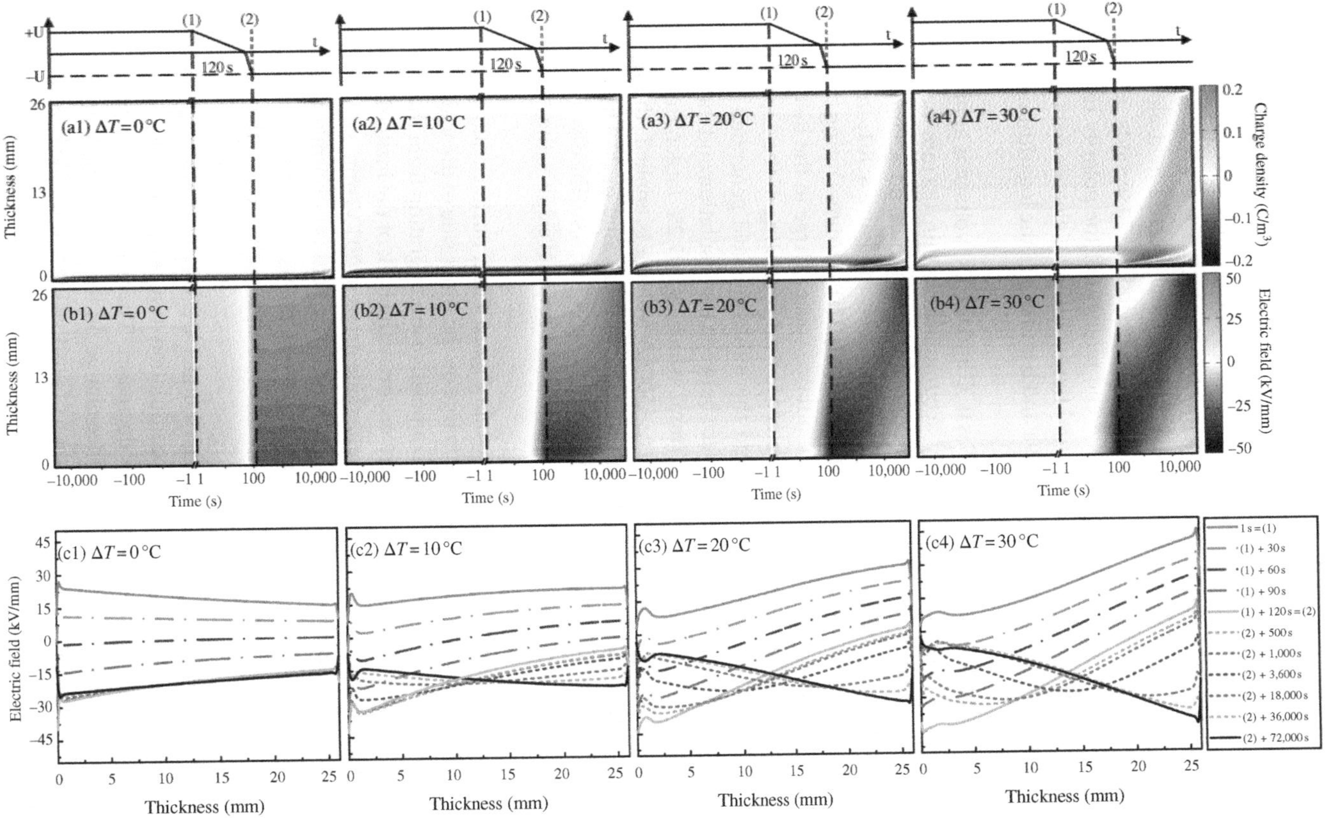

Figure 4.9 Effect of ΔT on (a) space charge and (b) electrical field evolution. (c) Electric field profiles during polarity reversal. One second is the moment when the polarity reversal voltage is applied, which is labeled (1) in (c1–c4). *Source:* Li et al. [33]/with permission from IEEE.

insulation drops to 1 kV/mm, while that on the outer side surges to 45.3 kV/mm, thus inducing the "electric field reversal" phenomenon. This phenomenon leads to a lag of the polarity reversal of the electric field on the outer side compared to that on the inner side during polarity reversal as shown in Figure 4.9b3 and b4. After the polarity reversal, the E_{max} reaches 46.1 kV/mm at the $\Delta T = 30\,°C$, representing an increase of 15.8% compared to that at the $\Delta T = 0\,°C$. Furthermore, due to the high temperature of the conductor, there are intense charge behaviors on the inner side during polarity reversal, which causes the local electric field to weaken as shown in Figure 4.9c4.

As shown in Figure 4.10, to further investigate the impact of the ΔT on the electric field, the E_{max} is compared under different ΔT. It can be observed that the E_{max} during the polarity reversal exhibited an increasing trend as the ΔT increases. Additionally, in Figure 4.9, with the increase of ΔT, the influence of ion in the high-temperature region on the electric field weakens, while that in the low-temperature region gradually strengthens.

In the process of reversal, the electric field consists of the capacitive field caused by the applied voltage and the induction field generated by the residual space charge. The space charge and the E_{max} mainly occur near the inner and outer semi-conductive layers. Therefore, Figure 4.11 further analyzes the evolution of space charge within 1 mm near the inner and outer semi-conductive layers under different ΔT. The average density of space charge within 1 mm near the inner and outer semi-conductive layers is calculated. The impact of the ΔT on the electric

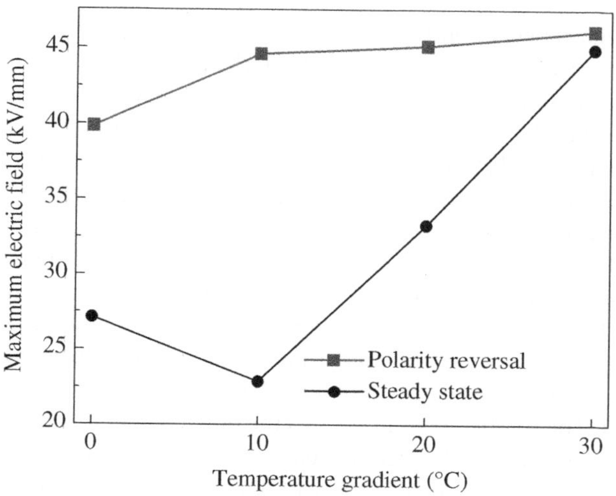

Figure 4.10 E_{max} before and during polarity reversal under different ΔT. *Source:* Li et al. [33]/with permission from IEEE.

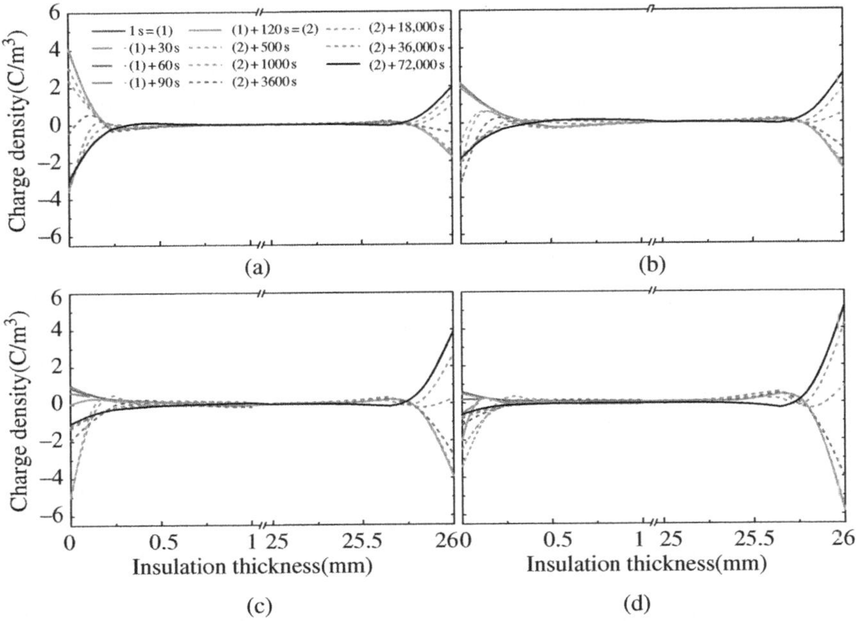

Figure 4.11 Space charge behaviors in the vicinity of the inner and outer semi-conductive layers under different ΔT. (a) ΔT = 0 °C. (b) ΔT = 10 °C. (c) ΔT = 20 °C. (d) ΔT = 30 °C. *Source:* Li et al. [33]/with permission from IEEE.

field is manifested through its effect on the distribution of space charge in the steady state and the charge behaviors in the process of polarity reversal. As shown in Figure 4.11a, in the steady state at the $\Delta T = 0\,°\text{C}$, the charge behaviors are not violent, but the small accumulation range leads to high space charge density, resulting in serious electric field distortion on both sides of the insulation. The average density of space charge on the interior side is 0.28 C/m^3 and that on the exterior side has a negative polarity with a density of 0.14 C/m^3. As the ΔT increases, the amount of hole on the inner side of the insulation increases more than that of negative ion, thereby weakening the influence of negative ion on the field. The density of the space charge on the inner side decreases. When the ΔT is 30 °C, the average density of space charge on the inner side drops to 0.11 C/m^3. Due to the enhancement of carrier mobility, the range of space charge accumulation is enlarged. The positive ion migrates to the outer side of the insulation, leading to the enhancement of the electric field. Therefore, the phenomenon of "uniform distribution of electric field" ($\Delta T = 10\,°\text{C}$) and the phenomenon of "electric field reversal" (ΔT = 20 and 30 °C) are observed in Figure 4.9. As shown in Figure 4.11d, due to the enhanced electric field on the outer side, the injection of electron increases, and the average density of the negative space charge reaches 0.36 C/m^3 when the ΔT rises to 30 °C.

During polarity reversal, the increase in conductor temperature aggravates the charge behaviors on the insulation layer's interior side. Meanwhile, the $E_{\max}$ appears in the proximity of the inner semi-conductive layer. Hence, Figure 4.12 analyzes the charge variation rate on the interior side of the insulation under different ΔT. The rate of carrier recombination (C/(m^2·s)) within 1 mm near the inner semi-conductive layer is calculated according to (4.7)–(4.9). Charge injection current density (A/m^2) is calculated according to Schottky injection. Charge extraction current density (A/m^2) is calculated according to Ohmic extraction. The dotted line indicates the end of the polarity reversal. The integration of the curve with respect to time represents the amount of charge variation:

$$R_{\mathrm{eh}} = S_{\mathrm{hte}\mu} n_{\mathrm{ht}} n_{\mathrm{e}\mu} + S_{\mathrm{htet}} n_{\mathrm{ht}} n_{\mathrm{et}} + S_{\mathrm{h}\mu\mathrm{e}\mu} n_{\mathrm{h}\mu} n_{\mathrm{e}\mu} + S_{\mathrm{h}\mu\mathrm{et}} n_{\mathrm{h}\mu} n_{\mathrm{et}} \tag{4.38}$$

$$R_{\mathrm{np}} = S_{\mathrm{ptn}\mu} n_{\mathrm{pt}} n_{\mathrm{n}\mu} + S_{\mathrm{p}\mu\mathrm{n}\mu} n_{\mathrm{p}\mu} n_{\mathrm{n}\mu} + S_{\mathrm{p}\mu\mathrm{nt}} n_{\mathrm{p}\mu} n_{\mathrm{nt}} + S_{\mathrm{ptnt}} n_{\mathrm{pt}} n_{\mathrm{nt}} \tag{4.39}$$

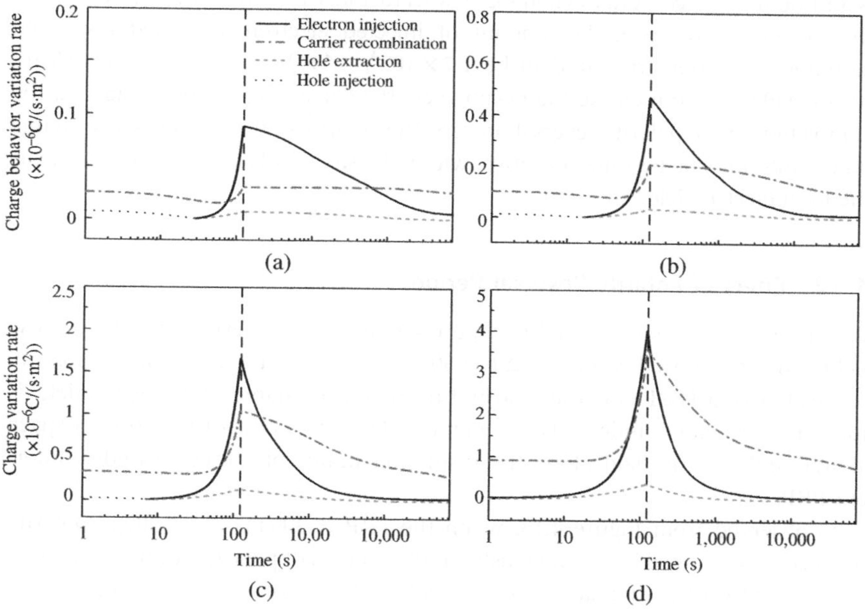

Figure 4.12 Charge variation rate on the inner side of the insulation under different ΔT. One second is the moment when the polarity reversal voltage is applied. (a) ΔT = 0 °C. (b) ΔT = 10 °C. (c) ΔT = 20 °C. (d) ΔT = 30 °C. *Source:* Li et al. [33]/with permission from IEEE.

$$R = \int_{r_d}^{r_d + l} \left(R_{eh} + R_{np}\right) dr \tag{4.40}$$

where subscripts h and e represent hole and electron, respectively. R_{eh} and R_{np} represent the rate of electron–hole recombination (C/(m^3·s)) and the rate of positive–negative ion recombination (C/(m^3·s)), respectively. R is the rate of carrier recombination within 1 mm near the inner semi-conductive layer (C/(m^2·s)). Take the hole as an example, n_{ht} is the density of trapped hole (C/m^3), $n_{h\mu}$ is the density of free hole (C/m^3), and $l = 1$ mm.

As shown in Figure 4.12, under different ΔT, electron injection and carrier recombination are the main charge behaviors during polarity reversal. With the increase of the ΔT, the injection time of holes during the polarity reversal decreases while the injection time of electrons increases. When the ΔT is up to 30 °C, the rates of electron injection, hole extraction, and carrier recombination are over 10 times greater than those at the ΔT of 0 °C. As shown in Figures 4.11a and 4.12a, when the $\Delta T = 0$ °C, the amount of injected electron within 1 mm in the vicinity of the inner semi-conductive layer is very small, which is 2.4×10^{-6} C/m^2 in the process of polarity reversal. At the end of the polarity reversal, the charge density is almost unchanged. As shown in Figures 4.11d and 4.12d, when the ΔT is 30 °C, in the process of reversal, the amount of injected electron, extracted hole, and recombined carrier are found to be 1.7×10^{-4}, 2.5×10^{-5}, and 1.9×10^{-4} C/m^2, respectively, which increase the negative charge density on the outer side. At the completion of the polarity reversal process, the polarity of the average space charge on the interior side becomes negative, with a density of 0.11 C/m^3, which weakens the local electric field.

4.4.2 Effect of Polarity Reversal Periods

Based on the results presented in Figure 4.9, it can be inferred that the electric field is most distorted when the ΔT is 30 °C. Therefore, the effect of the polarity reversal period (PRP) on the distribution of space charge and electric field is further investigated under this ΔT. Figure 4.13 displays the distribution of space charge and electric field along the insulation radius of the cable under different PRP.

As observed from Figure 4.13, when the PRP = 300 ms, the voltage polarity reversal is completed instantaneously. In the process of polarity reversal, the E_{max} appears on the insulation layer's interior side with a value of 51.1 kV/mm, which is 28.4% higher than that when the $\Delta T = 0$ °C. With the extension of the PRP to 10 seconds, the E_{max} decreases to 49.6 kV/mm due to the charge transport behaviors on the interior side of the insulation during the polarity reversal process. As shown in Figure 4.13c3, the PRP of 120 seconds results in a decrease of 9.8% in the E_{max}

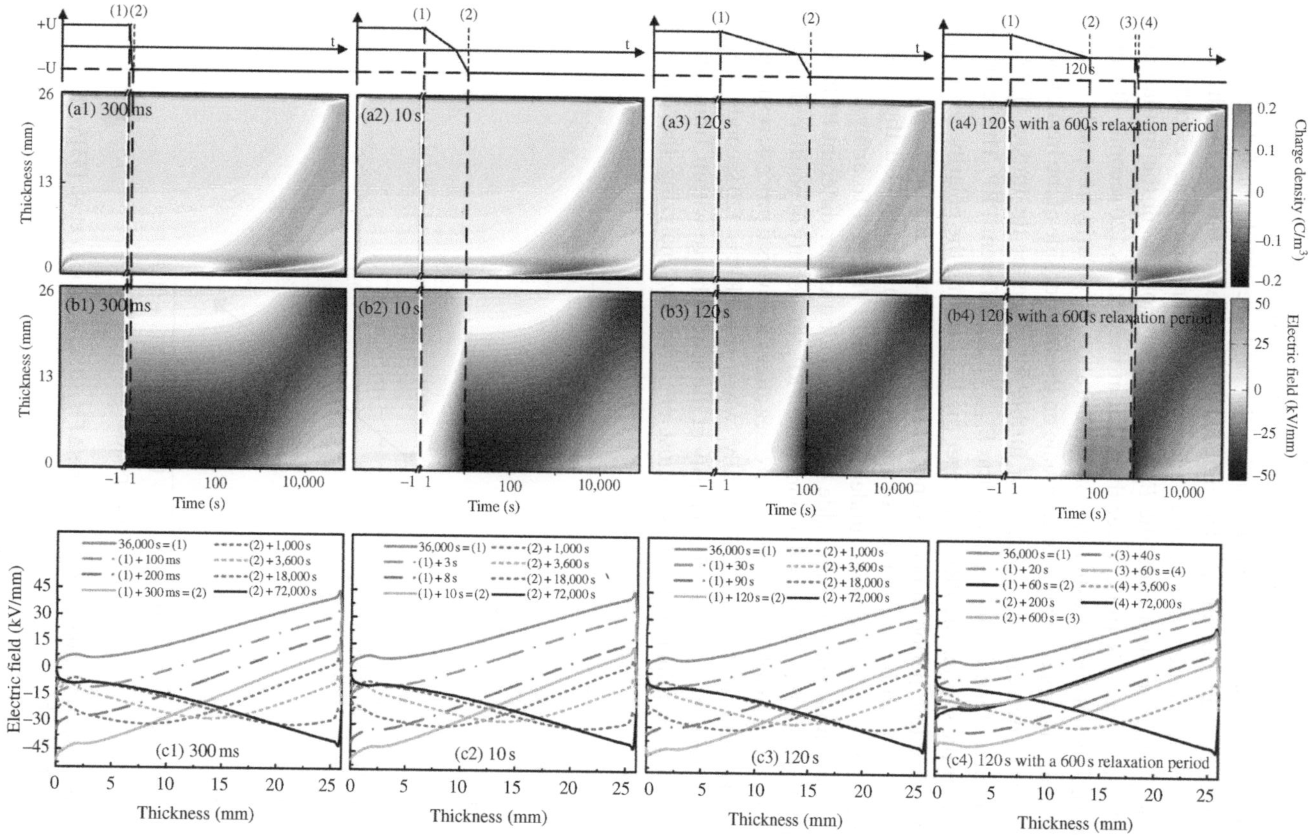

Figure 4.13 Effect of PRP on (a) space charge and (b) electric field evolution as the ΔT is 30 °C. (c) Electric field profiles during polarity reversal. One second is the moment when the polarity reversal voltage is applied, which is labeled (1) in (c1–c4). *Source:* Li et al. [33]/with permission from IEEE.

compared with the PRP of 300 ms. As illustrated in Figure 4.13b4 and c4, when the PRP is 120 seconds with 600 seconds relaxation period, during the relaxation process, the polarity of the electric field on the interior side is reversed. The field is generated solely by the space charge. The induced electric field promotes the charge behaviors on the inner side such as electron injection and hole extraction. When the polarity reversal ends, the E_{max} on the interior side of the insulation is 38.7 kV/mm, which is 16.1% lower than that at the PRP is 120 seconds. It can be observed that with the extension of the duration of polarity reversal, more and more negative space charges accumulate on the interior side of the insulation, resulting in the decrease of the E_{max}. For further investigating the impact of PRP on electric field, Figure 4.14 compares the E_{max} under different PRP. The results indicate that the shorter the PRP, the larger the E_{max} during the polarity reversal. And, the relaxation period of 600 seconds is effective in reducing the E_{max}.

Figure 4.15 presents the evolution of space charge within 1mm near the inner and outer semi-conductive layers under different PRP. Figure 4.16 analyzes the charge variation rate on the inner side of the insulation under different PRP. As shown in Figure 4.15, with the extension of the PRP, the density of positive charge on the insulation layer's interior side decreases after the reversal process, resulting in the weakening of the E_{max}. Especially, when the PRP is 120 seconds with 600 seconds relaxation period, the polarity of the space charge within 1mm near the inner semi-conductive layer is almost negative. By analyzing the charge

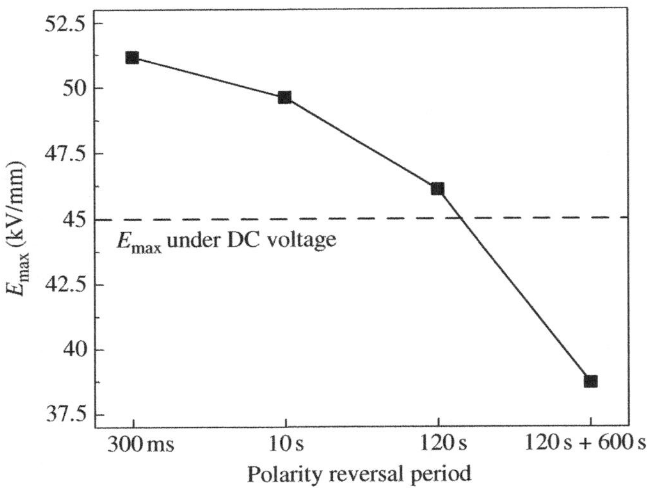

Figure 4.14 E_{max} under different PRP as the ΔT is 30 °C. *Source:* Li et al. [33]/with permission from IEEE.

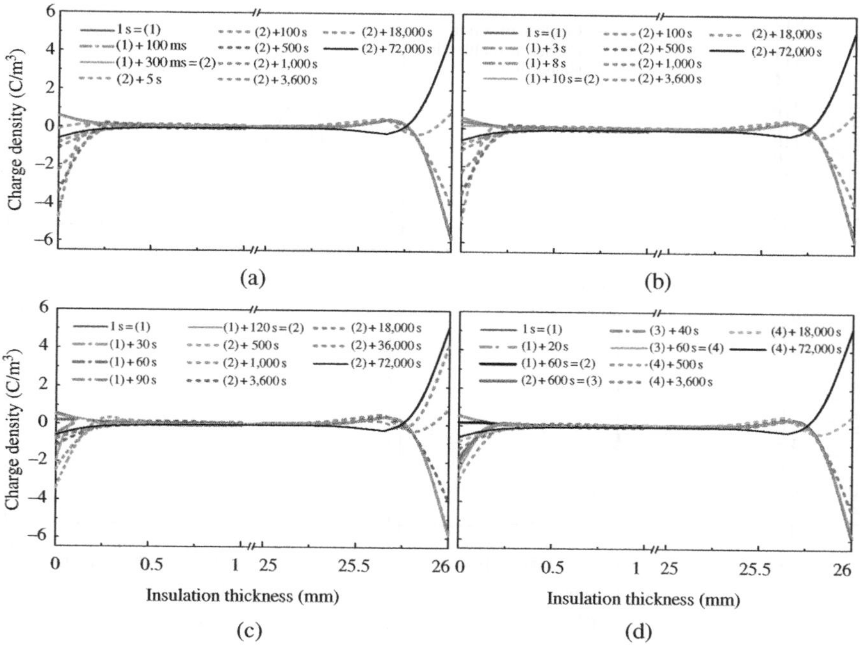

Figure 4.15 Space charge behaviors in the vicinity of the inner and outer semi-conductive layers under different PRP as the ΔT is 30 °C. (a) 300 ms. (b) 10 s. (c) 120 s. (d) 120 s with a 600 s relaxation period. *Source:* Li et al. [33]/with permission from IEEE.

variation rate in Figure 4.16, it can be observed that electron injection and carrier recombination are the main charge behaviors during the polarity reversal process. As illustrated in Figures 4.15a and 4.16a, when the PRP is set at 300 ms, the rate of charge variation increases sharply. However, due to the short PRP, the amount of injected electron, extracted hole, and recombined carrier within 1 mm in the proximity of the inner semi-conductive layer are 6.8×10^{-7}, 7.7×10^{-8}, and 5×10^{-7} C/m^2, respectively. As the PRP increases, there is a corresponding increase in the duration of electron injection and hole extraction but the rate of them decreases. When the PRP is 10 seconds, after the reversal progress, the average density of space charge on the insulation layer's inner side is 0.075 C/m^3 which is reduced by 31.8% compared to that at the PRP is 300 ms. When the PRP is 120 seconds, the average density of the negative space charge is 0.11 C/m^3.

In Figures 4.15d and 4.16d, when the PRP is 120 seconds with 600 seconds relaxation period, during the relaxation period (voltage is 0 V), the field generated solely by the space charge is relatively small, resulting in a reduction of the

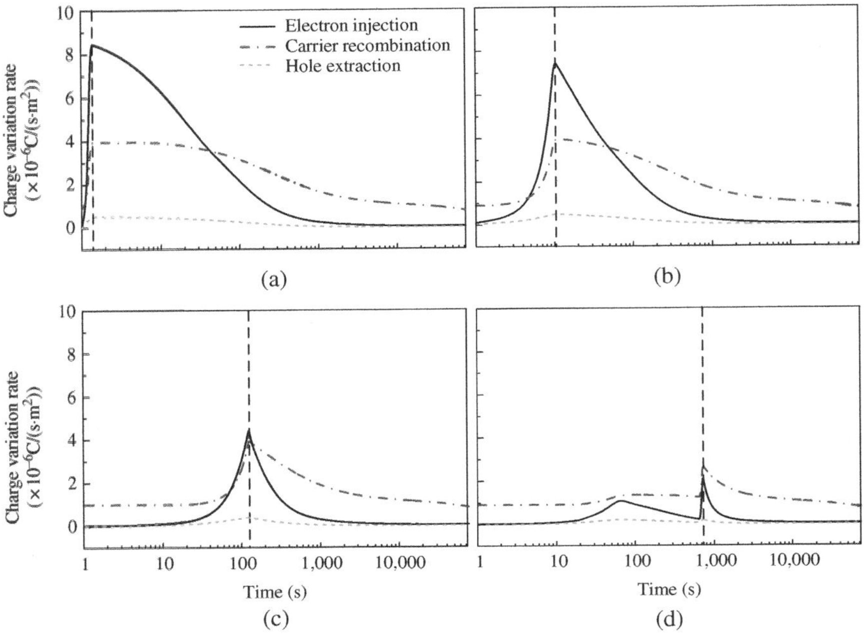

Figure 4.16 Charge variation rate on the inner side of the insulation under different PRP. One seconds is the moment when the polarity reversal voltage is applied. (a) 300 ms. (b) 10 s. (c) 120 s. (d) 120 s with a 600 s relaxation period. *Source:* Li et al. [33]/with permission from IEEE.

charge variation rate. Due to the longer PRP, the amount of injected electron, extracted hole, and recombined carrier increases to 3.7×10^{-4}, 8.1×10^{-5}, and 9.4×10^{-4} C/m^2, respectively, which is larger than those at the PRP is 120 s. Consequently, at the end of polarity reversal, the negative average space charge is present on the inner side with a density of 0.16 C/m^3, resulting in a significant reduction in the E_{max}.

4.5 Insulation Thickness Optimization for HVDC Cables

4.5.1 Theoretical Design and Verification of Insulation Thickness of DC Cable

The operation of the HVDC cable needs to withstand DC voltage and surge voltage. The insulation thickness of the cable is designed according to DC voltage and surge voltage, respectively, and the maximum value of the two is taken.

4.5.1.1 Design Method of Insulation Thickness of HVDC Cables

Insulation Thickness Design Under DC Voltage

The DC withstand voltage can be described as [25]:

$$V_{dc} = V_0 k_1 k_2 k_3 k_4 \tag{4.41}$$

where V_{dc} is the DC design voltage; V_0 is the rated voltage of the system; k_1 is the aging coefficient; k_2 is the temperature coefficient; k_3 is the safety factor; k_4 is the conversion coefficient of breakdown field strength.

The aging coefficient k_1 can be calculated according to the ratio of the cable design life to the trial voltage application time and by introducing the life index n [26]:

$$k_1 = \sqrt[n]{t_1/t_2} \tag{4.42}$$

where t_1 is the design life of the cable; t_2 is the step-up time interval of DC breakdown test.

The existing design method of DC cable insulation thickness does not consider the relationship between insulation strength and sample thickness. In this chapter, the conversion coefficient k_4 of breakdown field strength is introduced. For samples with different thicknesses, k_4 has the following relationship with insulation thickness [27]:

$$k_4 = (d/d_1)^{0.18} \tag{4.43}$$

where d is the thickness of the cable; d_1 is the thickness of the sample.

Insulation design thickness [28] under DC voltage is:

$$d_{dc} = V_{dc}/E_{dc,bd} \tag{4.44}$$

where d_{dc} is the insulation design thickness under DC voltage; $E_{dc,bd}$ are the breakdown electric field strength of impulse voltage.

Insulation Thickness Design Under Impulse Voltage

The impulse voltage of DC cable can be described as [18]

$$V_{imp} = \left(V_p M + KV_0\right) K_1 K_2 K_3 \tag{4.45}$$

where V_{imp} is the impulse design voltage; V_p is the residual voltage of lightning arrester; M is the safety level of lightning arrester; K is the Bahder coefficient; K_1 is the aging coefficient; K_2 is the temperature coefficient; K_3 is the safety margin.

Insulation design thickness [28] under impulse voltage is:

$$d_{imp} = V_{imp}/E_{imp,bd} \tag{4.46}$$

where d_{imp} is the insulation design thickness under impulse voltage; $E_{imp,bd}$ are the breakdown electric field strength of impulse voltage. Select the maximum value of d_{dc} and d_{imp} as the insulation design thickness:

$$d = \mathrm{Max}\{d_{dc}, d_{imp}\} \tag{4.47}$$

4.5.1.2 Analysis and Calculation of Insulation Thickness of HVDC Cables

Key Factors Affecting Insulation Thickness of HVDC Cable

From the above section, there are 21 variables and parameters related to the insulation thickness of DC cables, among which life index, breakdown field strength conversion coefficient and Bahder coefficient have a wide range of changes and have a great influence on the insulation thickness.

a) Effect of life index n on insulation thickness
   Life index n is related to insulation material, manufacturing process, and test method. It is considered that the life index of cable under DC voltage is between 13 and 20 [26]. Figure 4.17 shows the insulation thickness of cables with life index n ranging from 10 to 25. When the life index n is less than 16, the insulation thickness of the cable depends on the DC withstand voltage. When the life index n is greater than 16, the insulation thickness of the cable depends on the impulse voltage.
b) Effect of conversion coefficient k_4 of breakdown field strength on insulation thickness

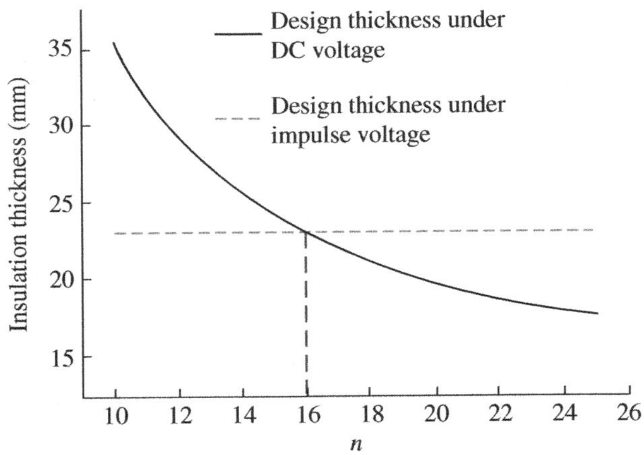

Figure 4.17 Relationship between n and thickness of the insulation. *Source:* Li et al. [34]/ Electric Power Engineering Technology Website Design.

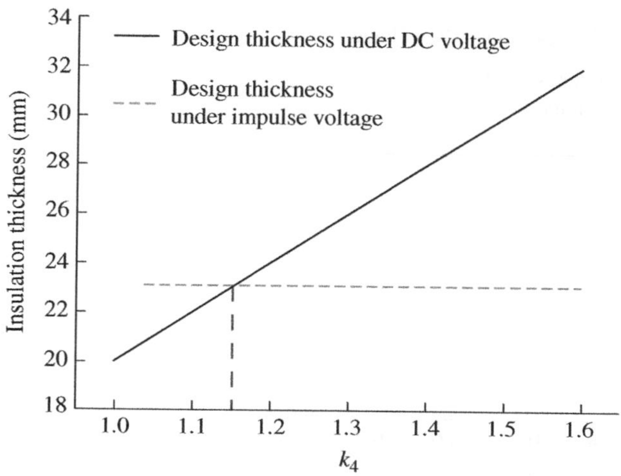

Figure 4.18 Relationship between k_4 and thickness of the insulation. *Source:* Li et al. [34]/ Electric Power Engineering Technology Website Design.

According to the theory of weak point breakdown, dielectric breakdown is caused by weak point breakdown of materials in uniform or nonuniform electric field, such as micropores and impurities in cable insulation, and the electric field intensity at these points is strong, which reaches the condition of breakdown at the earliest. The greater the insulation thickness, the more the number of weaknesses per unit length, the higher the insulation breakdown probability, and the lower the lowest breakdown voltage under the same conditions. The conversion coefficient k_4 of breakdown field strength reflects the relationship between the insulation thickness of the sample and the breakdown field strength, and the larger the value, the more obvious the dependence of the breakdown field strength on the insulation thickness.

Figure 4.18 shows the relationship between the breakdown field strength conversion coefficient k4 and the insulation design thickness. As can be seen from the figure, when k_4 is less than 1.15, the insulation thickness of the cable depends on the impulse withstand voltage; when k_4 is greater than 1.15, the insulation thickness depends on the DC withstand voltage; and with the increase of k_4, the required insulation thickness increases.

c) Effect of Bahder coefficient K on insulation thickness

Bahder coefficient K is the ratio of the breakdown voltage drop value caused by DC preloading and the DC voltage value. After DC voltage preloading, space charge accumulation causes electric field distortion, which leads to the decrease of lightning impulse breakdown voltage of insulation with opposite polarity. The larger Bahder coefficient K is, the more obvious the insulation breakdown

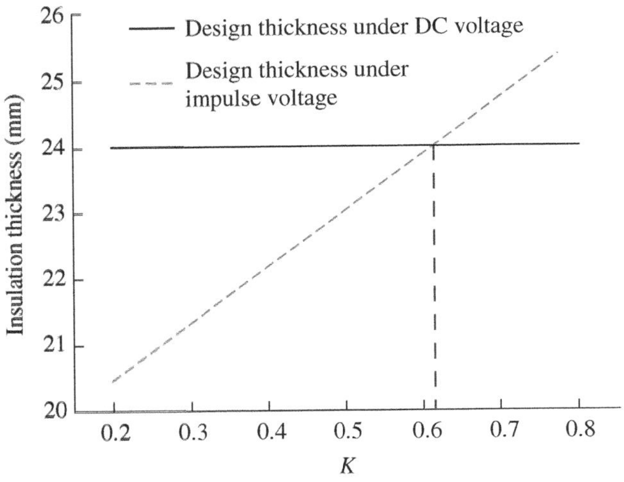

Figure 4.19 Relationship between K and thickness of the insulation. *Source:* Li et al. [34]/ Electric Power Engineering Technology Website Design.

field strength is affected by space charge. Figure 4.19 shows the relationship between K and insulation thickness. When K is less than 0.62, insulation thickness depends on DC withstand voltage. When K is greater than 0.62, the insulation thickness depends on the impulse withstand voltage.

Calculation of Insulation Thickness of HVDC Cables

Equations are established from Eqs. (4.1), (4.2), (4.4)–(4.7), and the 6 equations contain 6 unknown variables (V_{dc}, V_{imp}, k_1, d_{dc}, d_{imp}, and d) and 16 known parameters (V_0, V_p, k_2, k_3, k_4, t_1, t_2, n, d_1, K, K_1, K_2, K_3, M, $E_{dc,bd}$, and $E_{imp,bd}$).

According to the analysis in the previous section, take $n = 16$, $k_4 = 1.16$, and $K = 0.62$, and the insulation strength of the cable can be fully utilized at this time. It is known that the rated voltage of DC cable is $V_0 = 535$ kV, and the peak value of lightning surge overvoltage in Zhangbei Project is $V_p = 1020$ kV, $k_2 = 1$, $k_3 = 1.1$ [25], $t_1 = 40a$, and $t_2 = 3$ hours. Under impulse voltage, $K_1 = 1$, $K_2 = 1$, and $K_3 = 1.1$, safety level of lightning arrester $M = 1.15$; in reference [25], the lowest DC breakdown field strength and the lowest impact breakdown field strength of the model cable sample are 80 and 86 kV/mm at 363.15 K. Taking 80% of the lowest breakdown electric field as the design basis, $E_{dc,bd} = 64.0$ kV/mm and $E_{imp,bd} = 68.8$ kV/mm.

The calculated $V_{dc} = 1416$ kV, $V_{imp} = 1655$ kV, $k_1 = 2.07$, $d_{dc} = 22.12$ mm, $d_{imp} = 24.06$ mm, and $d = 24.06$ mm. The insulation thickness of cable shall be rounded by the one-in-one method, that is, the insulation design thickness of ±535 kV HVDC cable, $d = 25$ mm.

4.5.1.3 Verification of Insulation Thickness of DC Cable

In order to verify whether the insulation design method and insulation thickness of DC cable is reasonable, it is required that the electric field distortion rate of cable under different working conditions should not exceed the specified value. In this chapter, the finite-element simulation method is used to calculate the electric field distribution and electric field distortion rate of ±535 kV cable.

Cable Structure and Material Properties

The conductor section of ±535 kV DC cable is 3000 mm^2, the insulation inner diameter is 70 mm, and the insulation thickness is 25 mm. The electrical conductivity of insulation materials for commercial DC cables conforms to the following formula [29, 30]:

$$\sigma(E,T) = \sigma_0 e^{\alpha(T-273.15)} E^{\beta} \tag{4.48}$$

where σ_0 is the conductivity at reference temperature and electric field, S/m; α is the temperature coefficient of conductivity, 1/K; β is the electric field coefficient of conductivity; T is the temperature, K. In this chapter, the electrical conductivity of the material is measured, and the least square method is used to fit the test data, and the electrical conductivity $\sigma_0 = 3.51 \times 10^{-17}$ S/m, the temperature coefficient $\alpha = 0.05$ K^{-1}, and the electric field coefficient $\beta = 2.3$ are obtained.

Simulation Results of Electric Field Distribution

Assuming that the cable is empty, the temperature inside and outside the insulation layer is 313.15 K. At full load, the inner temperature of conductor insulation is 363.15 K and the outer temperature of insulation is 343.15 K. The radial electric field distribution curve of insulation is shown in Figure 4.4, and the maximum electric field appears inside the insulation at no load, which is 23.3 kV/mm; at full load, the maximum electric field appears outside the insulation, which is 22.8 kV/mm; the average electric field of insulation is 21.4 kV/mm.

a) Electric field distortion rate under different conductivity temperature coefficients

   The rated voltage of DC cable is $U_0 = 535$ kV, $\sigma_0 = 3.51 \times 10^{-17}$ S/m, electric field coefficient $\beta = 2.3$, and insulation temperature difference $\Delta T = 20$ K. The electric field distribution corresponding to different temperature coefficients is simulated and calculated, and the variation law of electric field distribution with temperature coefficient is obtained, as shown in Figure 4.5.

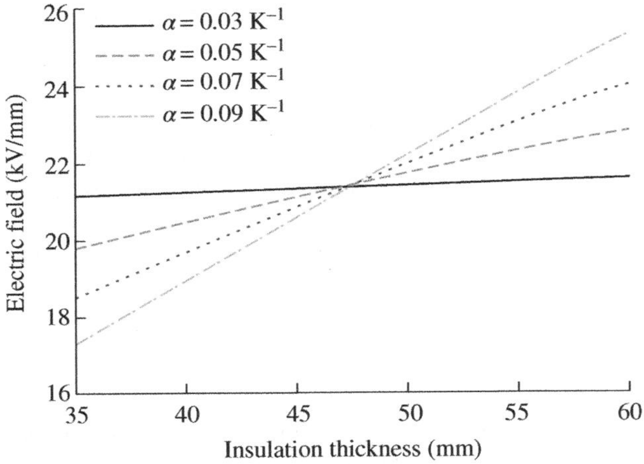

Figure 4.20 Relationship between temperature coefficient and electric field distribution. *Source:* Li et al. [34]/Electric Power Engineering Technology Website Design.

As can be seen from Figure 4.20, the influence of temperature coefficient on the electric field distribution is very significant. With the increase of temperature coefficient, the field strength of the insulation inner screen gradually decreases and the field strength of the insulation outer screen gradually increases.

To more clearly reflect the influence of various factors on the electric field distribution, the distortion rate η of field strength is defined here, and its definition formula is:

$$\eta = \frac{E_{max} - E_{mean}}{E_{mean}} \times 100\% \tag{4.49}$$

where E_{max} is the maximum field strength in the insulation layer; E_{mean} is the average field strength of insulation layer, and $E_{mean} = V_0/d = 21.4\,\text{kV/mm}$.

The statistical results of electric field intensity and electric field distortion rate at different temperature coefficients are shown in Table 4.1. With the increase

Table 4.1 Electric field distortion rate at different temperature coefficients. *Source:* Li et al. [34]/Electric Power Engineering Technology Website Design.

Temperature coefficient α/K^{-1}	Insulation internal screen electric field (kV/mm)	Insulation external screen electric field (kV/mm)	Electric field distortion rate (η/%)
0.03	21.2	21.6	0.1
0.05	19.8	22.8	6.5
0.07	18.5	24.1	12.6
0.09	17.3	25.3	18.2

of temperature coefficient, the electric field distortion rate increases gradually, and the electric field on the insulation surface is too high, which will bring difficulties to the matching of cable accessories and is not conducive to the safety and reliability of cable system in long-term operation.

b) Electric field distortion rate under different conductivity electric field coefficients
It is set that $\sigma_0 = 3.51 \times 10^{-17}$ S/m, $\alpha = 0.05$ K^{-1}, and insulation temperature difference $\Delta T = 20$ K. The electric field distribution with different electric field coefficients is simulated and calculated, as shown in Figure 4.21. The electric field distribution has a great dependence on the electric field coefficient. With the increase of the electric field coefficient, the electric field of the insulated inner screen increases and the electric field of the insulated outer screen decreases. The electric field distortion rate corresponding to Figure 4.21 is shown in Table 4.2. The electric field coefficient β actually plays an adaptive adjustment role between conductivity and electric field intensity. The larger β, the stronger the adjustment effect and the more uniform the electric field distribution.

c) Electric field distortion rate under different insulation temperature differences
The electric field distribution with insulation temperature difference of 0 ~ 30 K is calculated by setting $\sigma_0 = 3.51 \times 10^{-17}$ S/m, $\alpha = 0.05$ K^{-1}, electric field coefficient $\beta = 2.3$, and conductor temperature of 363.15 K, as shown in Figure 4.22. When the insulation temperature difference is low, with the increase of the temperature difference, the electric field of the insulation inner screen decreases, the electric field of the insulation outer screen increases, and the

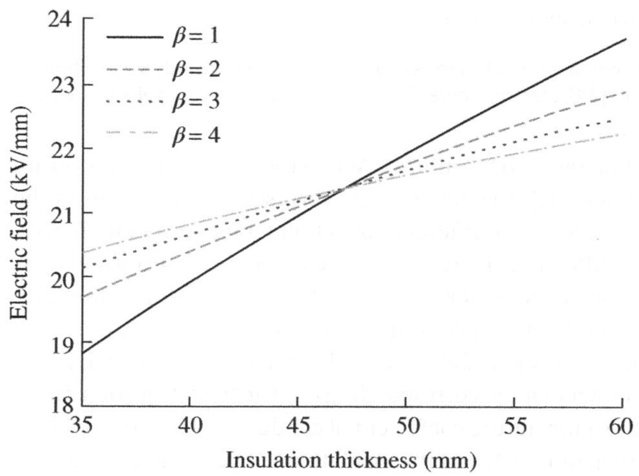

Figure 4.21 Relationship between electric field coefficient and electric field distribution. *Source:* Li et al. [34]/Electric Power Engineering Technology Website Design.

Table 4.2 Electric field distortion rate at different electric field coefficients. *Source:* Li et al. [34]/Electric Power Engineering Technology Website Design.

Electric field coefficient β	Insulation internal screen electric field (kV/mm)	Insulation external screen electric field (kV/mm)	Electric field distortion rate (η/%)
1	18.8	23.7	10.7
2	19.7	22.9	7.0
3	20.1	22.5	5.1
4	20.4	22.3	4.2

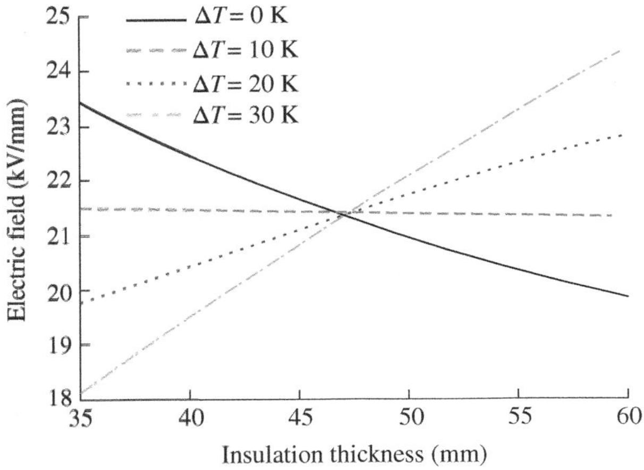

Figure 4.22 Relationship between insulation temperature difference and electric field distribution. *Source:* Li et al. [34]/Electric Power Engineering Technology Website Design.

electric field distribution tends to be uniform; When the insulation temperature difference reaches about 10 K, the electric field reverses. Table 4.3 shows the electric field distortion rate under different insulation temperature differences. As the temperature difference increases, the electric field distortion first decreases and then increases. When the insulation temperature difference reaches 30 K, the electric field distortion rate reaches 14%.

Through the calculation results of electric field distribution and electric field distortion rate, it is known that the corresponding electric field distortion rate is less than 20% when the temperature coefficient of conductivity is 0.03–0.09 K^{-1}, the electric field coefficient is 1–4, and the insulation temperature difference is less than 30 K, which meets the requirements of Section 6.4.10 in GB/T 31489.1 [31], and the calculation results verify the rationality of the design method.

Table 4.3 Electric field distortion rate at different insulation temperature differences.

Insulation temperature difference (K)	Insulation internal screen electric field (kV/mm)	Insulation external screen electric field (kV/mm)	Electric field distortion rate (η/%)
0	23.4	19.8	9.3
10	21.5	21.3	0.5
20	19.8	22.8	6.5
30	18.2	24.4	14.0

4.5.2 Insulation Thickness Optimization Based on Modified BEICT Model

According to the BEICT model proposed in Section 4.2, the simulation results for electric fields and thermal fields of cable models with various insulation thicknesses are calculated and shown in Figure 4.23. Among them, Figure 4.23a represents the distribution of temperature within the insulation layer at various thicknesses. It can be observed that the increased insulation thickness, on the one hand, increases the difficulty of the heat dissipation process, resulting in a relatively higher temperature environment for the insulation layer at the same radius distance location. On the other hand, it also leads to a broader range of cooling behavior on the distance scale, which results in lower temperatures on the outer side of the insulation compared to smaller thickness cables, that is, the larger the thickness, the greater the ΔT between the insulation layer's inner and outer sides. Figure 4.23b and c corresponds to the changes in charge behavior and electric field evolution within the cable during polarization under various insulation thickness conditions, respectively. It can be seen that even though there are differences in temperature and thickness inside the insulation layer, the distribution and change trend of charge during polarization are similar, so it is not much elaborated here.

With the decrease in insulation thickness, the injection of holes at the interface of the inner semiconducting layer gradually increases so that the weakening effect on the electric field at nearby locations gradually increases. This can also be confirmed by comparing the electric field change of the insulation layer and the semi-conductive layer interface before and after polarization of 22 and 30 mm in Figure 4.24. Compared with before polarization, the former decreased by 14.36 kV/mm, while the latter fell by 9.02 kV/mm.

Meanwhile, with the decrease of insulation thickness, the homocharges accumulated around the outer semi-conductive layer gradually increase, whereas the heterocharge amount on the same side gradually decreases. The resulting

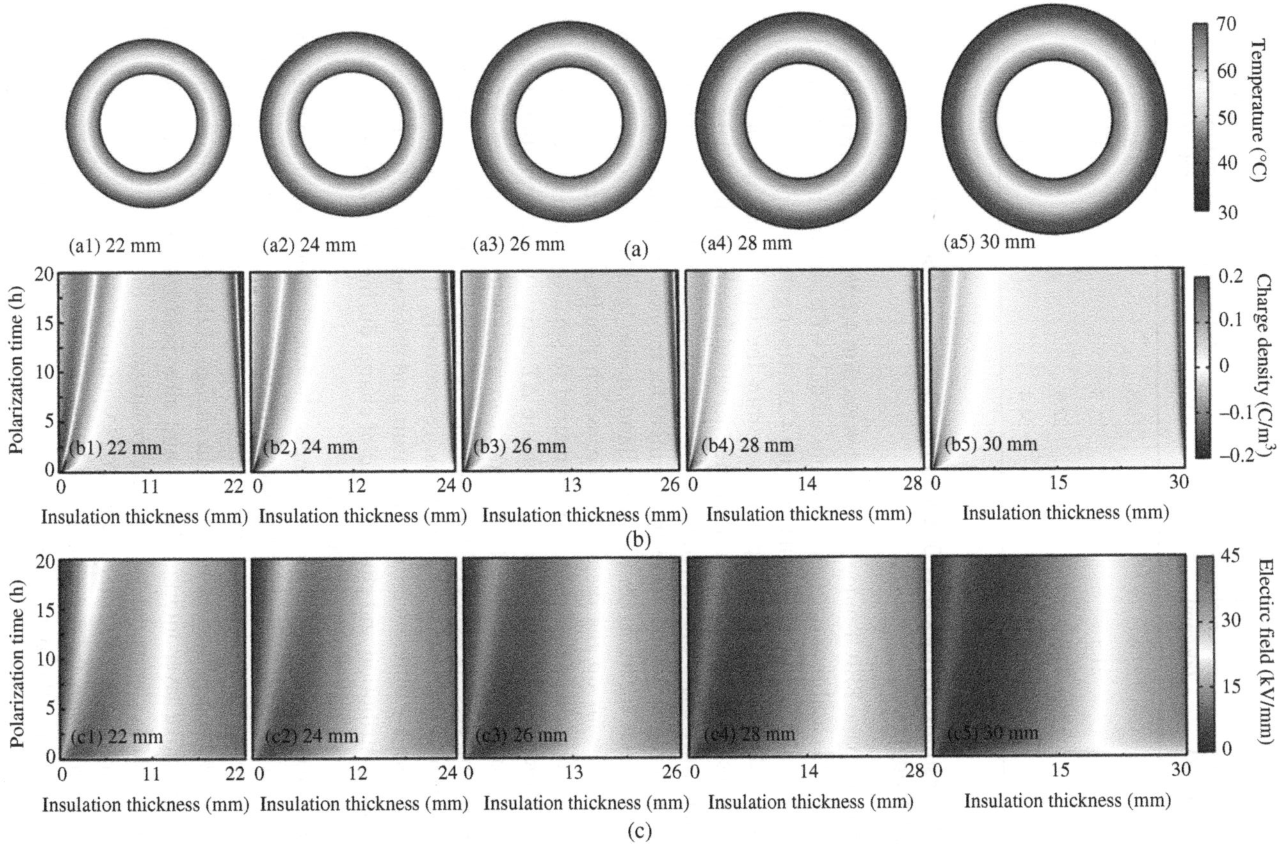

Figure 4.23 Effect of insulation thickness on (a) temperature gradient, (b) space charge, and (c) electric field distributions under +500 kV. *Source:* Li et al. [32]/with permission from IEEE.

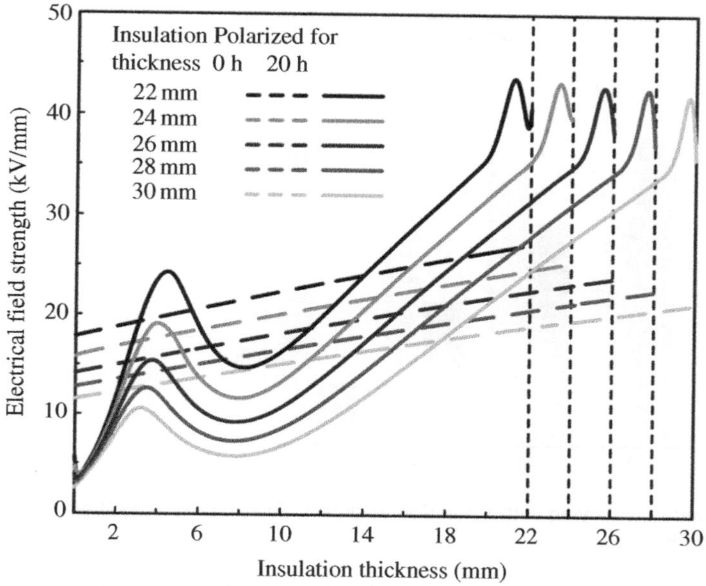

Figure 4.24 Electrical field distributions varying with insulation thicknesses. *Source:* Li et al. [32]/with permission from IEEE.

enhancement effect for the local field strength gradually decreased, of which the electric field strength at 30 mm increased 97% compared with that before polarization. In comparison, the increase in the electric field of 22 mm is only 62%.

Two parameters, E_{max} and C_E, are introduced here to quantify the effect of insulation thickness on the electric field evolution characteristics within the cable, and the results are presented in Figure 4.25. Among them, the nonuniformity coefficient of electric field C_E can be used to quantify the extent to which the distribution of electric field deviates from the average field strength caused by space charges, which can be written as

$$C_E = \frac{E_{max}}{E_{mean}} \tag{4.50}$$

where $E_{mean} = U/d$, which represents the average electric field; U represents the rated DC voltage, ±500 kV; d represents the insulation thickness.

As can be found from Figure 4.25, with the decrease in cable insulation thickness, E_{max} in the insulation layer increases slightly in value, but the C_E decreases significantly. The C_E is 2.51 with a thickness of 30 mm but decreases significantly to 1.91 with a thickness of 22 mm. The main reason leading to this result is that the local field strength increases due to the heterocharges accumulation on the

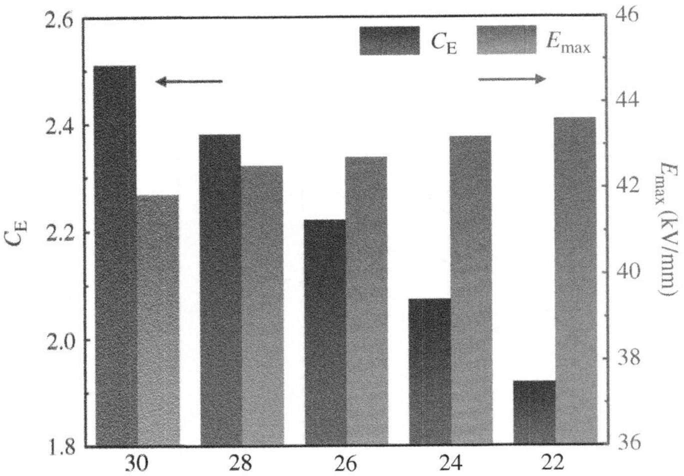

Figure 4.25 C_E and E_{max} varying with insulation thickness. *Source:* Li et al. [32]/with permission from IEEE.

insulation layer's outer side, forming a strong local electric field. By comparison, the conclusion can be drawn that the variation of insulation thickness has little influence on E_{max}, and the difference is only within 2 kV/mm. At this point, C_E becomes extremely important. The bigger it is, the more serious the electric field distortion problem is, which is not conducive to the reliable and efficient operation of the transmission. Therefore, appropriate insulation optimization becomes extremely important and necessary.

The difference in insulation thickness results in a synergistic variation of the ΔT and electric field evolution within the insulation layer. The thinner the insulation thickness, the greater the average strength field inside the cable. Meanwhile, the ΔT between the insulation layer's inner and outer sides decreases with the optimization of insulation thickness, which also causes the average temperature within the insulation to be higher.

Owing to the synergistic increase of temperature and electric field, the interface charge's injection process and the impurity molecules' dissociation process are simultaneously enhanced. For the insulation layer's inner side with higher temperature, with the decrease of thickness, the injected and accumulated electronic charges increase significantly in both quantity and range. The resulting electric field weakening effect at the inner semi-conductive layer interface is also more significant.

Meanwhile, the insulation layer's outer side, with a lower temperature, is still dominated by ionic charges. But, with the decrease of thickness, the number of impurity molecules participating in dissociation inside the insulation layer decreases, so the ionic charges that migrate and accumulate at the insulation layer's outer side with lower temperature also decrease. Meanwhile, the increase in temperature and electric field at the outer semi-conductive layer interface caused by the decrease in insulation thickness enhances the electron injection process. With the combined effect of increasing homocharges and decreasing heterocharges, the electric field distortion degree at the insulation layer's outer side gradually decreases.

Figure 4.26 shows the effect of insulation thickness on space charge and electric field evolution under the polarity reversal voltage with a polarity reversal period of 120 seconds, as the ΔT is 30 °C. Figure 4.27 shows the radial electric field distribution of insulation at the beginning of polarity reversal, the middle time of polarity reversal (voltage is 0 V), and the end time of polarity reversal. At the beginning of polarity reversal, due to the influence of ΔT, the E_{max} appears outside the insulation layer. The maximum value of the electric field decreases with the increase in thickness. In the process of polarity reversal, as the applied voltage decreases, the polarity of the electric field near the wire core will reverse, and the internal holes will gradually become heteropolar charges, thus distorting the electric field. The smaller the insulation thickness, the more holes accumulated in the steady state, and the greater the distortion of the transient electric field. When the polarity reversal is completed, the E_{max} the insulation thicknesses from 22 to 30 mm are 46.85, 44.17, 42.40, 40.96, and 39.63 kV/mm, respectively.

In order to better analyze the influence of thickness on electric field distribution during polarity reversal, the E_{max} and distortion rate (C_E) under different thicknesses are compared, as shown in Figure 4.28.

The conclusion can be drawn that the increase in insulation thickness does not achieve the goal of lowering the strong local electric field within the cable, but further aggravates the electric field distortion problem on the insulation layer's outer side by enhancing the dissociation process of impurity molecules and weakening the injection process of interfacial charges. It can be seen that with the decrease of the insulation thickness, the more charges of the same polarity accumulated near the semi-conductive layer at the beginning of polarity reversal, resulting in the greater maximum electric field at the end of polarity reversal. At the end of polarity reversal, smaller insulation thickness causes greater electric field distortion, which is contrary to the steady-state effect. Therefore, when optimizing the insulation thickness, too thin or too thick is not conducive to the uniform electric field.

Therefore, the design of the cable insulation structure should be based on the consideration of the E_{max} and C_E of the electric field within the insulation layer and be achieved by appropriate insulation thinning. Based on the above experimental research, 26 mm is considered the most suitable insulation thickness design parameter for ±500 kV HVDC-extruded clad cable.

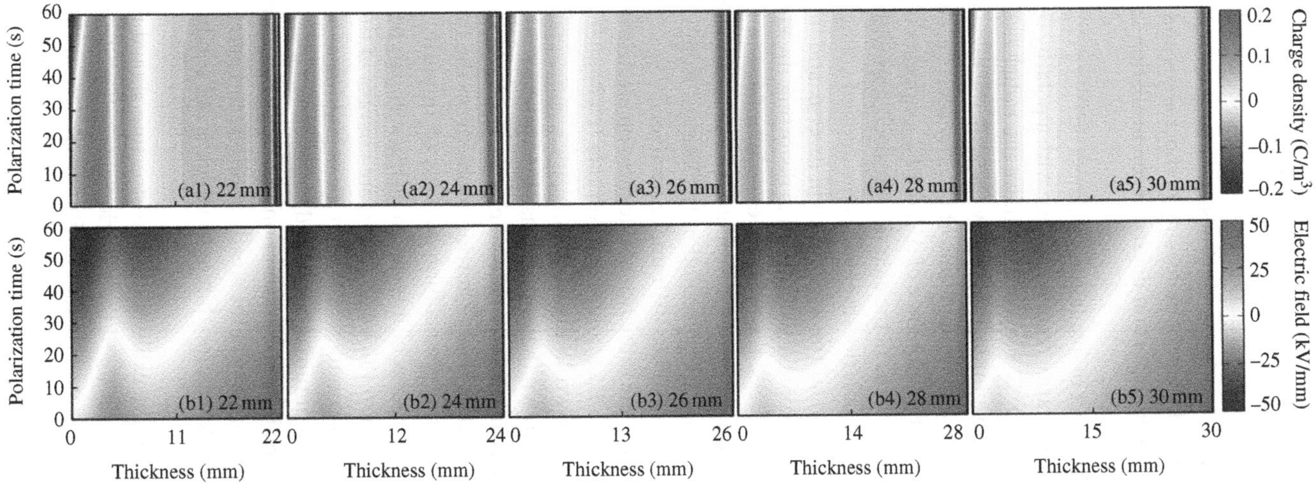

Figure 4.26 Effect of insulation thickness on (a) space charge and (b) electric field evolution under the polarity reversal voltage with a polarity reversal period of 120 seconds, as the ΔT is 30 °C. *Source:* Zheng et al. [35]/with permission from IEEE.

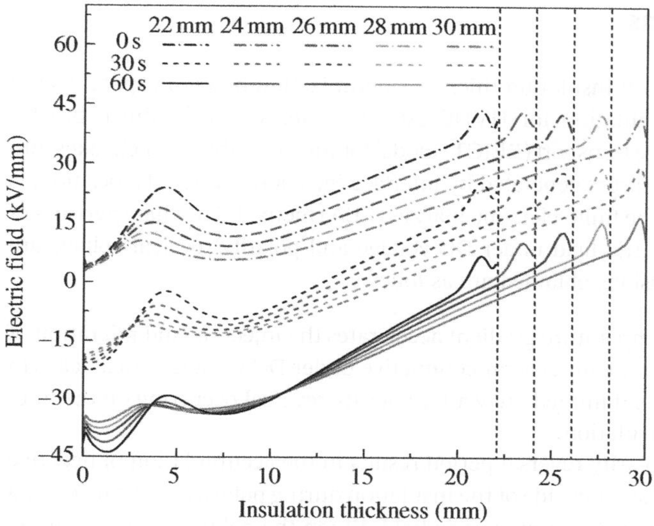

Figure 4.27 Electric field distribution at 0, 60, and 120 seconds. *Source:* Zheng et al. [35]/ with permission from IEEE.

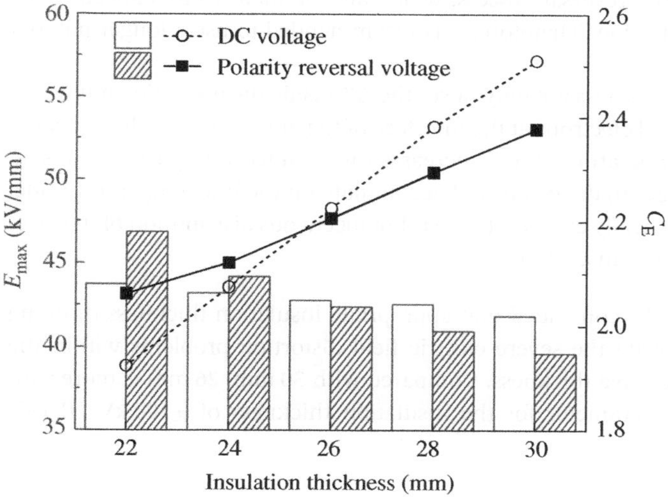

Figure 4.28 E_{max} and C_E under different thicknesses. *Source:* Zheng et al. [35]/with permission from IEEE.

4.6 Conclusions

This chapter focuses on insulation thickness optimization based on space charge and electric field simulation for HVDC-extruded cables. The modified bipolar electronic–ionic charge transport (BEICT) model for full-size cables, which takes into account both space charges generated by interface injection and ion dissociation, is established based on the finite-element analysis (FEA) method. The characteristics of space charge and electric field under DC voltage and polarity reversal voltage are analyzed. The conclusions obtained are as follows:

1) The increase of temperature gradient accelerates the injection and migration of homocharges near the inner semi-conductive under DC voltage, which leads to a higher electric field immediately after polarity reversal occurring on the interior side of the insulation.
2) Prolonging the polarity reversal period results in the accumulation of negative space charge on the inner side of the insulation during polarity reversal, leading to a weaken of the maximum electric field. When the polarity reversal period increases from 300 ms to 120 seconds, the maximum electric field decreases by 9.8%. Moreover, a relaxation period of 600 seconds when the voltage drops to 0 kV will further reduce the maximum electric field by 16.1%.
3) Electron injection and carrier recombination dominate the charge behaviors during the polarity reversal process, which are the main factors affecting the electric field distortion. Therefore, it is recommended to use a longer polarity reversal period.
4) The increasing thickness not only raises the ΔT inside the insulation but weakens the injection of electrons at the interface of the outer semi-conductive layer with the low temperature. It also increases the number of impurities, causing more ionic charges to dissociate and accumulate on the low-temperature side. The two synergistically exacerbate the inhomogeneous distribution of the electric field within the insulation.

Ultimately, it can be concluded that appropriate insulation thickness thinning can effectively alleviate the severe electric field distortion problems within the cable caused by excessive thickness. Compared with 30 mm, 26 mm is more suitable as the design parameter for the insulation thickness of ±500 kV HVDC-extruded cable.

References

1 Igi, T., Asai, S., Mashio, S et al. (2019). Qualification, installation and commissioning of world's first DC 400 kV XLPE cable system. *10th International Conference on Insulated Power Cables*, Paris, France, 23–27 June 2019.

2 Wu, S., Zhang, X., Jia, W. et al. (2021). A modular multilevel converter with integrated energy dissipation equipment for offshore Wind VSC-HVDC system. *IEEE Transactions on Sustainable Energy* 13 (1): 353–362.

3 Alassi, A., Bañales, S., Ellabban, O. et al. (2019). HVDC transmission: technology review, market trends and future outlook. *Renewable and Sustainable Energy Reviews* 112: 530–554.

4 Li, C., Yang, Y., Xu, G. et al. (2022). Insulating materials for realising carbon neutrality: opportunities, remaining issues and challenges. *High Voltage* 7 (4): 610–632.

5 Mazzanti, G., Chen, G., Fothergill, J.C. et al. (2015). A protocol for space charge measurements in full-size HVDC extruded cables. *IEEE Transactions on Dielectrics and Electrical Insulation* 22 (1): 21–34.

6 Mirebeau, P., Frohne, C. and Larsen, V. (2015). Review of HVDC insulated transmission cables technologies. *9th International Conference on Insulated Power Cables*, Paris, France, from 21 to 25 June 2015.

7 Su, J., Du, B., Li, J., and Li, Z. (2020). Electrical tree degradation in high-voltage cable insulation: progress and challenges. *High Voltage* 5: 353–364.

8 Tanaka, T. and Greenwood, A. (1978). Effects of charge injection and extraction on tree initiation in polyethylene. *IEEE Transactions on Power Apparatus and Systems* 97 (5): 1749–1759.

9 Zhao, X., Liu, Y., Wu, J. et al. (2020). Technical and economic demands of HVDC submarine cable technology for Global Energy Interconnection. *Global Energy Interconnection* 3 (2): 120–127.

10 Lv, Z., Cao, J., Wang, X. et al. (2015). Mechanism of space charge formation in cross linked polyethylene (XLPE) under temperature gradient. *IEEE Transactions on Dielectrics and Electrical Insulation* 22 (6): 3186–3196.

11 Chen, X., Wang, X., Wu, K. et al. (2010). Space charge measurement in LPDE films under temperature gradient and DC stress. *IEEE Transactions on Dielectrics and Electrical Insulation* 17 (6): 1796–1805.

12 Hanley, T.L., Burford, R.P., Fleming, R.J., and Barber, K.W. (2003). A general review of polymeric insulation for use in HYDC Cables. *IEEE Electrical Insulation Magazine* 19 (1): 14–24.

13 Vu, T.T.N., Teyssedre, G., Vissouvanadin, B. et al. (2015). Correlating conductivity and space charge measurements in multi dielectrics under various electrical and thermal stresses. *IEEE Transactions on Dielectrics and Electrical Insulation* 22 (1): 117–127.

14 Qin, S. and Boggs, S. (2012). Design considerations for high voltage DC components. *IEEE Electrical Insulation Magazine* 28 (6): 36–44.

15 Boggs, S., Damon, D.H., Hjerrild, J. et al. (2001). Effect of insulation properties on the field grading of solid dielectric DC cable. *IEEE Transactions on Power Delivery* 16 (4): 456–461.

16 Montanari, G.C., Seri, P., Lei, X.Z. et al. (2011). Next generation polymeric high voltage direct current cables – a quantum leap needed? *IEEE Electrical Insulation Magazine* 34 (2): 24–31.

17 Du, B.X., Kong, X.X., Cui, B., and Li, J. (2017). Improved ampacity of buried HVDC cable with high thermal conductivity LDPE/BN insulation. *IEEE Transactions on Dielectrics and Electrical Insulation* 24 (5): 2667–2676.

18 Chen, J., Huang, X.Y., Zhu, Y., and Jiang, P.K. (2016). Cellulose nanofiber supported 3D interconnected BN nanosheets for epoxy nanocomposites with ultrahigh thermal management capability. *Advanced Functional Materials* 27 (5): 1604754.

19 Huang, X.Y., Jiang, P.K., and Tanaka, T. (2011). A review of dielectric polymer composites with high thermal conductivity. *IEEE Electrical Insulation Magazine* 27 (4): 8–16.

20 Kumara, S., Serdyuk, Y.V., and Jeroense, M. (2021). Calculation of electric fields in HVDC cables: comparison of different models. *IEEE Transactions on Dielectrics and Electrical Insulation* 28 (3): 1070–1078. https://doi.org/10.1109/TDEI.2021.009371.

21 Han, C., Du, B., Li, J. et al. (2020). Investigation of charge transport and breakdown properties in XLPE/GO nanocomposites part 2: Effect of polarity reversal. *IEEE Transactions on Dielectrics and Electrical Insulation* 27 (4): 1213–1221. https://doi.org/10.1109/TDEI.2020.008931.

22 Tanaka, Y., Fujitomi, T., Kato, T. et al. (2017). Packet-like charge formation in cable insulating materials at polarity reversal. *IEEE Transactions on Dielectrics and Electrical Insulation* 24 (3): 1372–1379. https://doi.org/10.1109/TDEI.2017.006132.

23 Chen, J., Gao, Y., Zhu, M. et al. (2020). Space charge dynamics in double-layered insulation cable under polarity reversal voltage. *IEEE Transactions on Dielectrics and Electrical Insulation* 27 (2): 622–630. https://doi.org/10.1109/TDEI.2019.008457.

24 Huang, M., Zhou, Y., Chen, W. et al. (2014). Influence of voltage reversal on space charge behavior in oil-paper insulation. *IEEE Transactions on Dielectrics and Electrical Insulation* 21 (1): 331–339. https://doi.org/10.1109/TDEI.2013.004010.

25 Diban, B., Mazzanti, G., and Seri, P. (2023). Life-based geometric design of HVDC cables. Part 2: effect of electrical and thermal transients. *IEEE Transactions on Dielectrics and Electrical Insulation* 30 (1): 97–105. https://doi.org/10.1109/TDEI.2022.3212974.

26 Maekawa, Y., Watanabe, K., Maruyama, S. et al. (2002). Research and development of DC ±500 kV extruded cables. *CIGRE Session Papers & Proceedings*, 21–203, 1–8.

27 Xia, L., Qinxue, Y., Lisheng, Z. et al. (2015). Review on research of V-t characteristics of HV XLPE cables. *Southern Power System Technology* 9 (10): 57–63.

28 Kubota, T., Takahashi, Y., Sakuma, S. et al. (1994). Development of 500 kV XLPE cables and accessories for long distance underground transmission line-Part I: insulation design of cables. *IEEE Transactions on Power Delivery* 9 (4): 1741–1749.

29 Terashima, K., Murata, Y., Muto, H. et al. (1999). Study of design method of XLPE cable for DC voltage. *Transactions Institute of Electrical Engineers of Japan B* 119 (2): 212–222.

30 Hampton, R.N. (2008). Some of the considerations for materials operating under high-voltage, direct-current stresses. *Electrical Insulation Magazine IEEE* 24 (1): 5–13.

31 Standardization Administration of China (2015). *DC Extruded Cable Systems for Power Transmission at a Rated Voltage Up to and Including 500 kV – Part 1: Test Methods and Requirements: GB/T 31489.1 – 2015.* Beijing: China Standard Press.

32 Li, Z., Wu, Y., Zheng, Z., & Du, B. (2024). Temperature-Gradient-Dependent Space Charge and Electric Field Evolutions of ±500 kV HVDC Cables With Different Thicknesses. *IEEE Transactions on Dielectrics and Electrical Insulation*, 31 (3), 1620–1627. https://doi.org/10.1109/tdei.2023.3343671.

33 Li, Z., Zheng, Z., Wu, Y., & Du, B. (2024). Space Charge and Electric Field Dependent on Polarity Reversal of HVDC Cable Insulation. *IEEE Transactions on Dielectrics and Electrical Insulation*, 31 (1), 58–65. https://doi.org/10.1109/tdei.2023.3335189.

34 Li, D., Zhu, Z., Yang, L., Chen, L., Zeng, H., & Gao, K. (2020). Theoretical design and verification of ±535 kV DC cable insulation thickness. *Electric Power Engineering Technology* 39 (1): 151–156. doi: 10.12158/j.2096-3203.2020.01.021 (in Chinese with English abstract).

35 Zheng, Z., Li, Z., Wu, Y., & Du, B. (2023, septembre 24). Effect of Insulation Thickness on Space Charge Behavior and Electric Field Evolution in HVDC Cable Under Polarity Reversal Voltage. 2023 International Symposium on Electrical Insulating Materials (ISEIM). https://doi.org/10.23919/iseim60444.2023.10329060.

Part II

Polypropylene Insulation for HVAC Cables

5

Polarization and Dielectric Relaxation

5.1 Introduction

Dielectric relaxation is usually considered to be one of the most important properties of a polymer [1]. The polarization forms of dielectric include electronic polarization, ionic polarization, orientational polarization, and space charge polarization. Electronic polarization refers to the deformation or translation of the originally symmetrical distribution of the electron clouds of atoms or molecules under the action of an external electric field, resulting in the geometric center of the positive charge and the electron cloud no longer being aligned. This polarization process belongs to instantaneous polarization, which mainly occurs in the frequency range of 10^{14}–10^{16} Hz. Ionic polarization is the polarization effect of molecules or crystals caused by an externally applied electric field. When an electric field is applied to a substance, ions are displaced, causing the geometric centers of positive and negative ions to no longer coincide. Ionic polarization has a permanent dipole moment and is accompanied by a slight energy loss, and this polarization process is also a kind of instantaneous polarization, mainly occurring in the frequency range of 10^{9}–10^{13} Hz. Orientational polarization occurs only in materials consisting of molecules or particles with a permanent dipole moment. The electric field causes the reorientation of the dipoles toward the direction of the field, and the polarization is a relaxation polarization, which generally occurs in the frequency range of 10^{3}–10^{8} Hz. At higher fields, carrier injection becomes important. For materials with a high concentration of carriers, polarization due to the migration of carriers to form space charges at interfaces or grain boundaries becomes important. This type of polarization is called space charge polarization, which mainly occurs in the lower frequency range of 10^{-3}–10^{2} Hz.

Polypropylene Cable Insulation, First Edition. Boxue Du and Zhonglei Li.

Published 2025 by John Wiley & Sons, Inc.

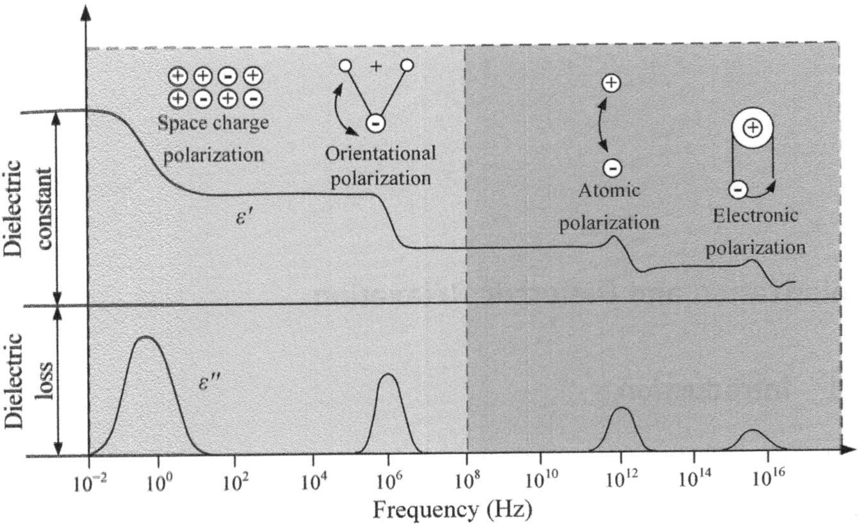

Figure 5.1 The variation of different types of polarization with frequency under AC electric field. *Source:* Lai et al. [38]/Walter de Gruyter GmbH / CC BY 4.0.

The variation of different types of polarization with frequency under an external electric field is shown in Figure 5.1.

The dielectric response at different frequencies and temperatures serves a function to characterize the molecule chain motion, dipole orientation, and carrier transport properties of the polymer [2]. These characteristics provide a novel method to investigate a macroscopic electrical property from the microstructure view. Fuse et al. observed the four relaxation processes in polyamide nanocomposites by analyzing their complex permittivity spectra: the dipole orientation, space charge polarization, interface polarization, and group rotation, and it was found that the longer average relaxation time indicates the restriction of molecular motion, which should in turn suppress carrier transport [3]. Huang et al. investigated the molecular chain movement and carrier hopping relaxation of an epoxy resin and concluded that the enhanced size of free volume can provide enough space for chain movement and energy for carrier hopping [4].

Currently, some models are available for the analysis of the dielectric relaxation process in insulating materials. These models include the Debye model, the Cole–Cole (C–C) model, the Cole–Davidson (D–C) model, and the Havriliak–Negami (H–N) model [1].

The Debye model can only describe a single relaxation process in simple materials and cannot accurately describe multiple relaxation behaviors in complex materials. The Debye equation can be expressed as

$$\varepsilon_{\mathrm{D}}^{*}(\omega) = \varepsilon_{\infty} + \frac{\Delta\varepsilon}{1 + i\omega\tau_{\mathrm{D}}} \tag{5.1}$$

In this case, a distribution of relaxation times is necessary to interpret the experimental data, thus forming the C–C model, the D–C model, and the H–N model.

The C–C equation can be expressed as

$$\varepsilon_{\mathrm{CC}}^{*}(\omega) = \varepsilon_{\infty} + \frac{\Delta\varepsilon}{1 + (i\omega\tau_{\mathrm{CC}})^{\beta}} \tag{5.2}$$

The D–C equation can be expressed as

$$\varepsilon_{\mathrm{DC}}^{*}(\omega) = \varepsilon_{\infty} + \frac{\Delta\varepsilon}{(1 + i\omega\tau_{\mathrm{DC}})^{\gamma}} \tag{5.3}$$

Compared to the Debye model, the H–N model can more accurately describe the relaxation process in complex materials and applies to a wider frequency range. In addition, the H–N model can transform into the C–C model and D–C model under certain conditions. The H–N equation can be expressed as

$$\varepsilon_{\mathrm{HN}}^{*}(\omega) = \varepsilon_{\infty} + \frac{\Delta\varepsilon}{\left(1 + (i\omega\tau_{\mathrm{HN}})^{\beta}\right)^{\gamma}} \tag{5.4}$$

where τ_{HN} is the relaxation time of the H–N equation, β and γ are the shape coefficients satisfying $0 < \beta \leq 1$ and $0 < \gamma \leq 1$. When $\beta = \gamma = 1$, the H–N equation can be simplified to the Debye model. When $\beta = 1$ and $\gamma \neq 0$, the H–N equation transforms to the D–C equation, and when $\gamma = 1$ and $\beta \neq 0$, the H–N equation transforms to the C–C equation [4].

In this chapter, the H–N equation is used to calculate the values of complex permittivity ε_{α}^{*} for α relaxation and ε_{δ}^{*} for δ relaxation. The equation of complex permittivity $\varepsilon_{\mathrm{r}}^{*}$ can be expressed as [5]

$$\varepsilon_{\mathrm{r}}^{*} = \varepsilon_{\infty} + \frac{\Delta\varepsilon_{\alpha}}{\left(1 + (i\omega\tau_{\alpha})^{\beta_{\alpha}}\right)^{\gamma_{\alpha}}} + \frac{\Delta\varepsilon_{\delta}}{\left(1 + (i\omega\tau_{\delta})^{\beta_{\delta}}\right)^{\gamma_{\delta}}} + \frac{\sigma}{i\varepsilon_{0}\omega} \tag{5.5}$$

where $\Delta\varepsilon_{\alpha}$ and τ_{α} are the relaxation strength and relaxation time of the α relaxation, respectively, γ_{α} and β_{α} are the shape coefficients of α relaxation, while $\Delta\varepsilon_{\delta}$, τ_{δ}, γ_{δ}, and β_{δ} are the similar parameters for δ relaxation. σ is the direct-current conductivity, and ε_{0} is the vacuum dielectric constant. The frequency domain spectroscopy (FDS) fitting of grafted-PP samples based on the H–N model is shown in Figure 5.2.

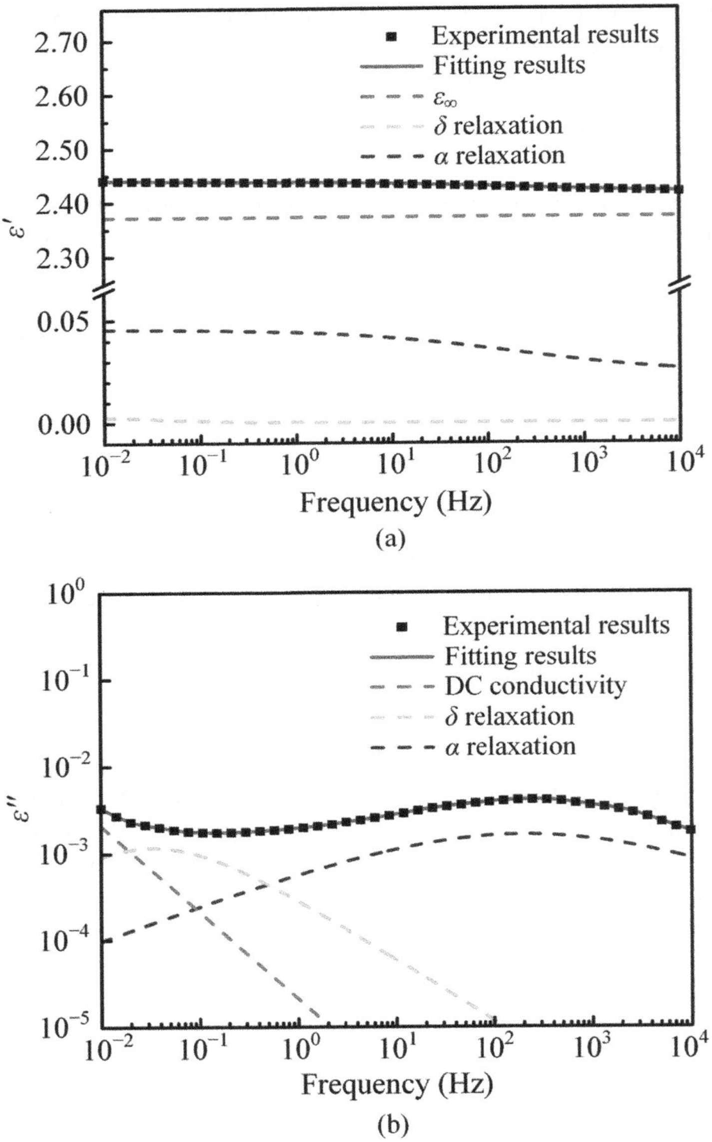

Figure 5.2 Example for FDS fitting of grafted-PP samples based on the H–N model: (a) real permittivity and (b) imaginary permittivity. *Source:* Li et al. [39]/with permission from IEEE.

5.2 Effect of Blending Modification

5.2.1 FDS of PP Blend Insulation

Blending iPP with syndiotactic polypropylene (sPP) has demonstrated a large enhancement in the breakdown performance and flexibility of iPP [6]. Atactic polypropylene (aPP) is a subsidiary product of the PP industry and has poor electrical performance. However, iPP/aPP blends exhibit better ductility with aPP acting as a softening agent [7]. Exploiting PP blends with aPP at relatively low contents for high-voltage cable insulation is of great significance for both economic and environmental factors. Therefore, this subsection employs the iPP/sPP/aPP blends as the test samples. The descriptions of the five PP blends, with different contents, studied in this investigation are shown in Table 5.1.

Figures 5.3 and 5.4 show the real part (ε') and imaginary part (ε'') of the complex permittivity (ε^*) of the samples at frequencies ranging from 10^{-2} to 10^4 Hz and temperatures from 50 to 90 °C. In the FDS measurement, both sides of the samples are sputtered with aluminum electrodes, and the diameter of the electrodes is 90 mm. The complex permittivity is measured by a broadband dielectric spectrometer (Concept 40, Novocontrol) under 1 kV alternating voltage. The frequency in the test varies from 10^{-2} to 10^4 Hz and the temperature is at 50, 70, and 90 °C. In Figure 5.3, there are labeled two monotone rising trends of dielectric constant as the frequency decreases, representing the growth of the α and δ relaxation strengths, respectively. However, in Figure 5.4, two relaxation peaks are observed in the relaxation spectra. First, the dielectric loss spectroscopy shows a relaxation peak above 10^2 Hz, and it can be identified as α relaxation process. The α relaxation is associated with the motion of segmental chains [4]. Moreover, as the frequency decreases, an increase in ε'' is observed. These behaviors can be determined as δ relaxation processes that are involved with carrier hopping in localized states [4]. The peaks of α and δ relaxations both shift to higher frequency as the temperature rises, indicating that more thermal energy is conducive to lowering the average

Table 5.1 Index for PP blend samples investigated in this subsection.

Index	iPP content (wt%)	sPP content (wt%)	aPP content (wt%)
sPP0	95	0	5
sPP5	90	5	5
sPP15	80	15	5
sPP30	65	30	5
sPP45	50	45	5

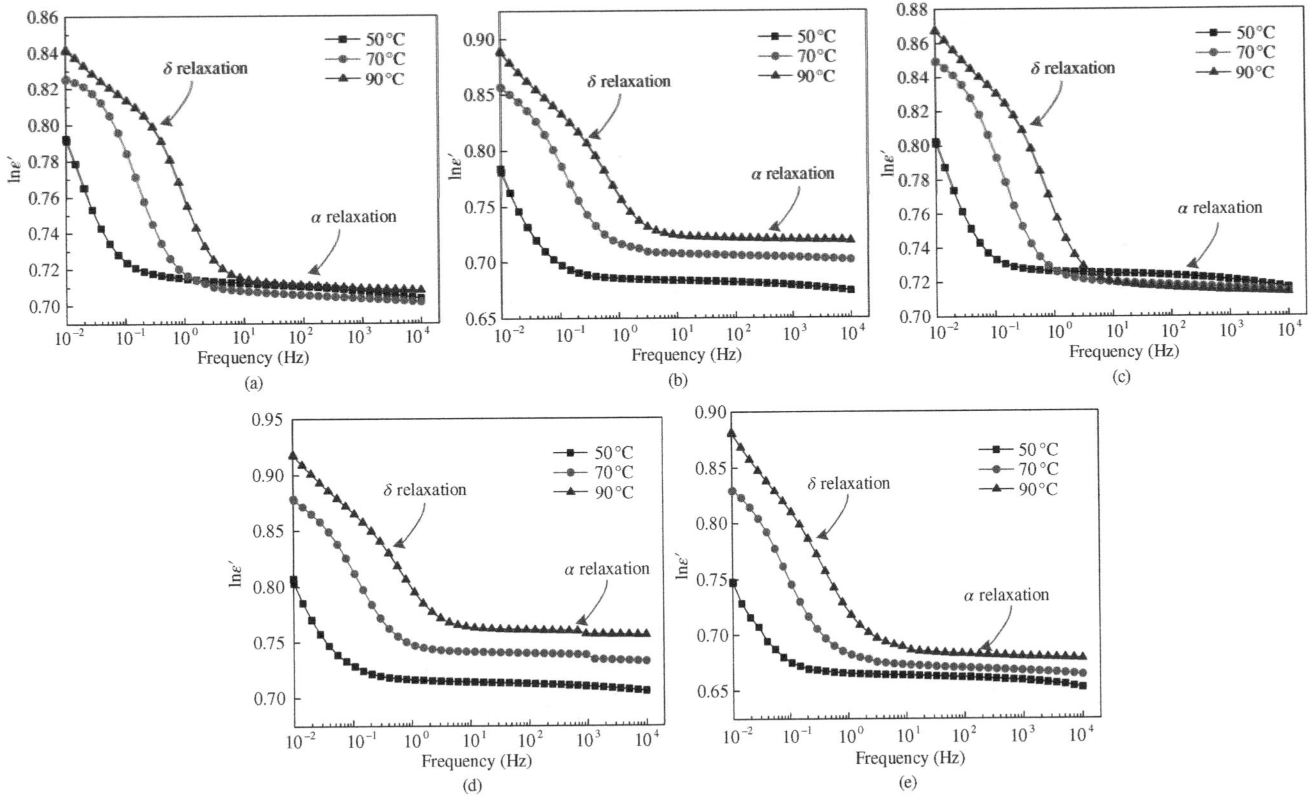

Figure 5.3 Real part of permittivity ln(ε') as a function of frequency at 50, 70, and 90 °C for (a) sPP0, (b) sPP5, (c) sPP15, (d) sPP30, and (e) sPP45. *Source:* Li et al. [5]/with permission of Elsevier.

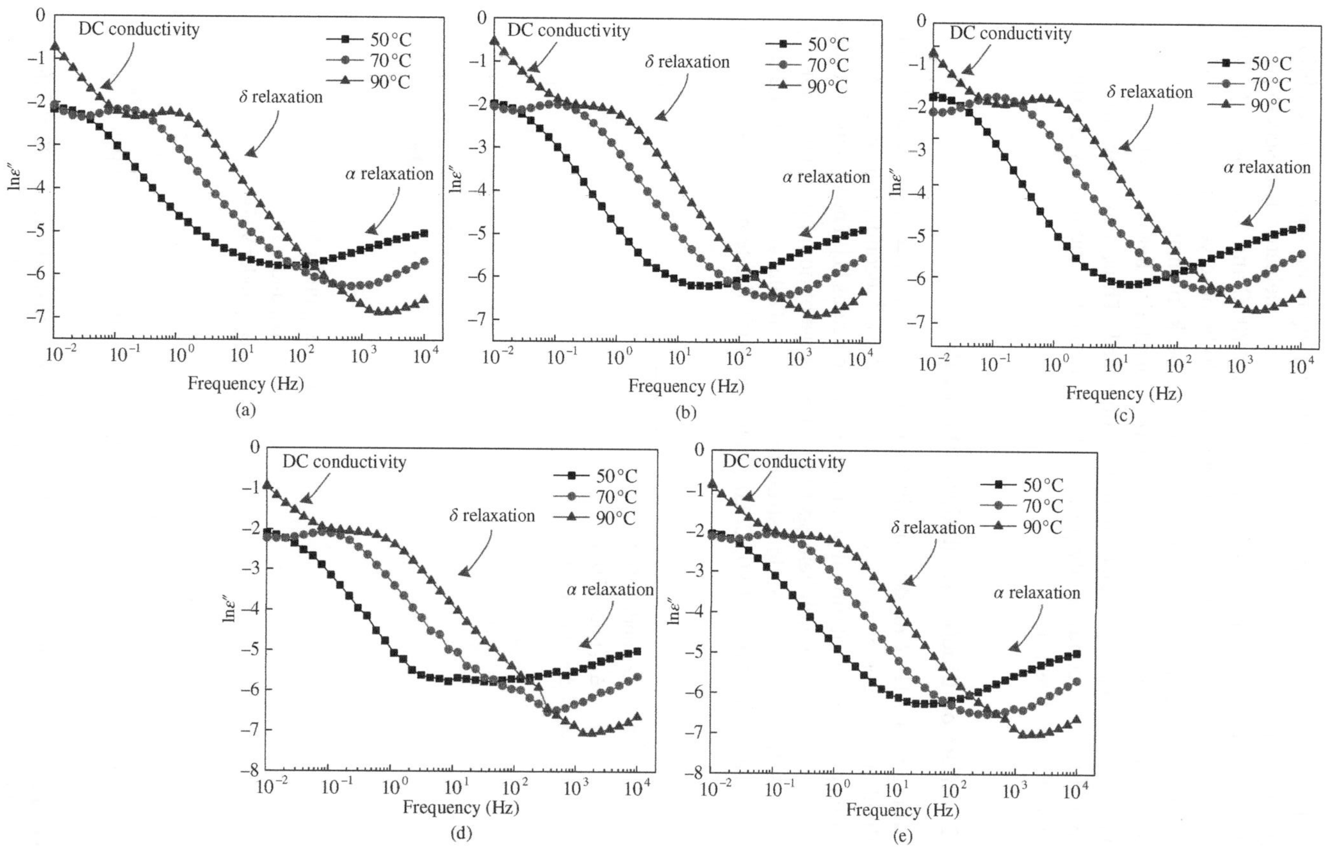

Figure 5.4 Imaginary part of permittivity ln(ε'') as a function of frequency at 50, 70, and 90 °C for (a) sPP0, (b) sPP5, (c) sPP15, (d) sPP30, and (e) sPP45. *Source:* Li et al. [5]/with permission of Elsevier.

time of segmental chains' motion and carrier hopping in PP samples. In addition, when the temperature rises to 70 °C and 90 °C, the dielectric loss spectroscopy shows a new transition process below 10^{-1} Hz for 70 °C and 1 Hz for 90 °C. This new process also exists at 50 °C but is too weak to be observed or covered by the adjacent δ relaxation peak. For this process, the peak intensity increases dramatically with temperature, and a linear negative correlation between $\ln(\varepsilon'')$ and frequency in the logarithm distribution is observed. Thus, this new process is attributed to a DC conductivity process at low frequency and mainly contributes to the dielectric loss [1].

5.2.2 Effect on Dipole Orientational Polarization

According to the H–N model, the fitting results for τ_α, τ_δ, $\Delta\varepsilon_\alpha$, $\Delta\varepsilon_\delta$, and σ_{DC} are obtained. The values of τ_α are between 10^{-9} and 10^{-4} seconds, which are in the time range for the orientational polarization of molecules [1]. The τ_δ is in the range of 10^{-1}–10^2 seconds, which belongs to the time range of hopping polarization [1]. Figure 5.5 shows the α relaxation strength ($\Delta\varepsilon_\alpha$) and δ relaxation strength ($\Delta\varepsilon_\delta$) of five samples at 50, 70, and 90 °C. $\Delta\varepsilon_\delta$ increases with the temperature since the

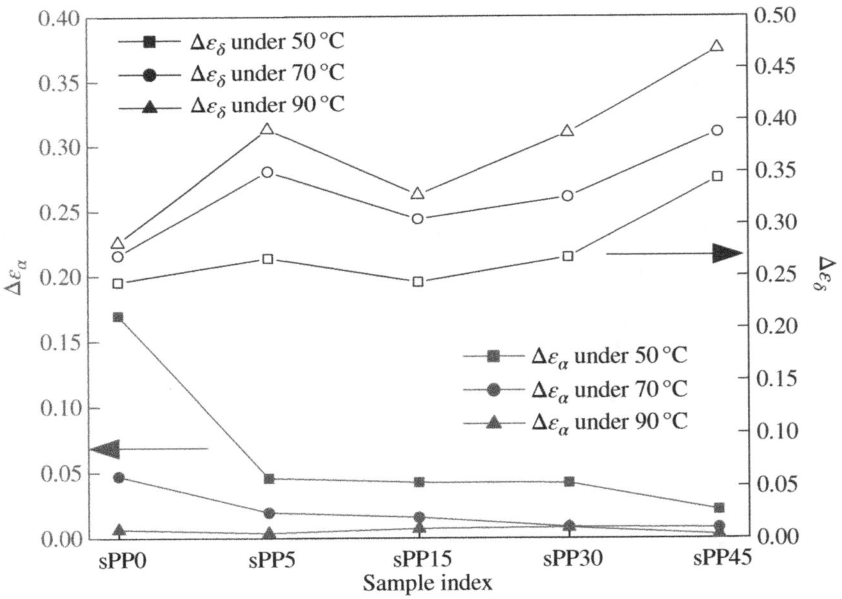

Figure 5.5 α Relaxation strength ($\Delta\varepsilon_\alpha$) and δ relaxation strength ($\Delta\varepsilon_\delta$) of PP blends at 50, 70, and 90 °C. *Source:* Li et al. [5]/with permission of Elsevier.

thermal motion of charges is more intense, and the carrier mobility is enhanced in the PP bulk.

The α relaxation results from segmental motions in the amorphous phase of iPP, sPP, and aPP [8], since the polarity of iPP, sPP, and aPP molecules is similar. From the dipole point of view, α relaxation is a large-range dipolar motion relaxation contributed to by micro-Brownian segmental motions of polymer chains [9]. As the temperature rises, more dipoles gain mobility to orient with the electrical field and contribute to the increase in permittivity. Therefore, $\Delta\varepsilon_\alpha$ increases with the temperature. However, further increasing of the temperature will lead to even stronger Brownian motion, causing the dipoles involved to undergo random thermal motion, thus preventing their complete orientation with the applied electrical field. Therefore, $\Delta\varepsilon_\alpha$ decreases with temperature as illustrated in Figure 5.8. Moreover, in Figure 5.8, it can be found that $\Delta\varepsilon_\alpha$ decreases as the amount of sPP increases. This phenomenon can be further investigated by the Kirkwood–Fröhlich equation [10]

$$\Delta\varepsilon_\alpha = \frac{1}{3\varepsilon_0} g \frac{\mu^2}{k_B T} \frac{N}{V} = \frac{1}{3\varepsilon_0} g \frac{\mu^2}{k_B T} \frac{\rho_A N_A}{M} \tag{5.6}$$

where ε_0 is the dielectric constant under vacuum; g is the Kirkwood–Fröhlich correlation factor, which describes the molecular interaction between dipoles involved in the relaxation; μ is the mean dipole moment related to the process under consideration; N/V is the number density of dipoles contributing to the concerned dielectric relaxation; ρ_A is the mass density; M is the molecular weight of the material involved in the relaxation process; and N_A is Avogadro's constant. Since the α process is associated with dipoles in the amorphous region, ρ_A is the mass density of the amorphous PP. $\Delta\varepsilon_\alpha$ decreases as the temperature rises based on Eq. (5.6), which is coincident with the above conclusions in this work.

Considering all other factors to be constant in Eq. (5.6), $\Delta\varepsilon_\alpha$ can be simplified as

$$\Delta\varepsilon_\alpha \propto g \frac{\mu^2}{T} \tag{5.7}$$

Therefore, $\Delta\varepsilon_\alpha$ is proportional to the mean dipole moment and intermolecular interaction at the same temperature. In addition, $\Delta\varepsilon_\alpha$ can also be decided by the amount of dipole moments in the amorphous regions [9]. The samples with lower crystallinities contain more dipole moments in the amorphous regions and have higher $\Delta\varepsilon_\alpha$. It is also assumed that the mean dipole moment (μ) of PP molecules does not vary in the PP blend systems, thus $\Delta\varepsilon_\alpha$ is proportional to the product of the number of dipole moments and correlation factor g [9, 10]. There is no significant change in the crystallinity of PP and PP blend systems, therefore there is no specific coincidence between $\Delta\varepsilon_\alpha$ and the number of dipole moments. Thus, the amount of dipole moments can be concluded as a minor factor, and the intermolecular interaction represented by g mainly decides $\Delta\varepsilon_{\alpha}$. In this way, further investigations of g are made, as shown below. Previous work [11] showed that the value of the interaction factors for iPP and sPP in an iPP/sPP blend is essentially zero. These interaction factors are calculated from the Flory–Huggins theory, which is used to estimate the intermolecular interaction. Another work [12] also demonstrates that the effect of the intermolecular interaction between iPP and sPP exists only when the amount of sPP is small, and the intermolecular interaction effect will generally become zero with the addition of sPP. Accordingly, with the increasing sPP content, the interaction factor (g) of PP blends reduces and contributes to the decrease in $\Delta\varepsilon_{\alpha}$.

5.2.3 Effect on Carrier Hopping Polarization

The dielectric relaxation process δ is ascribed to hopping polarization [4]. In a dielectric material, localized charges (ions and vacancies, or electrons and holes) can hop out from site A and be trapped in neighboring site B, leaving a positive charge in site A, and creating a negative charge in site B, by this way a dipole is formed [1]. A large number of deep traps concentrate in the interfaces between amorphous and crystalline phases of PP [13–15]. When the trapping of carriers takes place, especially in the interfaces between iPP and sPP regions or amorphous and crystalline phases, this process can be named interfacial polarization or Maxwell–Wagner–Sillars polarization [1].

To calculate the activation energy of hopping polarization, τ_{δ} as a function of temperature is well fitted by the Arrhenius equation [16]:

$$\tau_{\delta}(T) = \tau_0 \exp\left(-\frac{E_{\tau}}{kT}\right) \tag{5.8}$$

where τ_0 is the pre-exponential factor, E_{τ} is the activation energy, and k is Boltzmann's constant. The Arrhenius plot of the temperature dependence of the relaxation time (τ_{δ}) is shown in Figure 5.6. The fitting parameters E_{τ} of sPP0, sPP5, sPP15, sPP30, and sPP45 are 0.86, 0.83, 0.92, 1.05, and 0.54 eV, respectively.

The trap energy level indicates the potential barrier for the charge to hop from a localized state. It also can be measured by the ISPD method used in Chapter 2. The shallow and deep trap level and their density are presented in Figure 5.7. It is shown that the deep trap level and its density of sPP5 are lower than sPP0. Then, the deep trap level, as well as its density, increases with the further increase of sPP content from 5 to 30 wt%. As the sPP content is up to 45 wt%, the deep trap level and its density decrease dramatically. The shallow trap level has similar behaviors varying with the content of sPP. Nevertheless, the density of shallow traps

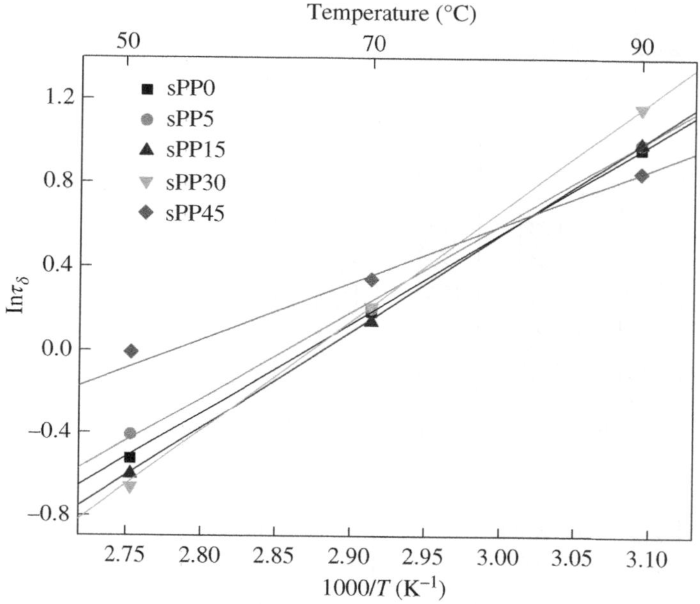

Figure 5.6 Relationship between the relaxation time constant ($\ln(\tau_\delta)$) of the δ process and the reciprocal of the temperature for PP blends. *Source:* Li et al. [5]/with permission of Elsevier.

generally does not change with the content of sPP. The trap level characteristics are closely related to the transport of carriers and play an important role in the breakdown process, which will be discussed in the following section.

There might be a correlation between hopping polarization and the trap characteristics. However, the effects of deep traps and shallow traps on hopping polarization can be significantly different. The deeper the trap, the longer the time that charges are detained in the trap. The difference in the trap filling time (τ_{trap}) might be several orders of magnitude, and this phenomenon is described by the following equation [17]:

$$\tau_{\text{trap}} = \upsilon_{\text{ATE}}^{-1} \exp\left(\frac{E_{\text{t}}}{k_{\text{B}}T}\right) \tag{5.9}$$

where υ_{ATE} is the escape frequency of trapped charges, $\upsilon_{\text{ATE}} = k_{\text{B}}T/h$; E_{t} is the trap energy level, e.g. the trap filling time constant is approximately 10^{-13} seconds when the trap energy is 0.1 eV, while the time constant is approximately 500 seconds in 1.0 eV traps [18]. Therefore, it is barely possible for charges to escape from deep traps due to the extremely long relaxation time [19]. The carriers

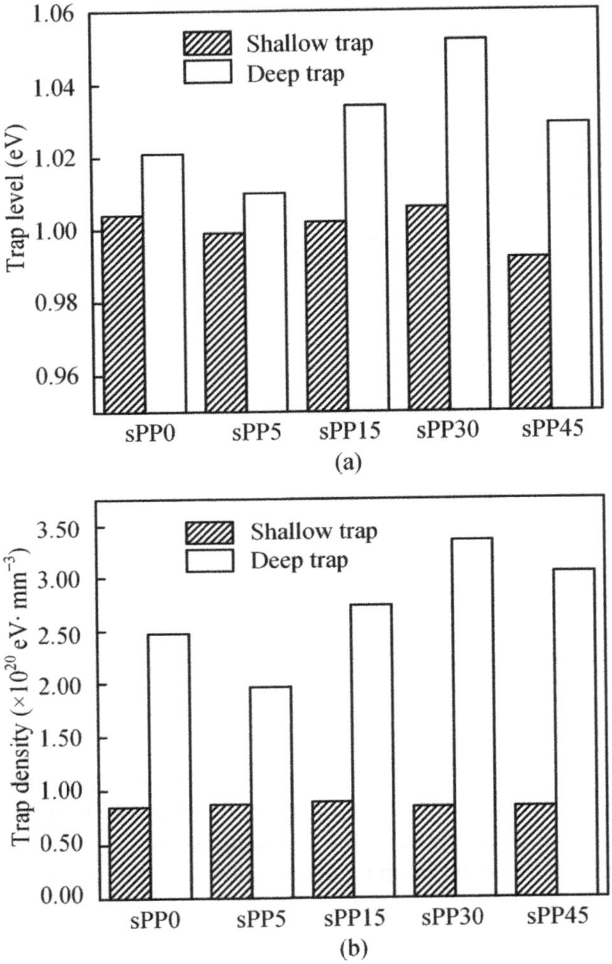

Figure 5.7 Trap center densities (a) and energy levels (b) of deep traps and shallow traps of PP blends. *Source:* Li et al. [5]/with permission of Elsevier.

can hop out from shallow traps, but carriers in deep traps are unlikely to hop or tunnel between deep trap sites [20]. Accordingly, the correlation between shallow trap energy and activation energy of δ relaxation is investigated, as shown in Figure 5.8.

As shown in Figure 5.8, the activation energy of the δ relaxation increases with the shallow trap energy. According to the hopping conductivity model, the

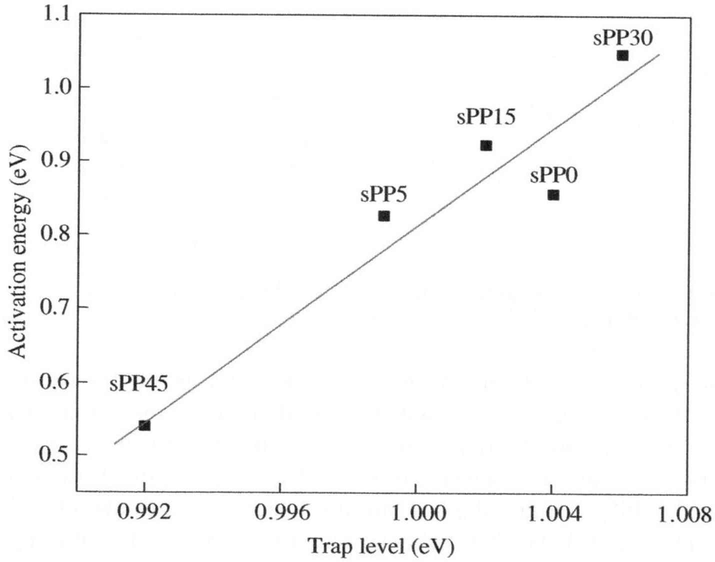

Figure 5.8 Activation energy of the δ relaxation as a function of the shallow trap level. *Source:* Li et al. [5]/with permission of Elsevier.

effective carrier de-trapping energy level (Φ_{eff}) in a dielectric is proportional to the trap level of the hopping states (Φ_t) [21]:

$$\Phi_{eff} = \Phi_t - \frac{1}{2} e E_{ext} \lambda \tag{5.10}$$

where E_{ext} is the applied field, and λ is the mean free path of electrons. Namely, Φ_{eff} is the potential barrier for charges to surmount when they hop from localized site A. In Eq. (5.10), a deeper trap level (Φ_t) indicates a higher potential barrier (Φ_{eff}) for the hopping process, thus, more activation energy is required for hopping polarization.

5.3 Effect of Monomer Grafting

The grafting monomers are dibenzylidene acetone, 2′-hydroxychalcone, and 2-*tert*-butyl-6-(3-*tert*-butyl-2-hydroxy-5-methylphenyl) methyl-4-methylp-henyl acrylate (hereinafter referred to as AO3052), which can be grafted to the PP molecular chains by initiator dicumyl peroxide (DCP), and the molecular structures are represented in Figure 5.9.

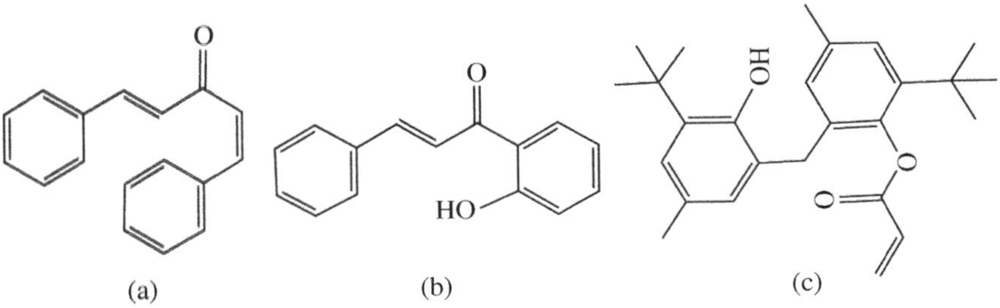

Figure 5.9 Molecular structures of grafting monomers: (a) dibenzylidene acetone, (b) 2′-hydroxychalcone, and (c) AO3052.

Dibenzylidene acetone with a content of 0.5 wt%, PP, and dicumyl peroxide (DCP) with a content of 0.3 wt% are melted and mixed at 180 °C for 10 minutes using a double roller machine. Subsequently, the resulting blend is hot-pressed using a plate vulcanizer to create sheet samples named PP-g-1. This process is replicated to make PP-g-2 and PP-g-3, with the sequential introduction of 2′-hydroxychalcone and AO3052. The schematic representation of the grafting process for PP is shown in Figure 5.10. Furthermore, reference samples named PP1, PP2, and PP3 are prepared by blending PP with the same mass fraction of the grafting monomer.

Figure 5.11a shows the FT-IR curves of PP, grafted-PP samples, and PP blend samples. When compared to PP, the grafted samples exhibit a new absorption peak

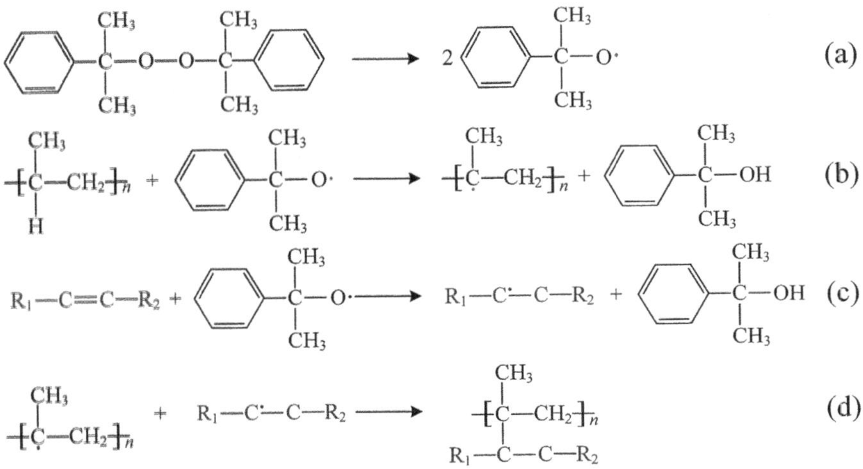

Figure 5.10 Schematic of free radical initiated grafting of PP: (a) the decomposition of DCP, (b) the generation of PP chain free radical, (c) the generation of grafted monomers free radical, (d) the grafting reaction between PP chain free radical and grafted monomers. *Source:* Li et al. [39]/with permission from IEEE.

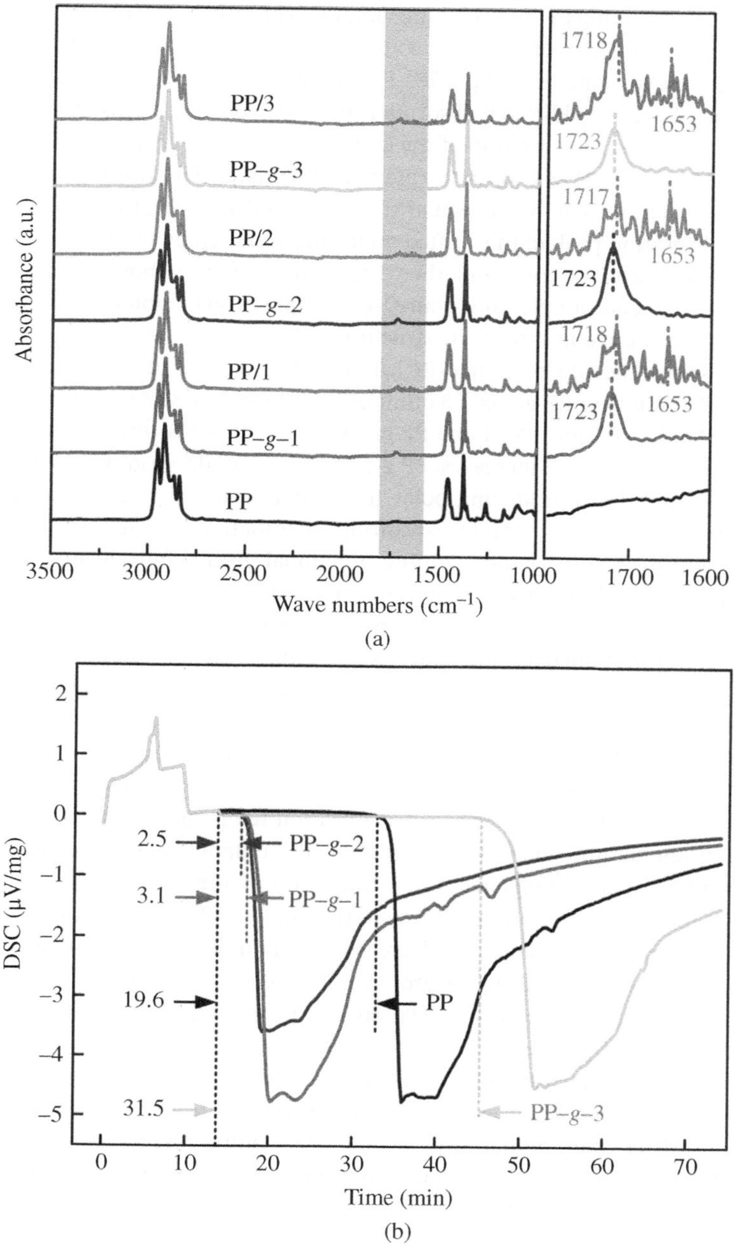

Figure 5.11 (a) FT-IR spectra of grafted-PP samples. (b) Oxidation induction times of grafted-PP samples. *Source:* Li et al. [39]/with permission from IEEE.

at 1723 cm^{-1}, corresponding to the stretching vibrations of carbonyl (C=O) within the grafted monomers. Furthermore, by comparing the local amplified FT-IR curves of the PP blend samples and grafted-PP samples, it can be seen that the PP blend samples have an additional absorption peak at 1653 cm^{-1}. This absorption peak corresponds to the vibration of the carbon–carbon double bond (C=C), signifying the disappearance of C=C in the grafted-PP samples. The grafting modification induces alterations in the molecular structure of PP, masking the benzene ring vibration absorption peak of the grafted molecule, and thereby rendering the FT-IR curves of grafted-PP samples smoother. The position of the carbonyl absorption peak in the PP blend samples and grafted-PP samples also exhibits some differences due to the change in molecular structure.

Figure 5.11b shows the OIT of grafted-PP samples. The OIT for PP is 19.6 minutes, PP-g-1 is 3.1 minutes, PP-g-2 is 2.5 minutes, and PP-g-3 is 31.5 minutes. In comparison to PP, the OIT for PP-g-1 and PP-g-2 decreases, whereas the oxidation induction time (OIT) for PP-g-3 increases. PP-g-1 and PP-g-2 undergo grafting with ordinary polar molecules, introducing polar functional groups that can readily interact with oxygen molecules, thereby expediting the initiation of oxidation reactions. Simultaneously, the oxygen-containing functional groups introduce additional reaction sites, enabling oxygen molecules to bind and react more readily with PP molecules. Because AO3052 contains a hindered phenolic group and the hydroxyl group is limited by the space barrier, the hydrogen atoms easily detach from the original molecular structure and combine with peroxide radicals and alkyl radicals, resulting in the cessation of the oxygen aging reaction.

The differential scanning calorimetry (DSC) melting and crystallization curves are shown in Figure 5.12. In comparison to PP, the grafted-PP samples exhibit a lower melting temperature and a higher crystallization temperature. The reduction in melting temperature may be attributed to the entanglement of grafted chains within them. This entanglement constrains the movement of molecular chains, leading to a slower grain growth rate in the grafted-PP samples. Furthermore, this entanglement contributes to decreased crystallinity in the grafted-PP samples. Moreover, the sluggish grain growth rate prevents certain molecular chains from achieving complete crystallization within the specified timeframe, further reducing crystallinity in the grafted-PP samples. The rise in crystallization temperature can be attributed to heterogeneous nucleation involving the grafted monomers. These monomers act as foreign impurities within the PP, with the potential to adsorb PP molecular chains during the melting phase, promoting the orderly arrangement of adsorbed chains and facilitating crystal nucleus formation. This process occurs at elevated temperatures, resulting in higher crystallization temperatures.

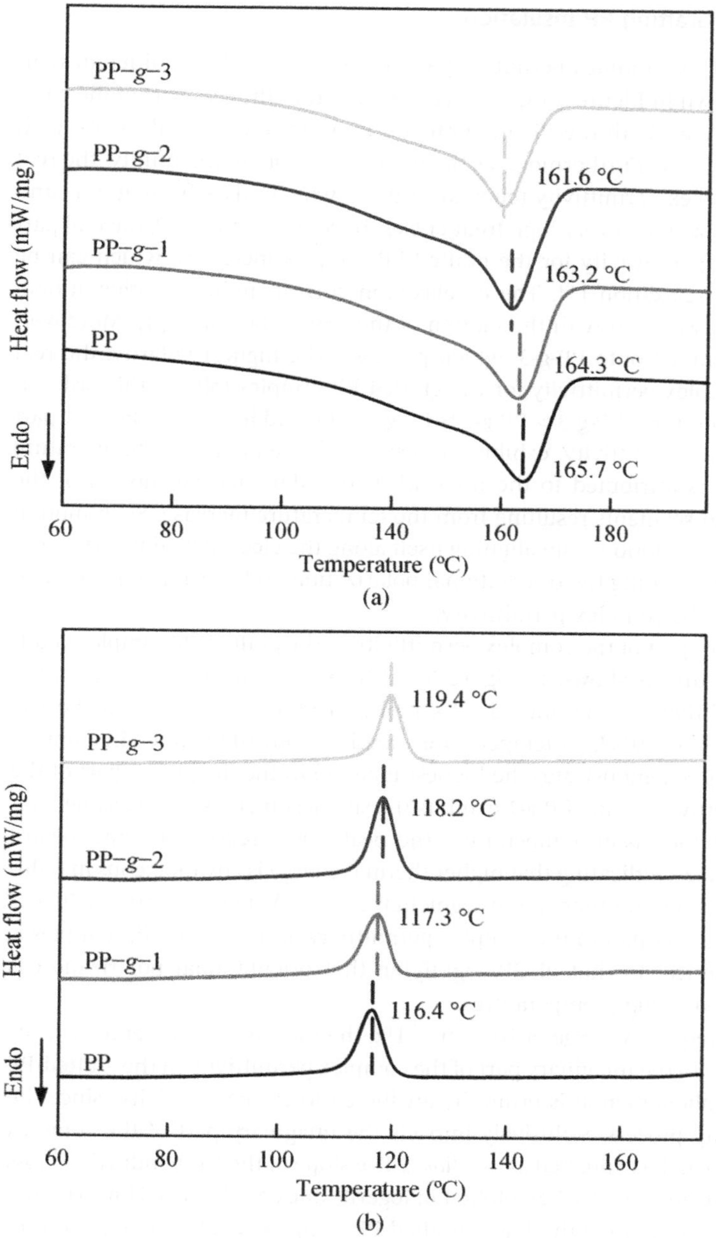

Figure 5.12 DSC curves of grafted-PP samples: (a) DSC melting curves and (b) DSC crystallization curves. *Source:* Li et al. [39]/with permission from IEEE.

5.3.1 FDS of Grafting PP Insulation

The real part of the complex permittivity of grafted-PP samples at different temperatures is shown in Figure 5.13a–d. It can be seen that the real part of the complex permittivity for both the PP and grafted-PP samples tends to decrease with increasing frequency. Furthermore, as the frequency approaches infinity, the real part of the complex permittivity tends toward ε_∞, which arises from atomic and electronic polarization at very high frequencies. In comparison to PP, the real part of the complex permittivity for the grafted-PP samples increases, which can be attributed to α relaxation [1]. The α relaxation corresponding to orientational polarization is associated with the motion of the segmental chain [4]. Moreover, as AO3052 is grafted onto PP-g-3, which possesses the highest polarity, the real part of the complex permittivity for the grafted-PP samples follows this order at the same temperature: PP-g-3 > PP-g-2 > PP-g-1 > PP. Additionally, the real part of the complex permittivity exhibits a decreasing trend as the temperature increases. This is attributed to the intensified irregular thermal motion of the molecular chain segments resulting from the temperature increase. Such motion inhibits the dipole motion from aligning itself along the electric field polarization, consequently weakening the orientational polarization and causing a decrease in the real part of the complex permittivity.

The imaginary part of the complex permittivity of the grafted-PP samples at different temperatures is shown in Figure 5.13e–h. As a result of α relaxation, the complex permittivity of the grafted-PP samples displays a peak within the frequency range of 10–10^4 Hz at temperatures ranging from 10 to 50 °C. In comparison to PP, PP-g-3 demonstrates the highest increase in the imaginary part of the complex permittivity, while PP-g-1 exhibits the smallest increase at the same temperature. With increasing temperature, the peaks of α relaxation shift toward higher frequencies, indicating that higher thermal energy is advantageous in reducing the average motion time of the segmental chains. At temperatures of 70 and 90 °C, the imaginary part of the complex permittivity in the grafted-PP samples is essentially identical to that of PP, signifying that orientational polarization is largely inhibited at high temperatures.

Within the frequency range of 10^{-2}–10^{-1} Hz, the increase in temperature leads to an escalation in the imaginary part of the complex permittivity in the grafted-PP samples. This phenomenon is primarily attributed to DC conductivity. Since the DC conductivity process exclusively impacts the imaginary part of the complex permittivity, it can be expressed as $\sigma_{DC}/i\omega\varepsilon_0$. The slope of the DC conductivity loss curve should ideally be -1 when plotted in logarithmic coordinates. However, the observed slope values in both PP and grafted-PP samples within the temperature range of 10–90 °C are not -1. This discrepancy indicates the existence of an

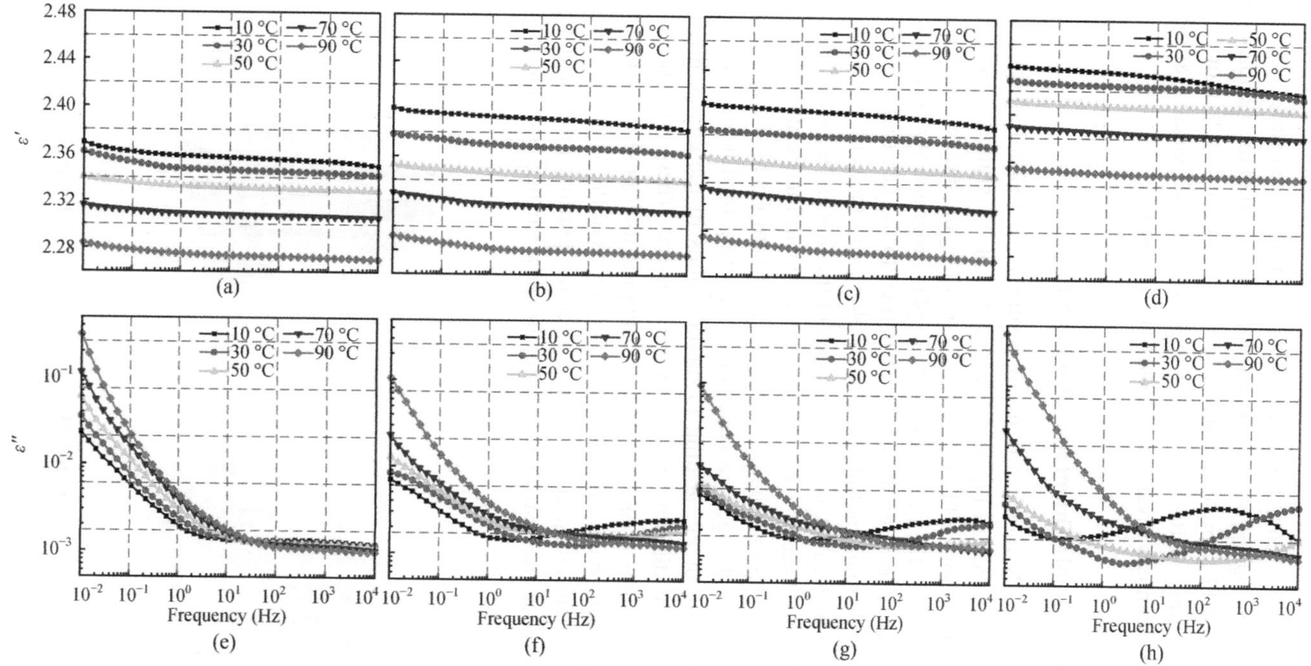

Figure 5.13 Real part and imaginary part of permittivity of grafted-PP samples at 10, 30, 50, 70, and 90 °C: (a) and (e) PP, (b) and (f) PP-*g*-1, (c) and (g) PP-*g*-2, and (d) and (h) PP-*g*-3. *Source:* Li et al. [39]/with permission from IEEE.

additional δ relaxation corresponding to hopping polarization within this frequency range, which is linked to carrier hopping in localized states [4].

In the frequency range of 10^{-2}–10 Hz, the imaginary part of the complex permittivity in the grafted-PP samples consistently remains lower than that of PP across the temperature range of 10–70 °C. This suggests that the chemical grafting modification method can effectively inhibit carrier transport.

5.3.2 Effect on Dipole Orientational Polarization

According to the H–N model, the fitting results for α relaxation are shown in Figure 5.14. As observed in Figure 5.14a, the $\Delta\varepsilon_\alpha$ of the grafted-PP samples has increased in comparison to PP, with PP-g-3 exhibiting the largest $\Delta\varepsilon_\alpha$ at the same temperature. The $\Delta\varepsilon_\alpha$ for the grafted-PP samples progressively decreases as the temperature rises. This phenomenon is attributed to the fact that α relaxation is linked to the Brownian motion of dipoles along the molecular chains. When the temperature is elevated, the arrangement of the dipoles becomes increasingly irregular, making it difficult to be polarized under the influence of an electric field. Figure 5.14b illustrates that the τ_α of the grafted-PP samples has increased in comparison to PP. This change is a result of the grafting modification altering the molecular structure of PP, leading to modifications in the charge distribution and dipole moment distribution within the material. These alterations reduce the rate of α relaxation, consequently increasing τ_α. Additionally, the interactions between the grafted chains and the molecular chains of PP, as well as among the grafted chains themselves, are enhanced after the grafting modification. These interactions hinder the rearrangement of dipoles, further contributing to the

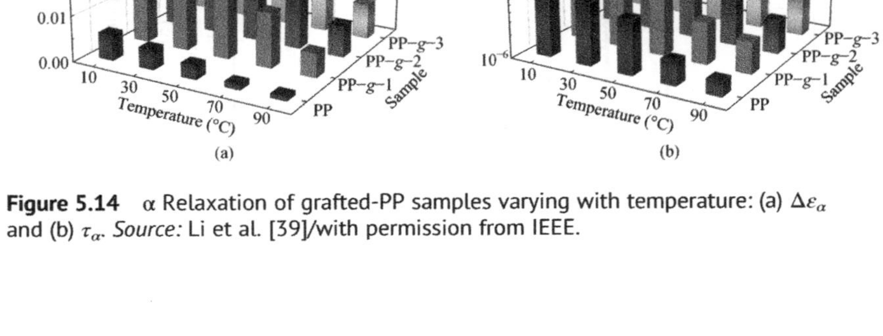

Figure 5.14 α Relaxation of grafted-PP samples varying with temperature: (a) $\Delta\varepsilon_\alpha$ and (b) τ_α. *Source:* Li et al. [39]/with permission from IEEE.

increase in τ_α. Furthermore, the τ_α of the grafted-PP samples decreases gradually with rising temperature, corresponding to the rightward shift of the α relaxation peaks, as shown in Figure 5.13.

According to Figure 5.14a, it can be seen that there is a large difference in $\Delta\varepsilon_\alpha$ for different grafted-PP samples, and this difference can be discussed by the Kirkwood–Fröhlich equation (Eqs. (5.6) and (5.7)). $\Delta\varepsilon_\alpha$ is proportional to g and μ at the same temperature. Figure 5.15 shows the relationship between $\Delta\varepsilon_\alpha$ and the temperature of the grafted-PP samples. The slopes of the fitted curves for the grafted-PP samples are larger than those for PP, indicating that the $g\mu^2$ of the grafted-PP samples surpasses that of PP. The enhancement in the interaction between dipoles within the grafted-PP samples can be attributed to the interaction between molecular chains, resulting in an increased g value for the grafted-PP samples. Furthermore, Gaussian16 serves as the platform for quantum chemistry computations. Molecular models of grafted-PP samples are created in GaussView, and subsequently optimized in Gaussian to obtain the conformation with the lowest potential energy. Finally, the dipole moment of both PP and grafted-PP samples can be calculated. The dipole moment calculation results for the grafted-PP samples, as shown in Figure 5.16, reveal the dipole moment of PP to be 0.48 *D*. In contrast, PP-g-1 has a dipole moment of 3.69 *D*, PP-g-2 is 3.81 *D*, and PP-g-3 reaches 4.14 *D*. This indicates that grafting modification can augment the dipole moment of PP. Notably, the dipole moment of PP-g-3 is the largest when compared to PP-g-1 and PP-g-2, resulting in a larger $\Delta\varepsilon_\alpha$ for PP-g-3 at the same temperature.

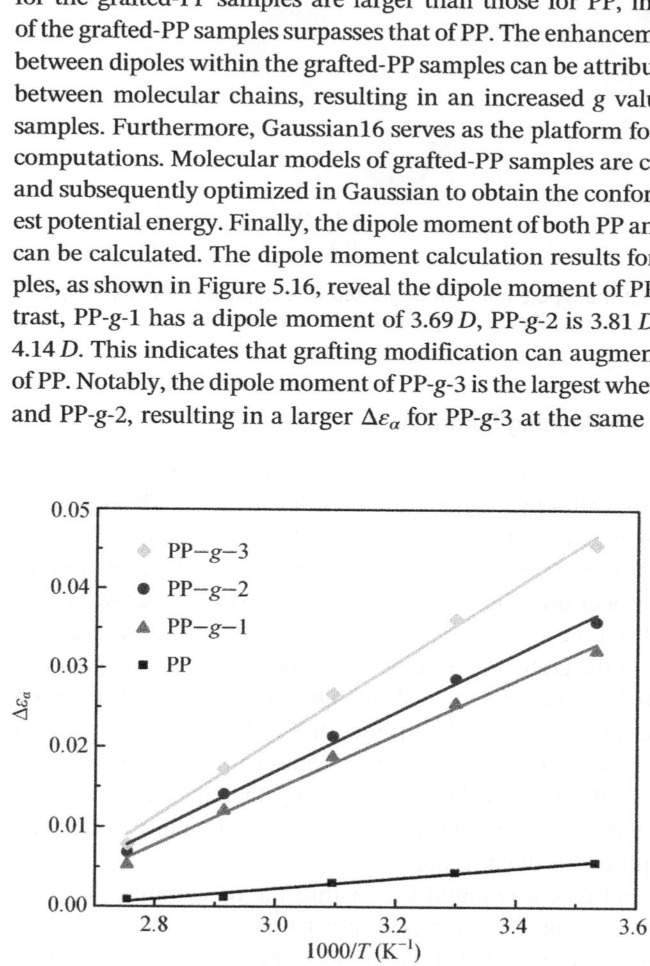

Figure 5.15 Relationship between $\Delta\varepsilon_\alpha$ and temperature of grafted-PP samples. *Source:* Li et al. [39]/with permission from IEEE.

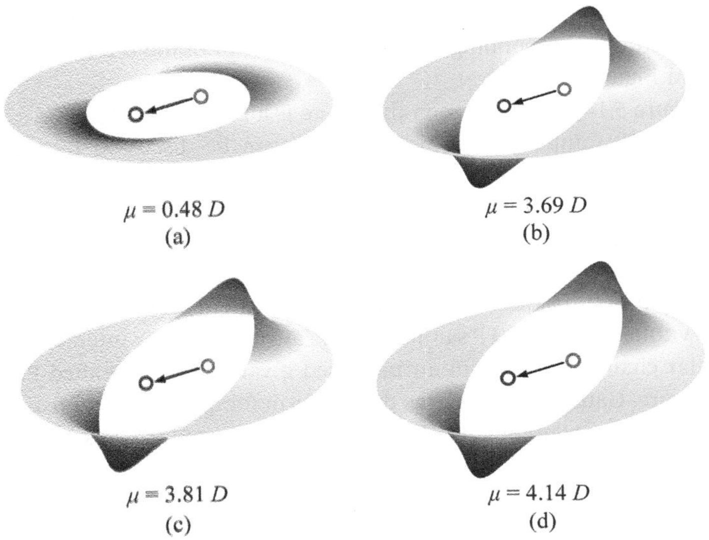

Figure 5.16 Schematic of the dipole moment of grafted-PP samples: (a) PP, (b) PP-*g*-1, (c) PP-*g*-2, and (d) PP-*g*-3. *Source:* Li et al. [39]/with permission from IEEE.

5.3.3 Effect on Carrier Hopping Polarization

Figure 5.17 shows the fitting results of δ relaxation. The $\Delta\varepsilon_\delta$ of grafted-PP samples is lower than that of PP within the temperature range of 10–90 °C. This indicates that the transportation of carriers within the grafted-PP samples is more challenging than in PP, resulting in the formation of δ relaxation being less pronounced. Notably, within the temperature range of 10–50 °C, PP-g-3 exhibits the lowest $\Delta\varepsilon_\delta$, suggesting that PP-g-3 is more effective in inhibiting carrier hopping. As the temperature rises, the $\Delta\varepsilon_\delta$ of the grafted-PP samples PP gradually increases. This phenomenon is attributed to the heightened mobility of carriers at elevated temperatures. At 90 °C, the $\Delta\varepsilon_\delta$ of PP-g-3 surpasses that of PP-g-1 and PP-g-2. This can be attributed to the thermal dissociation of molecules within PP-g-3, resulting in an increased number of carriers. Additionally, the τ_δ gradually decreases with rising temperature, which is a consequence of the increased mobility of carriers, reducing the time required for them to hop.

By comparing $\Delta\varepsilon_\alpha$ and $\Delta\varepsilon_\delta$, it is found that within the temperature range of 10–50 °C, the $\Delta\varepsilon_\alpha$ in the grafted-PP samples is larger than $\Delta\varepsilon_\delta$, while $\Delta\varepsilon_\alpha$ in PP is less than $\Delta\varepsilon_\delta$. This suggests that in this temperature range, α relaxation is stronger than δ relaxation in the grafted-PP samples, whereas δ relaxation is stronger than α relaxation in PP. As the temperature continues to rise, $\Delta\varepsilon_\delta$

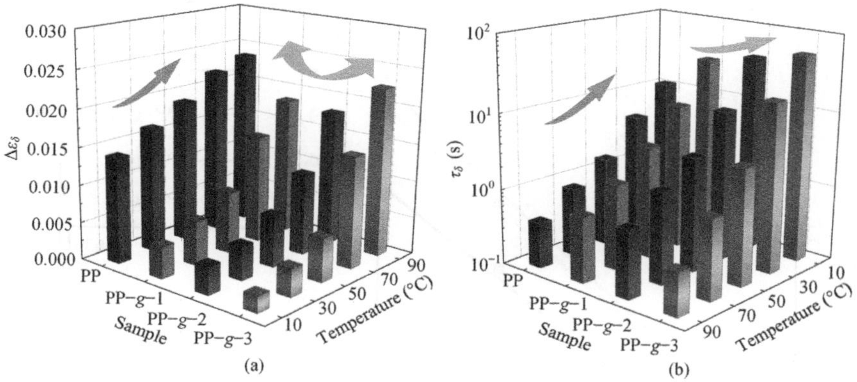

Figure 5.17 δ Relaxation of grafted-PP samples varying with temperature: (a) $\Delta\varepsilon_\delta$ and (b) τ_δ. *Source:* Li et al. [39]/with permission from IEEE.

gradually decreases, while $\Delta\varepsilon_\delta$ increases progressively. At 90 °C, $\Delta\varepsilon_\delta$ is larger than $\Delta\varepsilon_\alpha$ for both PP and grafted-PP samples.

In a dielectric material, localized charges (ions and vacancies, or electrons and holes) can hop from one site to the neighboring site, creating the δ relaxation [1]. The relationship between δ relaxation and barrier can be expressed as [34]

$$\tau = \tau_0 e^{\Phi/(kT)} \tag{5.11}$$

where τ_0 is the inherent relaxation time of δ relaxation, Φ is the barrier height that the carrier hop needs to overcome, and k is Boltzmann's constant.

The relationship between τ_δ and the temperature of the grafted-PP samples is shown in Figure 5.18. According to the slope of the fitted curve, the Φ of grafted-PP samples can be calculated. It can be seen that the Φ of PP is 0.3643 eV, PP-g-1 is 0.4137 eV, PP-g-2 is 0.4339 eV, and PP-g-3 is 0.5638 eV. Compared with PP, the Φ of the grafted-PP samples has increased, indicating that grafting modification can elevate the Φ of PP, leading to the inhibition of carrier hopping. Notably, PP-g-3 has the largest Φ among the grafted-PP samples, making it more effective in inhibiting carrier hopping and reducing $\Delta\varepsilon_\delta$.

The schematic of the double potential well model for δ relaxation is shown in Figure 5.19. In the absence of an electric field, the barrier height between points A and B in PP is denoted as Φ, while the barrier height in the grafted-PP samples is $\Phi + \Delta\Phi$, which exceeds that of PP. Under the action of the electric field, the barrier becomes tilted. During this period, the probability of a charge transitioning from A to B increases, while the probability of B to A decreases. It is worth noting that the barrier height that carriers must overcome for hopping in the grafted-PP

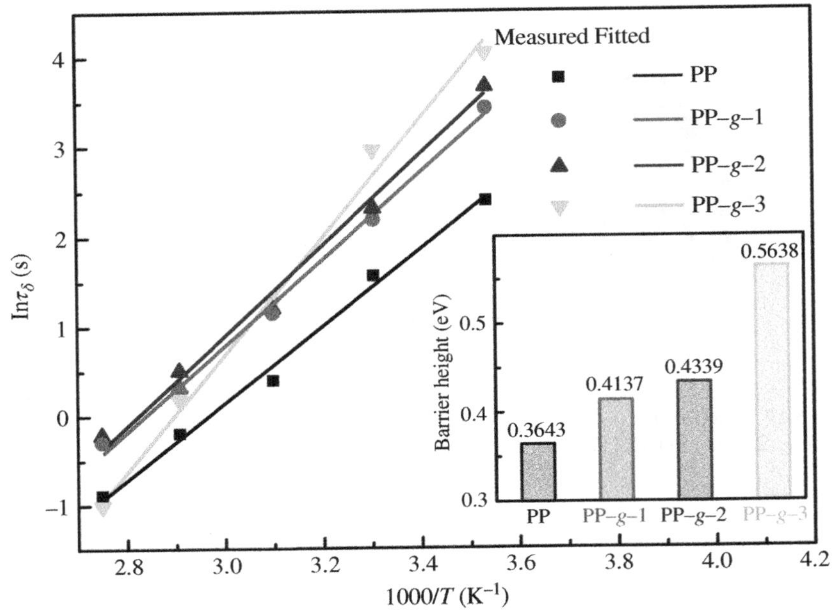

Figure 5.18 Relationship between τ_δ and temperature of grafted-PP samples. *Source:* Li et al. [39]/with permission from IEEE.

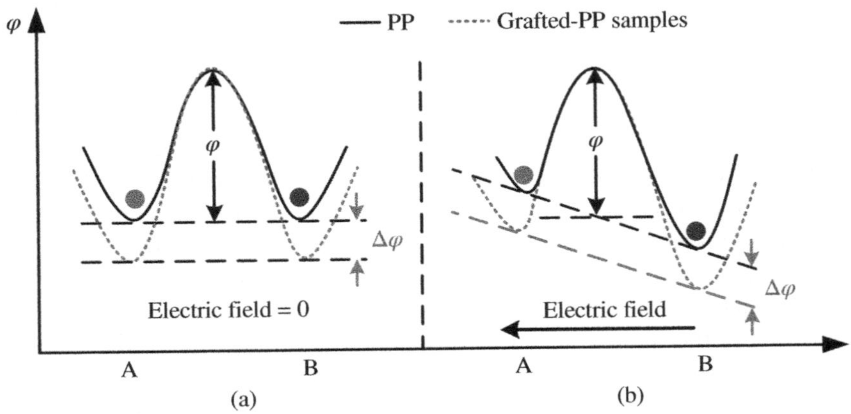

Figure 5.19 Relationship between τ_δ and temperature of grafted-PP samples: (a) PP, (b) grafted-PP samples. *Source:* Li et al. [39]/with permission from IEEE.

samples remains larger than that in PP. This signifies that the grafted-PP samples exhibit enhanced capabilities in inhibiting carrier hopping, reducing $\Delta\varepsilon_{\delta}$, and prolonging τ_{δ}.

5.4 Effect of Thermal Ageing

5.4.1 FDS of Thermal-Aged PP Insulation

The unused PP cable is used for experimentation, and the PP insulation layer is subjected to roll-cutting treatment, with a thickness set at 200 μm. The obtained samples are cut into rectangular samples with dimensions of 8 cm in length and width. These samples are suspended in a high-temperature test chamber with clips on it, with a distance of more than 20 mm between each sample. The temperature of the high-temperature test chamber is set at 135 °C according to the IEC60216 standard, with a ventilation frequency of 160 times per hour. After aging for 7, 14, 30, and 60 days, several samples are taken out and left to settle at room temperature for 12 hours before conducting the FDS test. The complex permittivity is measured using a broadband dielectric spectrometer (Concept 40, Novocontrol) under 1 kV alternating voltage. Circular electrodes with a diameter of 20 mm are chosen as the test electrodes. Prior to testing, a small amount of silver paste is daubed evenly on both sides of the sample using a brush with a size that matches the test electrodes. The samples are allowed to air dry before testing.

The changes in the real part of the complex permittivity of PP at different aging times and temperatures are shown in Figure 5.20a–e. It shows that the real part of the complex permittivity of PP increases with aging time. This indicates that the molecular chains of PP have undergone oxidation, resulting in the formation of polar groups or the breakage of molecular chains to produce polar small molecules [35]. It enhances the interaction between dipoles and shifts the positive and negative charge centers of PP, leading to an increase in the real part of the complex permittivity. As the frequency increases, the real part of the complex permittivity decreases, and when the frequency is very high, the real part of the complex permittivity tends to ε_{∞}, which arises from atomic and electronic polarization at very high frequencies [1]. With the increase in aging time, this decreasing trend becomes stronger, indicating an enhancement of the polarity of PP. The real part of the complex permittivity of PP decreases with increasing temperature. This is a result of the density of PP decreasing due to thermal expansion and contraction, resulting in a decrease in the real part of the complex permittivity [36]. When the test temperature is between 10 and 70 °C, the real part of the complex permittivity of PP shows little variation with frequency. However, when the test temperature exceeds 90 °C, the real part of the complex permittivity significantly increases at

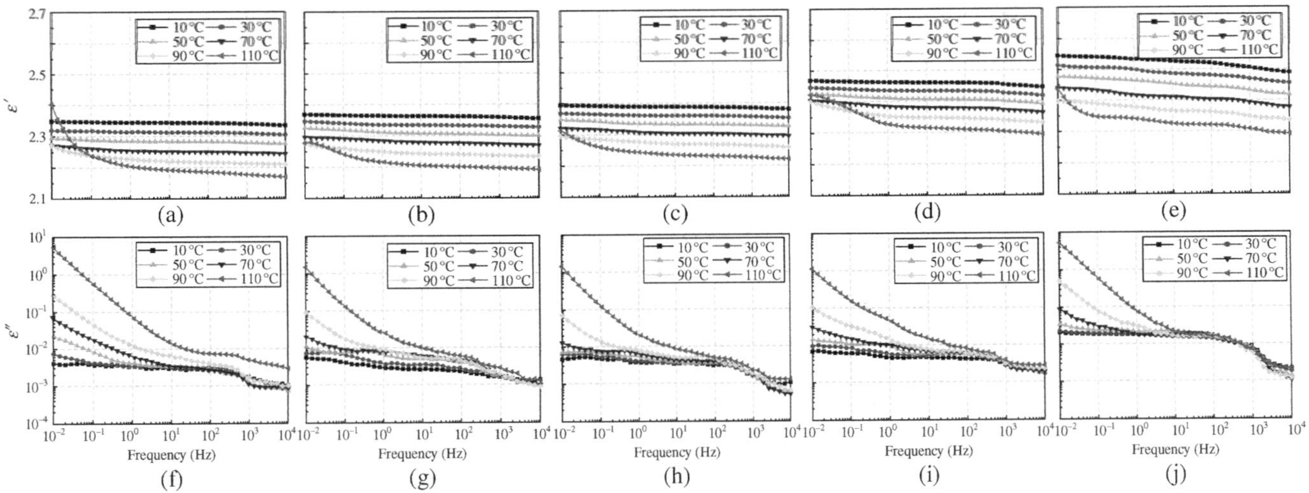

Figure 5.20 Real part and imaginary part of permittivity of thermal-aged PP samples at 10, 30, 50, 70, 90, and 110 °C: (a) and (f) unaged sample, (b) and (g) aged for 7 days, (c) and (h) aged for 14 days, (d) and (i) aged for 30 days, and (e) and (j) aged for 60 days. *Source:* Li et al. [40]/ with permission from IEEE.

lower frequencies (<1 Hz). This is a result of the higher temperature making it easier for carriers to transport, and at lower frequencies, some opposite polarity carriers accumulate near the electrodes, increasing the effective capacitance of the sample and leading to a significant increase in the real part of the complex permittivity of PP [37].

The variations of the imaginary part of the complex permittivity under different aging times and temperatures are shown in Figure 5.20f–j. The α relaxation can be observed in the frequency range of 10–10^4 Hz. The imaginary part of the complex permittivity in this frequency range significantly increases with aging time, which can be attributed to the enhancement of the α relaxation associated with segmental chain motion [1, 4]. In the frequency range of 10^{-2}–10^{-1} Hz, the imaginary part of the complex permittivity increases significantly with temperature, which can be attributed to the increase of DC conductivity. It can be expressed as $\sigma_{DC}/i\omega\varepsilon_0$. At this frequency, the imaginary part of the complex permittivity initially decreases and then increases. The reason is that in the early stage of aging, PP undergoes recrystallization, which limits the migration of carriers and leads to a decrease in DC conductivity. In the late stage of aging, the destruction of crystal regions and the formation of low-density regions lead to an increase in conductivity, increasing in the imaginary part of the complex permittivity. The slope of the DC conductivity loss curve should ideally be -1 when plotted in logarithmic coordinates. However, the experimental results show that the slope is not -1 within this frequency range. This indicates the existence of δ relaxation within this frequency range corresponding to hopping polarization, which is linked to carrier hopping in localized states [4].

5.4.2 Effect on Dipole Orientational Polarization

According to the H–N model, the fitting results for α relaxation are shown in Figure 5.21. As shown in Figure 5.21a, with the increase of aging time, the α relaxation intensity of PP significantly increases. This is a result of the polar groups and small molecules generated in PP after aging enhancing the polarity of PP, resulting in enhanced dipole polarization and an increase in $\Delta\varepsilon_\alpha$. As the temperature rises, the intensity of α relaxation decreases, which is related to the Brownian motion of the molecular chains. When the temperature rises, the enhanced Brownian motion increases the difficulty for dipoles to follow the direction of the electric field, resulting in a decrease in $\Delta\varepsilon_\alpha$. As shown in Figure 5.21b, with the increase of aging time, the τ_α of PP increases. This may be because the increase in polar groups strengthens the intermolecular interactions, hindering the rearrangement of dipoles, which is manifested as a left shift of the relaxation peak in the dielectric spectrum. As the temperature rises, the τ_α slightly decreases, resulting in a right shift of the relaxation peak in the dielectric spectrum.

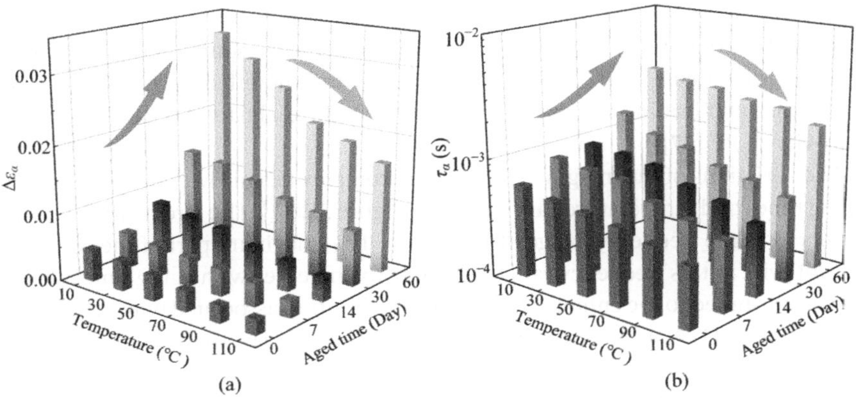

Figure 5.21 α Relaxation of thermal-aged PP samples varying with temperature: (a) $\Delta\varepsilon_\alpha$ and (b) τ_α. *Source:* Li et al. [40]/with permission from IEEE.

With the increase of aging time, $\Delta\varepsilon_\alpha$ of PP increases. This is because as aging progresses, the PP molecular chains undergo oxidation, resulting in the formation of polar groups such as carbonyl groups. Figure 5.22 shows the relationship between $\Delta\varepsilon_\alpha$ and temperature for samples with different aging times. It shows that as the aging time increases, the slope of the fitted line increases. According to Eqs. (5.6) and (5.7), the increase in slope indicates enhanced material polarity,

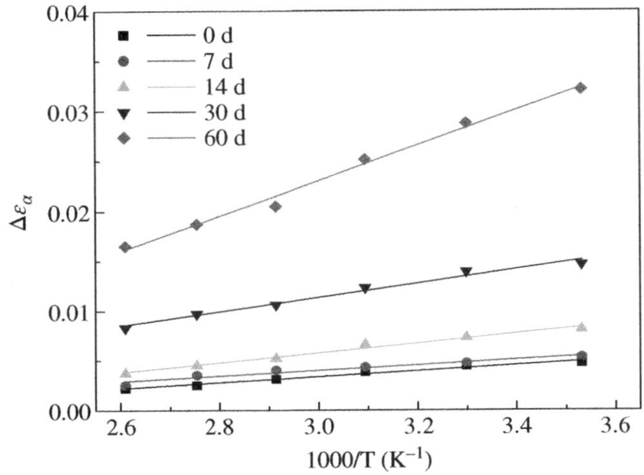

Figure 5.22 Relationship between $\Delta\varepsilon_\alpha$ and temperature of aged-PP samples. *Source:* Li et al. [40]/with permission from IEEE.

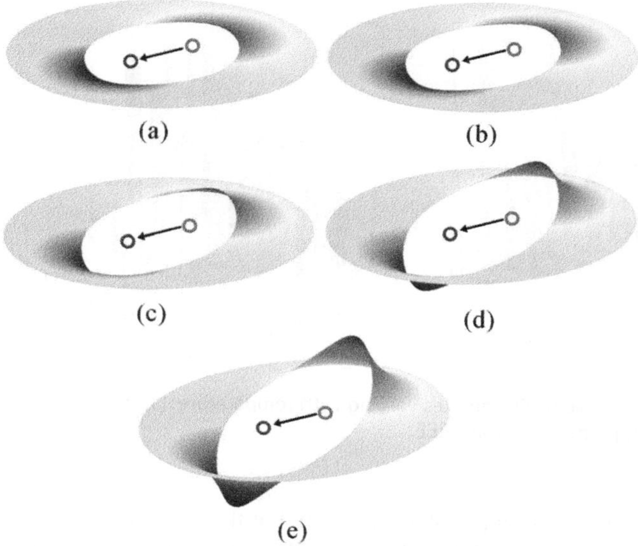

Figure 5.23 Relationship between $\Delta\varepsilon_\alpha$ and temperature of aged-PP samples: (a) unaged sample, (b) aged for 7 days, (c) aged for 14 days, (d) aged for 30 days, and (e) aged for 60 days. *Source:* Li et al. [40]/with permission from IEEE.

indicating an increase in PP polar groups. The formation of carbonyl groups can be observed in the infrared spectrum, which can be characterized using the carbonyl index. Molecular models of aged-PP samples are created in GaussView, and subsequently optimized in Gaussian to obtain the conformation with the lowest potential energy. Finally, the dipole moment of aged-PP samples can be calculated. The dipole moment calculation results for the aged-PP samples are shown in Figure 5.23. The results showed that after aging for 0, 7, 14, 30, and 60 days, the dipole moments of PP were 0.48 D, 0.62 D, 1.34 D, 2.35 D, and 3.41 D, respectively. This indicates that the polar functional groups or small molecules generated after aging will increase the dipole moment of the material, thereby increasing $\Delta\varepsilon_\alpha$.

5.4.3 Effect on Carrier Hopping Polarization

According to the H–N model, the fitting results for δ relaxation are shown in Figure 5.24. As shown in Figure 5.24a, with the increase in temperature, the strength of δ relaxation is enhanced, which also leads to a decrease of τ_δ in Figure 5.24b. With the increase in aging time, the strength of δ relaxation first decreases and then increases, indicating that the migration of carriers is limited in the early stage of aging. In the later stage of aging, the migration of carriers

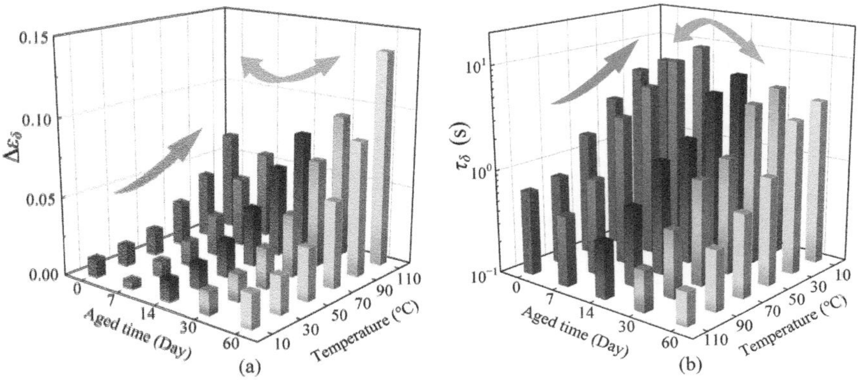

Figure 5.24 δ Relaxation of aged-PP samples varying with temperature: (a) $\Delta\varepsilon_\delta$ and (b) τ_δ. *Source:* Li et al. [40]/with permission from IEEE.

becomes easier, increasing the strength of δ relaxation. For the same reason, the τ_δ in Figure 5.24b initially increases and then decreases. With the increase in temperature, the strength of δ relaxation is enhanced, mainly because the higher temperature enhances the mobility of carriers, making the migration of carriers easier, and the increase in migration rate also leads to a decrease in the time constant. With the increase in aging time, the strength of δ relaxation exhibits a trend of initially decreasing and then increasing. This is because in the early stage of aging, imperfect spherulites in the PP undergo recrystallization under the thermal effect, making the spherulites in PP more perfect. These perfect spherulites restrict the migration of carriers, resulting in a decrease in the strength of δ relaxation and an increase of τ_δ. In the later stage of aging, as the molecular chains break, the perfect spherulites are destroyed, and the increased amorphous region and low-density region make the migration of carriers easier, increasing the strength of δ relaxation and a decrease of τ_δ.

The relationship between the activation energy (E_a) and temperature of PP after aging is shown in Figure 5.25a. According to Eq. (5.11), we can observe the variation of activation energy for PP after aging. The calculated values of activation energy for PP after aging for 0, 7, 14, 30, and 60 days are 0.261, 0.278, 0.272, 0.263, and 0.257 eV, respectively, as shown in Figure 5.25b. This indicates that the activation energy of PP initially increases and then decreases during aging. The schematic of the double potential well model for δ relaxation is shown in Figure 5.26. The barrier height between points A and B in PP is denoted as Φ, while the barrier height in the aged-PP samples is Φ-$\Delta\Phi$. According to Figure 5.26, higher activation energy limits the migration of carriers, leading to a decrease in the intensity of the δ relaxation. On the other hand, lower activation energy

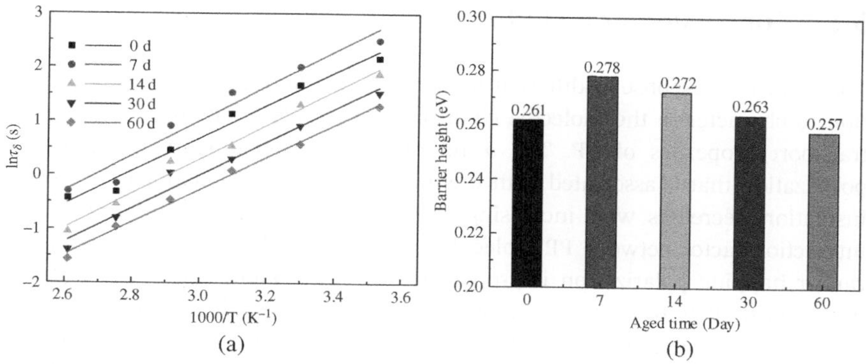

Figure 5.25 Relationship between τ_δ and temperature of aged-PP samples: (a) Arrhenius fitting curves, (b) barrier height. *Source:* Li et al. [40]/with permission from IEEE.

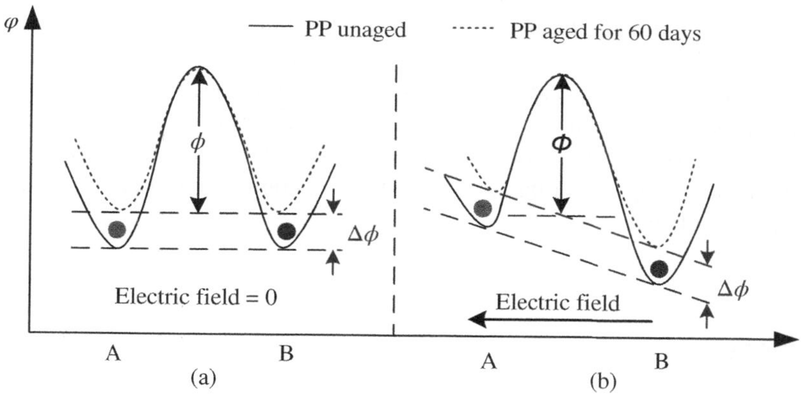

Figure 5.26 Schematic of the double potential well model of δ relaxation: (a) PP, (b) PP aged for 60 days. *Source:* Li et al. [40]/with permission from IEEE.

makes the migration of carriers in PP easier, increasing the intensity of the δ relaxation. This is consistent with the results shown in Figure 5.24.

In this subsection, by conducting FDS testing on aged PP and fitting it with the H–N model, parameters such as $\Delta\varepsilon_\alpha$ and $\Delta\varepsilon_\delta$ can be calculated. These parameters can be correlated with parameters that can characterize the aging state, such as carbonyl index and activation energy. Therefore, the aging state of the material can be evaluated by measuring the FDS of the material without causing damage to the insulation of the cable. This part will be introduced in Section 7.3.

5.5 Conclusion and Outlook

The dielectric response at different frequencies and temperatures serves a function to characterize the molecule chain motion, dipole orientation, and carrier transport properties of PP. The α relaxation corresponding to orientational polarization that is associated with the motion of segmental chains of PP blend insulation decreases with increasing content of sPP since the intermolecular interaction factor between PP molecules is weakened. The activation energy of carrier hopping polarization is proportional to the shallow trap level, and the increased shallow trap level of PP blend insulation poses a higher potential barrier for carrier hopping, resulting in the δ relaxation corresponding to hopping polarization being weakened. Grafting modification increases the intermolecular interaction factor and dipole moment, increasing α relaxation. At the same time, grafting modification increases the barrier height that the carrier hop needs to overcome, leading to a decrease in δ relaxation. PP insulation will produce polar groups such as carbonyl groups during the aging process, which increases the dipole moment and leads to an increase in α relaxation. The spherical crystal structure of PP insulation in the early aging period becomes more perfect, to better inhibit carrier hopping and reduce δ relaxation. However, with the further lengthening of the aging time, the PP molecular chain is broken, the spherical crystal structure is destroyed, the amorphous region increases, and the carriers are more likely to hop, resulting in increased δ relaxation and decreased insulation properties of PP.

References

1 Kao, K.C. (2004). *Dielectric Phenomena in Solids*. Elsevier.

2 Smith, R.C., Liang, C., Landry, M. et al. (2008). The mechanisms leading to the useful electrical properties of polymer nanodielectrics. *IEEE Transactions on Dielectrics and Electrical Insulation* 15 (1): 187–196.

3 Fuse, N., Sato, H., Ohki, Y., and Tanaka, T. (2009). Effects of nanofiller loading on the molecular motion and carrier transport in polyamide. *IEEE Transactions on Dielectrics and Electrical Insulation* 16 (2): 524–530.

4 Huang, Y., Min, D., Li, S. et al. (2017). Dielectric relaxation and carrier transport in epoxy resin and its microcomposite. *IEEE Transactions on Dielectrics and Electrical Insulation* 24 (5): 3083–3091.

5 Li, Z.L., Zhong, Z.Y., and Du, B.X. (2019). Dielectric relaxation and trap-modulated DC breakdown of polypropylene blend insulation. *Polymer* 185: 121935.

6 Hosier, I.L., Vaughan, A.S., and Swingler, S.G. (2011). An investigation of the potential of polypropylene and its blends for use in recyclable high voltage cable insulation systems. *Journal of Materials Science* 46 (11): 4058–4070.
7 Haylock, J.C., Phillips, R.A., and Wolkowicz, M.D. (2000). *Metallocene-Based Polyolefins: Preparation*. Properties and Technology.
8 Kurahashi, K., Matsuda, Y., Ueda, A. et al. (2002). The application of novel polypropylene to the insulation of electric power cable. *IEEE/PES Transmission and Distribution Conference and Exhibition* 2: 1278–1283.
9 Zhao, X., Liu, W., Jiang, X. et al. (2016). Exploring the relationship of dielectric relaxation behavior and discharge efficiency of P (VDF-HFP)/PMMA blends by dielectric spectroscopy. *Materials Research Express* 3 (7): 075304.
10 Yu, L. and Cebe, P. (2009). Effect of nanoclay on relaxation of poly(vinylidene fluoride) nanocomposites. *Journal of Polymer Science Part B: Polymer Physics* 47 (24): 2520–2532.
11 Woo, E.M., Cheng, K.Y., and Chen, Y.F. (2007). Experimental verification on UCST phase diagrams and miscibility in binary blends of isotactic, syndiotactic, and atactic polypropylenes. *Polymer* 48 (19): 5753–5766.
12 Zheng, C.X., Zhang, X.Q., Rottstegge, J. et al. (2007). Preliminary investigation on the miscibility of isotactic polypropylene (iPP) and syndiotactic polypropylene (sPP) blends. *Chinese journal of polymer science* 24 (06): 569–573.
13 Lin, Y.J., Du, W.C., Tu, D.M. et al. (2004). Space charge distribution and crystalline structure in low density polyethylene (LDPE) blended with high density polyethylene (HDPE). *Polymer international* 54 (2): 465–470.
14 Li, X., Du, Q., Kang, J., and Tu, D. (2002). Influence of microstructure on space charges of polypropylene. *Journal of Polymer Science Part B: Polymer Physics* 40 (4): 365–374.
15 Shen, W.W., Mu, H.B., Zhang, G.J. et al. (2013). Identification of electron and hole trap based on isothermal surface potential decay model. *Journal of Applied Physics* 113 (8): 083706.
16 Migahed, M.D. and Fahmy, T. (1994). Structural relaxation around the glass transition temperature in amorphous polymer blends: temperature and composition dependence. *Polymer* 35 (8): 1688–1693.
17 Simmons, J.G. and Taylor, G.W. (1971). Nonequilibrium steady-state statistics and associated effects for insulators and semiconductors containing an arbitrary distribution of traps. *Physical Review B* 4 (2): 502–511.
18 Teyssedre, G. and Laurent, C. (2005). Charge transport modeling in insulating polymers: from molecular to macroscopic scale. *IEEE Transactions on Dielectrics and Electrical Insulation* 12 (5): 857–875.
19 Skinner, D.E., Colombo, D.P., Cavaleri, J.J., and Bowman, R.M. (2005). Femtosecond investigation of electron trapping in semiconductor nanoclusters. *Journal of Physical Chemistry* 99 (20): 7853–7856.

20 Wang, W., Min, D., and Li, S. (2016). Understanding the conduction and breakdown properties of polyethylene nanodielectrics: effect of deep traps. *EEE Transactions on Dielectrics and Electrical Insulation* 23 (1): 564–572.

21 Ma, C., Min, D.M., Li, S.T. et al. (2017). Trap distribution and direct current breakdown characteristics in polypropylene/Al_2O_3 nanodielectrics. *Acta Physica Sinica* 66 (6): 067701.

22 Murakami, Y., Nemoto, M., Okuzumi, S. et al. (2008). DC conduction and electrical breakdown of MgO/LDPE nanocomposite. *IEEE Transactions on Dielectrics and Electrical Insulation* 15 (1): 33–39.

23 Kao, K.C. (2000). Electrical conduction and breakdown in insulating polymers. *Properties and Applications of Dielectric Materials* 1: 1–17.

24 Min, D.M. and Li, S.T. (2014). Simulation on the influence of bipolar charge injection and trapping on surface potential decay of polyethylene. *IEEE Transactions on Dielectrics and Electrical Insulation* 21 (4): 1627–1636.

25 Watson, P.K. (1995). The transport and trapping of electrons in polymers. *IEEE Transactions on Dielectrics and Electrical Insulation* 2 (5): 915–924.

26 Li, Z.L., Du, B.X., Yang, Z.R., and Han, C.L. (2018). Temperature dependent trap level characteristics of graphene/LDPE nanocomposites. *IEEE Transactions on Dielectrics and Electrical Insulation* 25 (1): 137–144.

27 Zhuang, Y., Chen, G., Chappell, P.H., and Rotaru, M. (2012). Surface potential decay: effect of different corona charging times. *Annual Report Conference on Electrical Insulation and Dielectric Phenomena* (14–17 October 2012). Montreal, QC, Canada, 620–623.

28 Du, B.X. and Li, Z.L. (2014). Surface charge and dc flashover characteristics of direct-fluorinated SiR/SiO_2 nanocomposites. *IEEE Transactions on Dielectrics and Electrical Insulation* 21 (6): 2602–2610.

29 Wang, Z.G., Phillips, R.A., and Hsiao, B.S. (2001). Morphology development during isothermal crystallization. I. Isotactic and atactic polypropylene blends. *Journal of Polymer Science Part B: Polymer Physics* 39 (16): 1803–1955.

30 Thyssen, A., Almdal, K., and Thomsen, E.V. (2017). Electret stability related to the crystallinity in polypropylene. *IEEE Transactions on Dielectrics and Electrical Insulation* 24 (5): 3038–3046.

31 Wu, Y.H., Zha, J.W., Li, W.K. et al. (2015). A remarkable suppression on space charge in isostatic polypropylene by inducing the β-crystal formation. *Applied Physics Letters* 107 (11): 112901.

32 Zhang, X.Q., Zhao, Y., Wang, Z. et al. (2005). Morphology and mechanical behavior of isotactic polypropylene (iPP)/syndiotactic polypropylene (sPP) blends and fibers. *Polymer* 46 (16): 5956–5965.

33 Zou, X.X., Yang, W., Zheng, G.Q. et al. (2007). Crystallization and phase morphology of injection-molded isotactic polypropylene (iPP)/syndiotactic polypropylene (sPP) blends. *Journal of Polymer Science Part B: Polymer Physics* 45 (21): 2948–2955.

34 Zhao, H., Mu, H., Zhang, D. et al. (2021). Analysis of FDS characteristics of oil-impregnated paper insulation under high electric field strength based on the motion of charge carriers. *IEEE Transactions on Dielectrics and Electrical Insulation* 28 (4): 1153–1161.

35 Singh, B. and Sharma, N. (2008). Mechanistic implications of plastic degradation. *Polymer degradation and stability* 93 (3): 561–584.

36 Min, D., Li, S., Hirai, N. et al. (2016). Dielectric spectroscopic analysis of degradation in ethylene-propylene-diene copolymer. *IEEE Transactions on Dielectrics and Electrical Insulation* 23 (6): 3620–3630.

37 Ohki, Y. (2022). Broadband complex permittivity and electric modulus spectra for dielectric materials research. *IEEJ Transactions on Electrical and Electronic Engineering* 17 (7): 958–972.

38 Lai, L., Bi, Y., Niu, B., Yu, G., Li, Y., Ding, G., & Xu, Q. (2024). Reinforcement mechanisms and current research status of silicon carbide whisker-reinforced composites: A comprehensive review. *REVIEWS ON ADVANCED MATERIALS SCIENCE*, 63 (1). https://doi.org/10.1515/rams-2024-0047.

39 Li, Z., Zhao, S., Yin, Y., & Du, B. (2024). Effect of Functional Grafting Modification on Dielectric Properties of Polypropylene Insulation—Part 1: Frequency Domain Spectroscopy. *IEEE Transactions on Dielectrics and Electrical Insulation*, 31 (4), 1748–1755. https://doi.org/10.1109/tdei.2024.3369001.

40 Li, Z., Zhang, M., Zhao, S., & Du, B. (2024). Condition Assessment of Thermally-Aged Polypropylene Cable Insulation Based on Frequency Domain Spectroscopy. *IEEE Transactions on Dielectrics and Electrical Insulation*, 1–1. https://doi.org/10.1109/tdei.2024.3389637.

6

AC Electrical Treeing and Dielectric Breakdown

6.1 Introduction

Polypropylene (PP) is considered a promising type of environmentally friendly and recyclable AC cable insulation because of its excellent insulation and heat resistance properties [1, 2]. As a typical semicrystalline polymer, PP itself has heterogeneous structural characteristics. The mesoscopic scale is the range of scales between the microscale and the macroscale, generally considered to be between nanometers and millimeters [3]. When its internal structure is observed at the mesoscopic scale, it can be found that the crystal and amorphous regions coexist. Recent research has found that the crystallization characteristic of PP is one of the most important factors affecting the inception and growth characteristics of the electrical tree [4, 5]. The crystallization temperature and post-treatment process are the important factors affecting the crystallization characteristics and modifying the electrical properties [6, 7]. It has been found that PP crystallized isothermally at 128 °C has large spherulites and a clearer boundary between spherulites, which leads to the breakdown strength and elongation at break lower [8]. The effects of quenching on electrical properties have been studied. Compared with PP crystallized isothermally at 120 °C, PP quenched in water with a temperature of 25 °C has small spherulites and higher breakdown strength. However, there are few studies on the effect of crystalline characteristics regulated by crystallization temperature on electrical tree initiation and growth characteristics. Recent studies also focused on the effects of α- and β-crystal forms on the electrical tree growth characteristics of isotactic polypropylene (iPP) insulation. Through the research of inducing montmorillonite (MMT) as an α-nucleating agent on electrical tree characteristics, it is found that uniformly distributed MMT plays the role of heterogeneous nucleation and makes the spherulites small and crystal structure dense [9, 10]. The electrical tree is shorter and denser in the PP/MMT composite insulation, and the electrical tree's initial voltage is higher in the sample with

Polypropylene Cable Insulation, First Edition. Boxue Du and Zhonglei Li.

Published 2025 by John Wiley & Sons, Inc.

denser spherulite distribution [11]. The molecular chains of PP are orderly folded to form lamellar crystals, and the lamellas are stacked to form spherulites while the molecular chains are disorderly arranged in the amorphous region resulting in the difference in dielectric properties and electrical tree growth characteristics [12, 13]. The influence of different contents of β-nucleating agents on electrical tree characteristics of PP insulation has been studied. The PP insulation modified by 0.1 wt% β-nucleating agent shows the most excellent resistance to electrical tree [14].

In 2021, Oriental Cable Company prototyped a batch of 35 kV PP insulation cables and conducted a voltage withstand test on the whole cable. During the 120 kV voltage withstand test, the PP cable broke down and obvious breakdown holes were observed in the cable body, as shown in Figure 6.1a. By dissecting the breakdown point, it was found that there were penetrating electrical dendritic channels in the PP insulation layer, as shown in Figure 1.1b and c, and the dendritic micro-channels ran through the entire insulation along the direction of the electric field, which led to the breakdown of the cable. Meanwhile, in the course of the cable voltage withstand test, the synchronous operation of the partial discharge monitor observed a more obvious local discharge signal, indicating that the growth process of electric dendrites is accompanied by the generation of local discharge. Dendrite deterioration and partial discharge are the root causes of PP cable insulation failure. It is of great significance for the application of PP-insulated cables to investigate the degradation mechanism of PP-insulated electrical dendrites and to suppress the growth process of electrical dendrites.

To analyze the cause of the PP cable insulation breakdown accident, we dissected the cable at the location of the breakdown. Through the dissection, we found that there was a light-colored ring in the middle insulation layer of the cable, see Figure 6.1b. Considering PP as a polymer material with semicrystalline

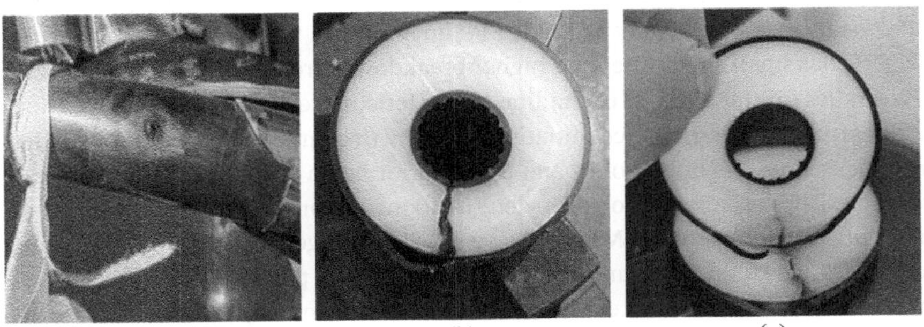

(a) (b) (c)

Figure 6.1 Electrical tree deterioration in polypropylene cable insulation. (a) Appearance, (b) cross-section, (c) slices.

structure, its crystalline structure is largely affected by the cooling rate during its processing. As far as the three-layer co-extruded cable manufacturing currently used in PP cables is concerned, after the cables are extruded at high temperatures, the outer layer of the cable insulation is cooled by water and the inner layer is cooled by a copper conductor, which is cooled at a relatively fast rate, resulting in the formation of a crystalline structure of the material with a greater number of smaller-sized spherical crystals. Limited by the poor thermal conductivity of the material itself, the cooling rate of the middle layer of cable insulation is slower, too slow cooling rate will result in the formation of large-sized spherical crystals. This difference in crystalline structure, caused by the difference in cooling rates, is visually apparent as a bright-colored circular phenomenon. Considering that the crystalline structure of PP has a large influence on its insulating properties, it is assumed that the difference in the crystalline structure of the material caused by the cooling rate is the main reason for the breakdown of the PP cable insulation.

In order to verify our conjecture, we rolled and cut the inner, middle, and outer layers of the PP cable insulation layer, respectively, obtained from the insulation sheet samples at each position, and conducted the breakdown test under AC voltage and plotted the data obtained from the test into a Weibull image, see Figure 6.2. By comparison, we can see that there is a significant difference between the breakdown strengths of the inner, middle, and outer layers of the insulation layer, and there is a positive correlation between the breakdown strengths of the inner, middle, and outer layers of the PP cable insulation layer, and there is

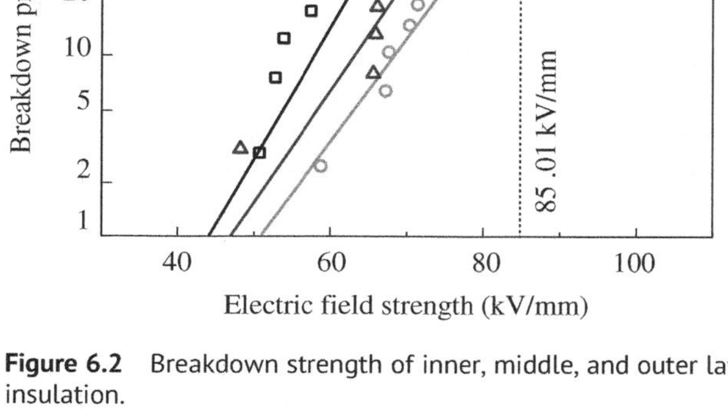

Figure 6.2 Breakdown strength of inner, middle, and outer layers in polypropylene cable insulation.

a positive correlation with the cooling rate which is positively correlated. The PP material on the outside of the insulation layer experiences the fastest cooling rate during the cooling process and has a higher breakdown strength. As the cooling rate decreases, the breakdown strength of the inner insulation layer, which has a larger spherical structure, decreases significantly. This test confirms, to some extent, our conjecture about the breakdown of PP cable insulation. That is, this difference in insulation properties at different locations of the insulation layer, caused by differences in crystalline structure, is believed to be the main cause of insulation breakdown.

Further, in order to establish the correlation between mesoscopic crystalline structure and macroscopic insulation performance of PP-insulating materials, on the basis of which to realize the improvement of cable insulation performance and to solve the hidden danger of cable insulation breakdown. In the next section of this chapter, a series of specific research work is carried out to address this issue, including the mechanistic investigation of the correlation relationship between electric dendrite and crystalline structure in Sections 6.2 and 6.3, and the study of electric dendrite suppression method in Section 6.4.

6.2 Electrical Treeing Dependent on Crystalline Morphology

6.2.1 Crystalline Morphology

This subsection aims to investigate the effect of the crystalline morphology on electrical tree characteristics of iPP/sPP blends for thermoplastic insulation. The iPP/sPP-blended samples were prepared with the weight ratio of syndiotactic polypropylene (sPP) by 0, 5, 15, 30, and 45 wt%. The catalog of blend samples investigated in this work is shown in Table 6.1. The sample prepared was heated to the melting state with needle (made of stainless steel) insertion in order to carry out the

Table 6.1 Index for PP samples investigated in this subsection.

Index	iPP content (wt%)	sPP content (wt%)
sPP0	100	0
sPP5	95	5
sPP15	85	15
sPP30	70	30
sPP45	55	45

electrical tree experiment. The protrusion of the tail of the needle contacted the edge of the model, thus ensuring a distance of 2 mm between the needle tip and the bottom. The diameter of the needle body was 300 μm and the radius of curvature of the needle tip was 3 μm. The unqualified samples with air cavities were removed before our experiment. The bottom surface of the sample was pasted with copper foil for the electrical tree experiment [15].

The crystal morphologies of blended insulation observed by polarizing optical microscope (POM) are shown in Figure 6.3. With the mass fraction of sPP increasing from 0 to 15 wt%, the volume of spherulites atrophies gradually with uniform distribution. With the sPP content increasing continuously, the volume of spherulites upsurges with sparse distribution. The crystallization characteristics of iPP/sPP blends were characterized by differential scanning calorimetry (DSC) test (Mettler Toledo DSC1 STARe system). Samples to be tested first rose from room temperature to 200 °C and then cooled to room temperature to eliminate the heat history. During the experiment, the temperature of samples increased from 20 to 200 °C at 10 K/min and then cooled to room temperature at the same rate. Melting and crystallization curves were recorded for analysis. The crystallization curve of blends tested by DSC is illustrated in Figure 6.4. It can be concluded that the crystallinity of sPP0, sPP5, sPP15, sPP30, and sPP45 is 36.3%, 42.5%, 46.4%, 31.1%, and 29.0%, respectively, indicating that a small amount of sPP can improve the crystallinity of the polymer.

The non-isothermal crystallization process of polymer can be roughly divided into two stages: primary crystallization and secondary crystallization. Primary

Figure 6.3 Crystal morphology of iPP/sPP blends: (a) sPP0, (b) sPP5, (c) sPP15, (d) sPP30, and (e) sPP45. *Source:* Shuofan Zhou et al. 2020/from MDPI/CC BY 4.0. [27].

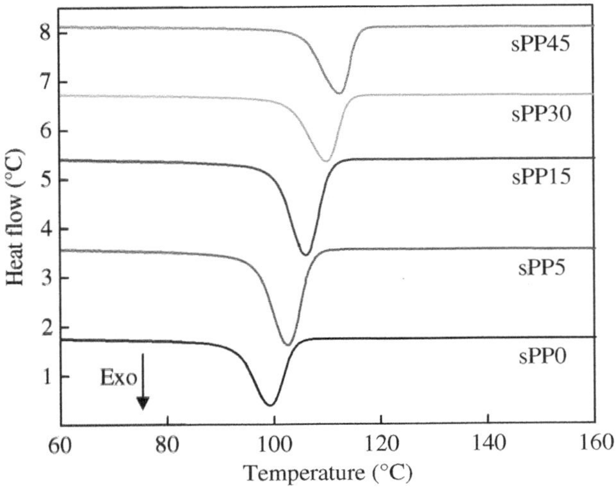

Figure 6.4 Differential scanning calorimetry (DSC) curves of iPP/sPP blends.
Source: Zhou et al. [27]/MDPI / CC BY 4.0.

crystallization is a process in which the crystal nuclei grow gradually until they collide with each other. The growth dimension of polymer crystal in primary crystallization is an important index to understand the crystallization dimension in polymer, in order to investigate the effect of crystallization on electrical properties. The values of the Avrami exponent can be calculated from the fitting of Jeziorny's improved Avrami equation in primary crystallization, which is shown as [16]:

$$\log[-\ln(1-x(t))] = n\log t + \log Z_t \tag{6.1}$$

where $X(t)$ represents the relative crystallinity, a function varying with crystallization time t, n is Avrami index, which is related to nucleation mechanism and growth mode, and is equal to the sum of space dimension of growth and time dimension of the nucleation process. Z_t represents non-isothermal crystallization rate constant. The values obtained for sPP0, sPP5, sPP15, sPP30, and sPP45 are 4.13, 3.70, 3.86, 3.77, and 4.36, respectively, as shown in Figure 6.5. Avrami index is generally considered as consisting of two parts: nucleation contribution and growth contribution. Considering that the crystal growth is identical, the difference in Avrami index is mainly attributed essentially to the difference in the nucleation process. Therefore, with the increase of sPP content, the crystallization gradually transforms from homogeneous nucleation to heterogeneous nucleation. When the content of sPP reaches 45 wt%, the heterogeneous nucleation effect weakens.

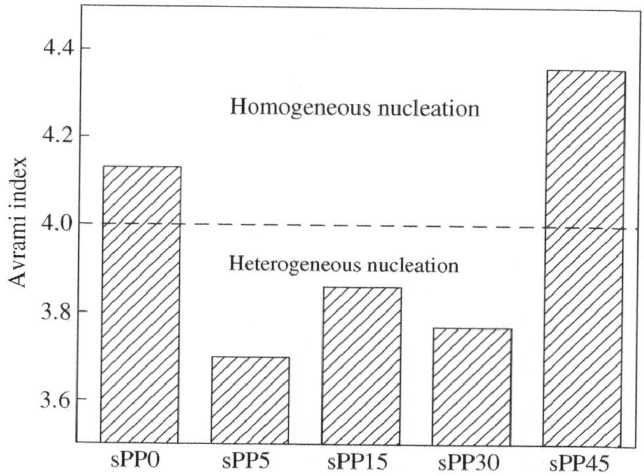

Figure 6.5 Avrami index of iPP/sPP blends. *Source:* Zhou et al. [27]/MDPI / CC BY 4.0.

6.2.2 Effect on Electrical Tree

Inception time is defined as the time from the beginning of voltage application to the beginning of the electrical tree. It can be observed and recorded when the length of the electrical tree is 10 μm. The inception time of the electrical tree for iPP/sPP blend insulation is shown in Figure 6.6. For the samples without the addition of sPP, the mean inception time of the electrical tree is 90 seconds under the ambient temperature of 50 °C. With the incorporation of sPP increasing from 0 to 15 wt%, the inception time rises by 8.2%. Further improvement of sPP content decreases the inception time of the electrical tree. The inception time decreases with the increase of temperature for all the samples and decreases rapidly at high temperature. When the temperature increased from 50 to 90 °C, the inception time of the electrical tree in sPP0 decreased by 85%. Meanwhile, the inception time of the electrical tree in sPP15 decreased by 41%, which is rapidly less than that of sPP0. At the temperature of 70 °C, the initiation time of sPP15 is 46% higher than that of sPP0. When the temperature rises to 90 °C, the initiation time of sPP15 is 403% higher than that of sPP0, indicating that the electrical tree resistance of pure iPP deteriorates rapidly at high temperature. The result indicates that comprising 15 wt% sPP in an iPP matrix restrains the inception of electrical tree effectively.

The structure of the electrical tree depicts the trend of growth and propagation of the electrical tree, and the research of blending content on the morphology of the electrical tree is helpful to analyze the mechanism of blending concentration and temperature on the propagation of the electrical tree. The morphology

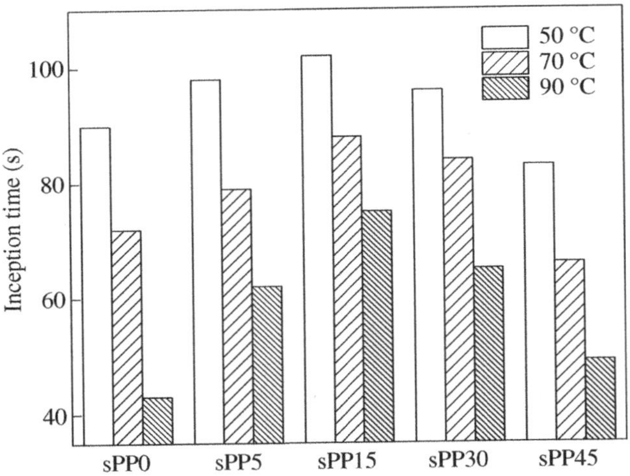

Figure 6.6 Inception time of electrical tree for isotactic polypropylene (iPP)/syndiotactic polypropylene (sPP) blends at 50, 70, and 90 °C. *Source:* Zhou et al. [27]/MDPI / CC BY 4.0.

characteristics of the electrical tree at different blending concentrations and temperatures are shown in Table 6.2. The brightness and contrast are adjusted to make the electrical tree clearer and easier to identify. At the temperature of 50 °C, sPP0 showed a branch-like tree. With the increase of temperature, more obvious vines are distinguished at the end of branches. Compared with the branches near the tip of needle, the density of vines decreases significantly and vine tracks exhibit extremely fast growth. When the temperature further rises to 90 °C, the electrical

Table 6.2 Electrical tree morphology of iPP/sPP blends at 50, 70, and 90 °C under AC voltage of 6 kV for 40 minutes. *Source:* Zhou et al. [27]/MDPI / CC BY 4.0.

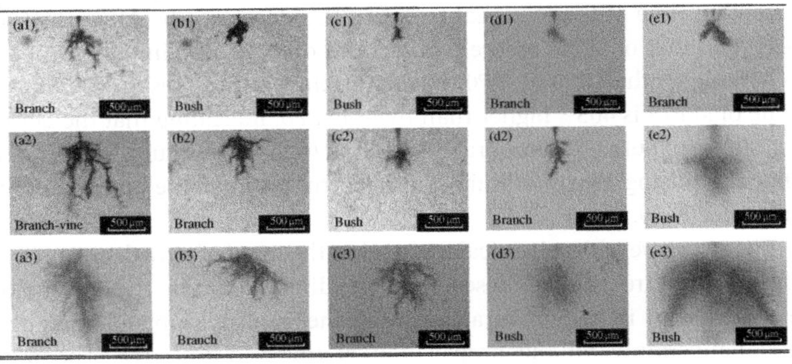

Temperature \ Samples	sPP0	sPP5	sPP15	sPP30	sPP45
50 °C	(a1) Branch 500 μm	(b1) Bush 500 μm	(c1) Bush 500 μm	(d1) Branch 500 μm	(e1) Branch 500 μm
70 °C	(a2) Branch-vine 500 μm	(b2) Branch 500 μm	(c2) Bush 500 μm	(d2) Branch 500 μm	(e2) Bush 500 μm
90 °C	(a3) Branch 500 μm	(b3) Branch 500 μm	(c3) Branch 500 μm	(d3) Bush 500 μm	(e3) Bush 500 μm

branches are concentrated with higher growth rate. Under the temperature of 50 °C, sPP5 and sPP15 are manifested as bush-like trees. With the increase of temperature, more discharge channels are generated and the structure of electrical tree gradually transformed into a branch-like tree. Electrical trees of sPP30 and sPP45 indicate branch trees at 50 °C, and the density of discharge channels gradually increases with the rise of temperature.

The electrical tree length is an essential index to characterize the degradation of the electrical tree, which is closely associated with the electrical tree structure. Figure 6.7 shows the relationship between the electrical tree length and the experimental time under different temperatures. It can be seen from the figure that the morphology of the electrical tree displays obvious influences on the growth speed of tree channels. Under the temperature of 50 and 70 °C, the morphology of the electrical tree for sPP0 terms is a branch or branch vine-mixed electrical tree. The growth rate of sPP0 suddenly increases after 15 minutes, reflecting the accelerated effect of the extension of the vine on the length of the electrical tree. At 90 °C, sPP0 terms branch tree with the tree length increased by 17% compared with 70 °C with the treeing time of 40 minutes. With the increase of the experimental temperature from 50 to 90 °C, the electrical tree morphology of sPP5 and sPP15 varies from bush to branch, and the electrical tree length rises enormously. Compared with sPP15, the tip dendrite of sPP5 tends to grow toward both sides at 90 °C, resulting in a relatively low growth length in the longitudinal direction. sPP30 and sPP45 display a trend of transition from branch to bush with the increase of temperature, and the growth speed of the electrical tree rises.

Fractal dimension and expansion coefficient are quantitative characterizations of electrical tree morphology, which is shown in Figure 6.8. With the increase of temperature from 50 to 70 °C, the morphology of the electrical tree of sPP0 changes from branch-like tree to branch vine-mixed tree. The fractal dimension of branch-like tree is smaller than that of the branch vine-mixed tree. When the temperature rises further to 90 °C, the structure of the electrical tree changes into branches with decreasing fractal dimension. However, the expansion coefficient keeps decreasing, which indicates a more obvious transverse growth trend at higher temperatures. With the increase of temperature, the fractal dimension of the electrical tree of sPP5 and sPP15 decreases and the expansion coefficient increases, with the morphology varying from bush to branch. With the increase of temperature, the fractal dimension of the electrical tree of sPP30 and sPP45 increases and the expansion coefficient increases, with the morphology varying from branch to bush. At the temperature of 50 °C, when the content of sPP is 15% or less, the electric tree appears bush-like tree. Further addition of sPP makes the electric tree a tendency to change into branches. When the temperature rises to 90 °C, the variation trend of the tree structure is opposite. Further addition of sPP makes the electric tree a tendency to change into bushes. The different tendency of the

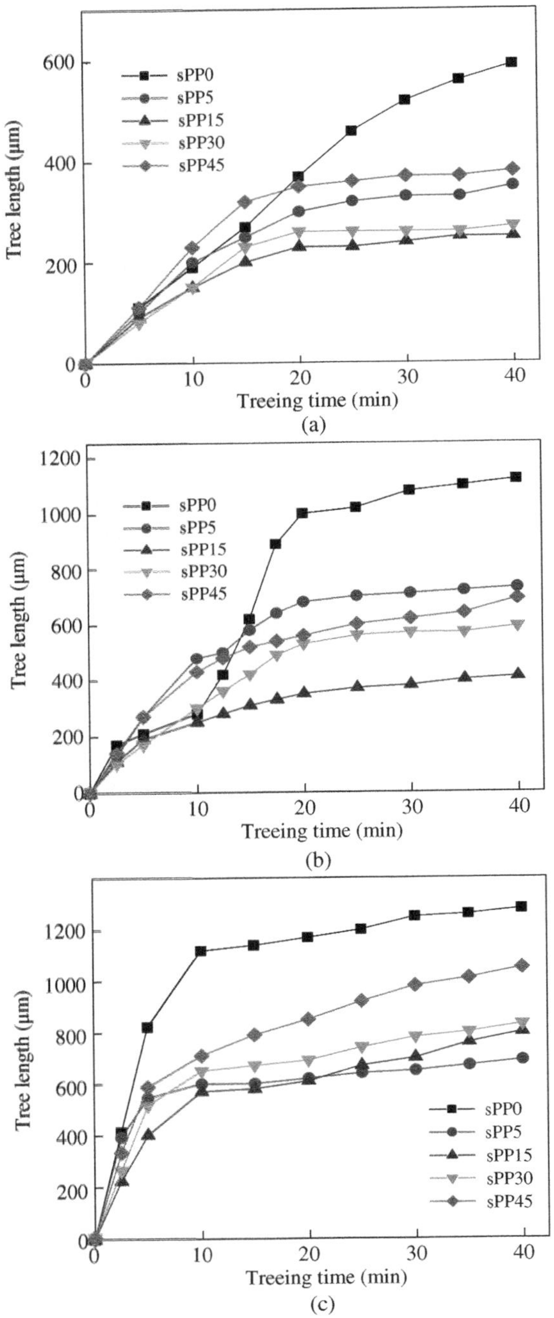

Figure 6.7 Variation of electrical tree length of different sPP contents at (a) 50 °C, (b) 70 °C, and (c) 90 °C under AC voltage of 6 kV. *Source:* Zhou et al. [27]/ MDPI / CC BY 4.0.

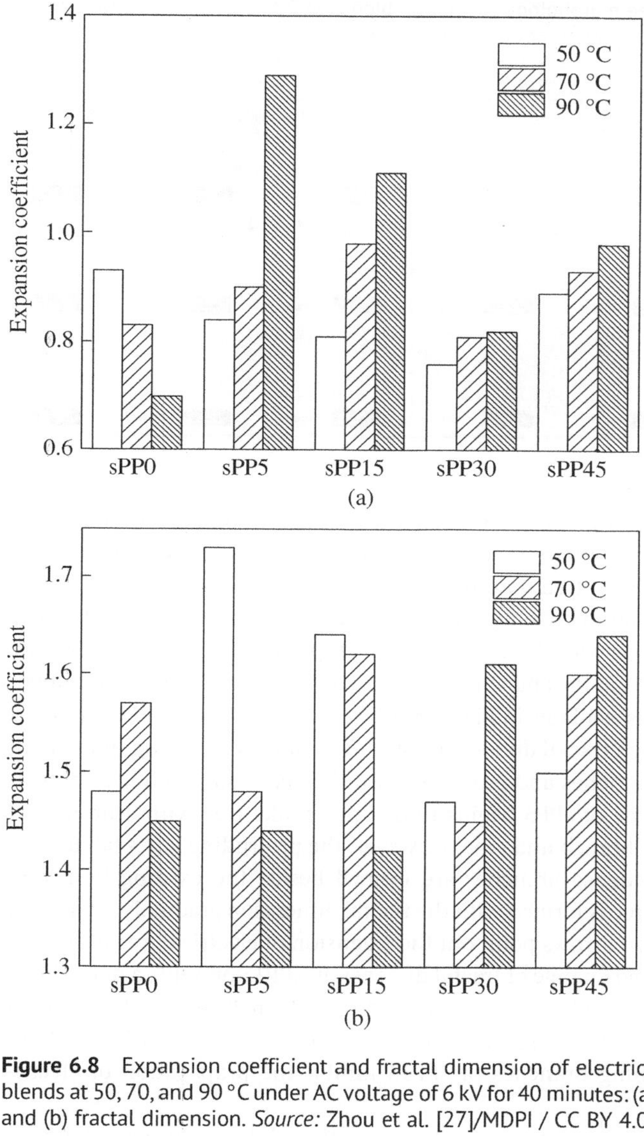

Figure 6.8 Expansion coefficient and fractal dimension of electrical tree in iPP/sPP blends at 50, 70, and 90 °C under AC voltage of 6 kV for 40 minutes: (a) expansion coefficient and (b) fractal dimension. *Source:* Zhou et al. [27]/MDPI / CC BY 4.0.

variation of the tree morphology at high and low temperatures makes the addition of sPP indicate an inflection point in the regulation of the electrical tree morphology at different temperatures. sPP15 distinguishes an excellent performance to inhibit electrical treeing, which benefits from the ability to promote the fractal

Table 6.3 Electrical tree morphology of iPP/sPP blends at 70 °C under AC voltage of 6 kV for 40 minutes. *Source:* Zhou et al. [27]/MDPI / CC BY 4.0.

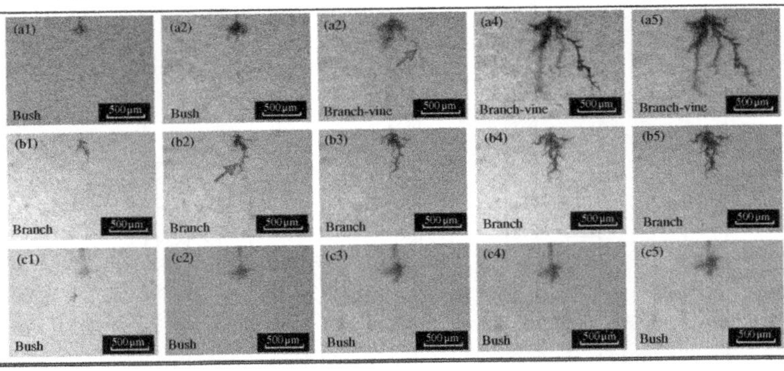

Time / Samples	5 min	10 min	15 min	20 min	40 min
sPP0	(a1) Bush 500 μm	(a2) Bush 500 μm	(a2) Branch-vine 500 μm	(a4) Branch-vine 500 μm	(a5) Branch-vine 500 μm
sPP5	(b1) Branch 500 μm	(b2) Branch 500 μm	(b3) Branch 500 μm	(b4) Branch 500 μm	(b5) Branch 500 μm
sPP15	(c1) Bush 500 μm	(c2) Bush 500 μm	(c3) Bush 500 μm	(c4) Bush 500 μm	(c5) Bush 500 μm

and lateral growth of branches compared with other samples at 50 and 70 °C. When the temperature rises to 90 °C, the blends with 5 wt% sPP exhibit a better inhibition of the electrical tree than that of 15 wt%.

The structure description not only depends on the final state of electrical tree but also varies and develops with the electrical tree propagation. Table 6.3 shows the growth process of sPP0, sPP5, and sPP15 at 70 °C, and time-varying fractal dimension of three samples is available in Figure 6.9. The electrical tree of sPP0 starts as a bush-like tree with high fractal dimension. With 15 minutes of treeing time, small vines extend out of branches and grow fast with the decreasing of fractal dimension. The electrical tree of sPP5 begins to be branch-like tree with a promising trend of mixed tree of branch and vine. However, the protruding vines fail to continue to grow. More dense branches are formed beside the existing branches. A dense branch-like tree is formed with the fractal dimension gradually increasing from 1.4 to 1.5. The red arrows point out the extension of electrical tree branches, which leads to the rapid change of fractal dimension. sPP15 presents a dense bush tree from the beginning to the end, and stretching out small twigs with a treeing time of 60 minutes, which is longer than any other sample. Dense discharge channels with less stretching branches of sPP15 inhibit the propagation of electrical tree.

The relationship between the electrical tree length and the experimental time at 70 °C can be referred to in Figure 6.7. It can be seen from the figure that the morphology of the electrical tree displays obvious influences on the growth speed of tree channels. The morphology of the electrical tree for sPP0 terms a branch vine-mixed electrical tree. The growth rate of sPP0 suddenly increases after 10 minutes, reflecting the accelerated effect of the extension of vine on the length

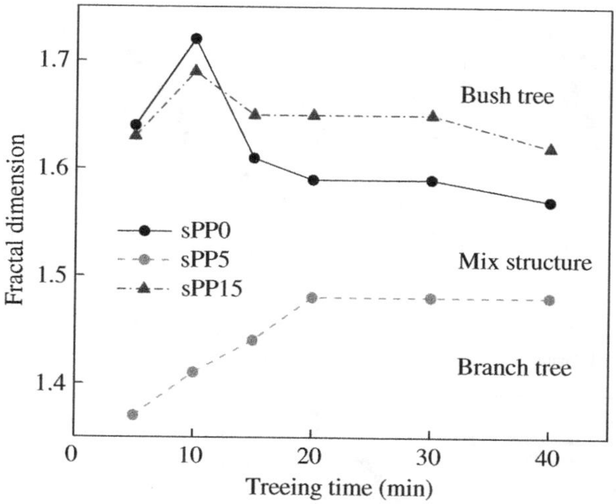

Figure 6.9 Time-varying fractal dimension of sPP0, sPP5, and sPP15 at 70 °C under AC voltage of 6 kV. *Source:* Zhou et al. [27]/MDPI / CC BY 4.0.

of electrical tree. It tends to stagnate after 20 minutes. The extension of vine of the electrical tree of sPP5 occurs earlier than sPP0, which exhibits a longer electrical tree length in the first 15 minutes. The electrical tree of sPP15 keeps a high fractal dimension during the experiment, which avoids the rapid change of the electrical tree length.

6.2.3 Effect on AC Breakdown

AC breakdown field strength is a significant method to evaluate the AC withstand ability of insulating materials. Weibull distribution of AC breakdown strength of iPP/sPP blended samples is revealed in Figure 6.10. All the experiments maintain the same rate of voltage rise, which is 1 kV/s. It can be observed that the AC breakdown strength of the three samples shows a downward trend with the temperature rising from 50 to 90 °C, in which the breakdown field strength of PP-neat decreases from 136.8 to 129.0 kV/mm, the data of sPP5 decreases from 146.9 to 142.2 kV/mm, the data of sPP15 decreases from 152.1 to 133.4 kV/mm, the data of sPP30 decreases from 149.6 to 136.1 kV/mm and of sPP45, the breakdown strength decreases from 145.5 to 137.0 kV/mm.

With the rising of the experimental temperature, the decrease rate of breakdown field strength is different. When the temperature increases from 50 to 70 °C, the breakdown field strength of the polymer decreases only within 5%, which shows the temperature-insensitive characteristic of PP under AC voltage. When the

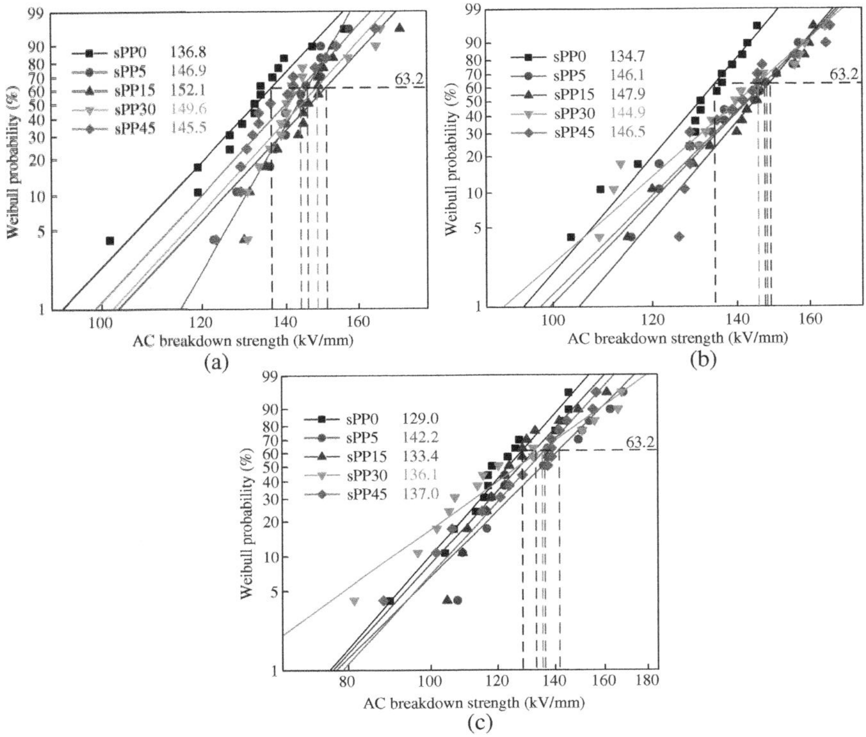

Figure 6.10 Weibull distribution of AC breakdown strength in iPP/sPP blends: (a) 50 °C, (b) 70 °C, and (c) 90 °C. *Source:* Zhou et al. [27]/MDPI / CC BY 4.0.

content of sPP blends increases from 0 to 15 wt%, the breakdown field strength is enhanced with the rise of the sPP content. With the content of sPP blends further increasing to 45 wt%, the breakdown field strength declines. When the temperature rises to 90 °C, the breakdown field strength decreases by about 10%, and the shape parameter of Weibull distribution decreases, which indicates that the breakdown field strength of blend insulation decreases greatly at high temperature, with a more decentralized distribution.

The change of crystal morphology affects localized states distributed in amorphous regions and interface between crystalline and amorphous regions, which are accessible capturing states for charge carriers. Figure 6.11 represents the trap energy level distribution characteristics based on IDC. It can be seen from the figure that the addition of sPP makes no difference to the trap energy level of the blend insulation. The shallow trap energy level measured by IDC remains

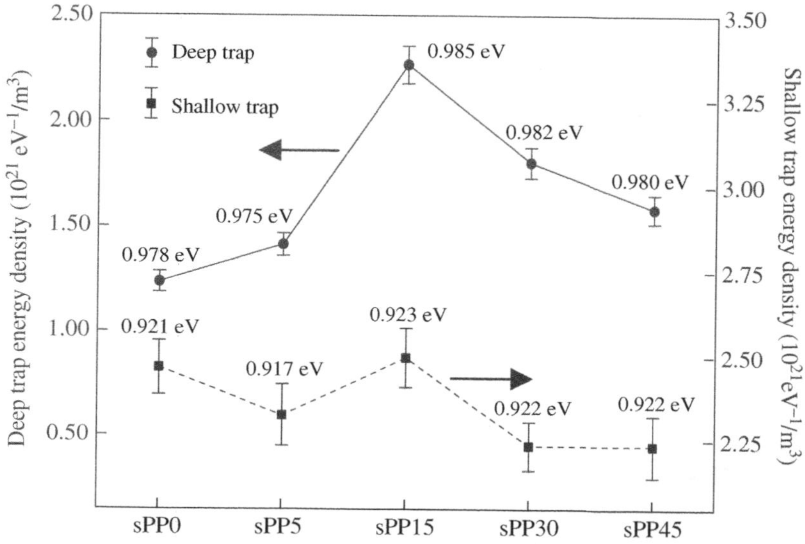

Figure 6.11 Trap level characteristics of iPP/sPP blends. *Source:* Zhou et al. [27]/MDPI / CC BY 4.0.

0.92 eV while the deep trap energy level is around 0.98 eV. The density of shallow trap is higher than deep trap. With the upsurge of sPP blending concentration, the density of shallow and deep traps of polymer insulation ascends. When the content of sPP is more than 15%, the density of shallow and deep traps declines with the rise of the blending concentration.

The density of charge traps, especially that of the deep traps, is closely related to the breakdown strength. The addition of sPP effectively improves the crystallization of the iPP matrix. The increase of crystallinity and heterogeneous nucleation facilitates the distribution of the interface between crystalline and amorphous regions, which limits the free transport of charge [17]. The incorporation of sPP with 15% content displays excellent crystallization with heterogeneous nucleation, thus inhibiting the generation of the electrical tree.

The distortion of the electric field is the primitive driving force of electric tree propagation. The electric field varies periodically with time under AC voltage, accompanied by the electric field of space charge. At the same time, space charge will also migrate under the action of the electric field, which will conversely affect the distribution of the electric field [18]. The analysis of the variation of the electric field under AC voltage is a necessary process in the investigation of electrical tree growth.

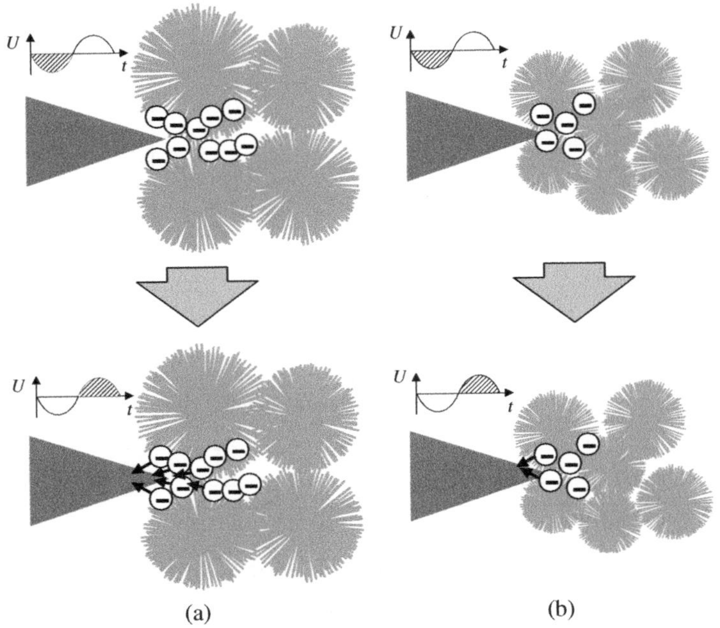

Figure 6.12 Schematic diagram of charge accumulation under AC voltage: (a) sPP0 and (b) sPP15. *Source:* Zhou et al. [27]/MDPI / CC BY 4.0.

As shown in Figure 6.12, in the negative half cycle of AC voltage, electronics are injected into the polymer promoted by the electric field and trapped by traps, forming a space charge limiting the electric field to inhibit the further injection of charges. When the voltage transits to the next half cycle, the injected charge will be drawn out with charge recombination. The released energy can produce energetic electrons through the Auger effect, which may destroy the molecular chain [19, 20]. The residual space charges form an electric field shielding for the injected charge in the next negative half cycle.

The charge injection and extraction processes are iteratively continued, resulting in continuous energy release to promote the growth of electrical tree. The incorporation of sPP in 15 wt% introduces more deep traps into the polymer, resulting in hard extraction in reverse electric field. In the next half cycle, a stronger shielding effect is formed, which inhibits the further charge injection. Intensified density of deep traps reduces kinetic energies of electrons accelerated in free volumes [10, 21, 22], thus inhibiting the further development of the electrical tree.

6.3 An Insight into Electrical Tree Growth Within Heterogeneous Crystalline Structure

6.3.1 Mechanism of Heterogeneous Crystalline Structure

The last subsection has proved that the crystalline structure has a significant impact on the degradation of electrical dendrites in PP cable insulation. This subsection further designs an experiment of the mesoscopic electrical treeing observed by POM, to demonstrate its mechanism. The film sample with a thickness of 280 μm is a stack of 2 thin films. An ultrafine silver with a diameter of 20 μm is placed between two films as a needle electrode. The curvature radius of the ultrafine silver tip is 2 μm and the distance between the needle tip and the bottom of the films is 2 mm. The preparation and test setup for the mesoscopic electrical tree test are shown in Figure 6.13. Two PP films were wiped with alcohol to remove impurities, then heated to 200 °C to melt and a pressure of 10 MPa was applied to remove bubbles to form a complete sample and cooled with the same methods as in Table 6.1. Finally, the samples have a thickness of 280 μm. The samples are subjected to tree degradation at the AC voltage of 8 kV and the temperature of 30 °C for 5 minutes. The samples were immersed in silicone oil and plexiglass was sandwiched up and down to prevent side breakdown of the samples caused by the electric field distortion. Images of mesoscopic electrical tree growth characteristics of PP samples with different crystalline morphologies are taken by the polarized microscope (POM) and saved in the computer for later analysis.

In order to clearly observe the crystal morphology of PP and the process of electrical tree development, iPP with an isotacticity greater than 98% was used as the experimental sample. The crystalline morphology of samples is altered by regulating the cooling rate and the isothermal crystallization temperature during the crystalline process, as shown in Table 6.4.

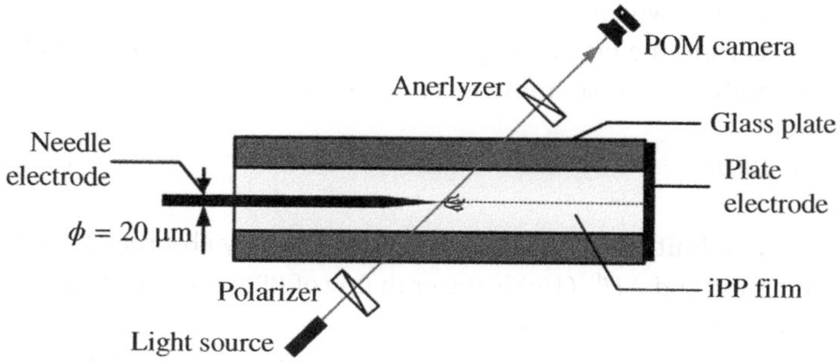

Figure 6.13 Setup of the mesoscopic electrical tree. *Source:* Li et al. [28]/with permission from IEEE.

Table 6.4 Index for PP samples with different crystallization process. *Source:* Li et al. [28]/ with permission from IEEE.

Samples	Crystallization process
A	Crystallized at a cooling rate of 10.0 °C/min
B	Crystallized at a cooling rate of 5.0 °C/min
C	Crystallized at a cooling rate of 2.5 °C/min
D	Crystallized at the isothermal temperature of 128 °C for 30 min
E	Crystallized at the isothermal temperature of 132 °C for 30 min

Samples A, B, and C are non-isothermally crystallized by cooling to room temperature at the rate of 10, 5, and 2.5 °C/min, respectively. Samples D and E are isothermally crystallized at the temperature of 128 and 132 °C, respectively. After isothermally crystallizing for 30 minutes, samples are cooled to room temperature. Figure 6.14 shows the crystalline process of samples with different cooling methods. When the temperature drops to 125 °C, there are a few nucleation sites in the sample crystallized at a cooling rate of 10 °C/min, while a lot of spherulites appear in the sample crystallized at a cooling rate of 2.5 °C/min as shown in Figure 6.14a and c. When the temperature drops to 95 °C, the spherulites fill the whole field of POM vision, as shown in Figure 6.14a–6.14c. As shown in Figure 6.14d and e, there are a large number of imperfect microcrystals between the interface of large spherulites when the samples crystallized at the isothermal temperature for 30 minutes. The average diameters of the five spherulites of samples are calculated. The average spherulite diameters of samples A, B, and C are 52.2, 58.9, and 101.4 μm, respectively. It can be found that the average diameters of spherulites of samples increase as the cooling rate decreases. The average diameter of spherulites of the sample crystallized at the isothermal temperature of 132 °C for 30 minutes is the largest, which is up to 129.6 μm.

DSC is used in measuring the crystallinity and melting point of samples with different cooling methods. The crystallinity of samples is calculated by Eq. (6.2):

$$X_c = \frac{\Delta \mathrm{H_m}}{\Delta \mathrm{H_m^0}} \times 100\% \tag{6.2}$$

where X_c (%) is the crystallinity of iPP, ΔH_m (J/g) is the enthalpy absorbed by iPP insulation during melting, and ΔH_m^0 (J/g) is the enthalpy of PP with a crystallinity of 100%. In this paper, ΔH_m^0 is 209 J/g [12].

Figure 6.15 shows the DSC curves. The melting enthalpy, melting point, and crystallinity of samples are calculated. It can be found that the melting point and crystallinity of samples increase at different degrees as the cooling rate

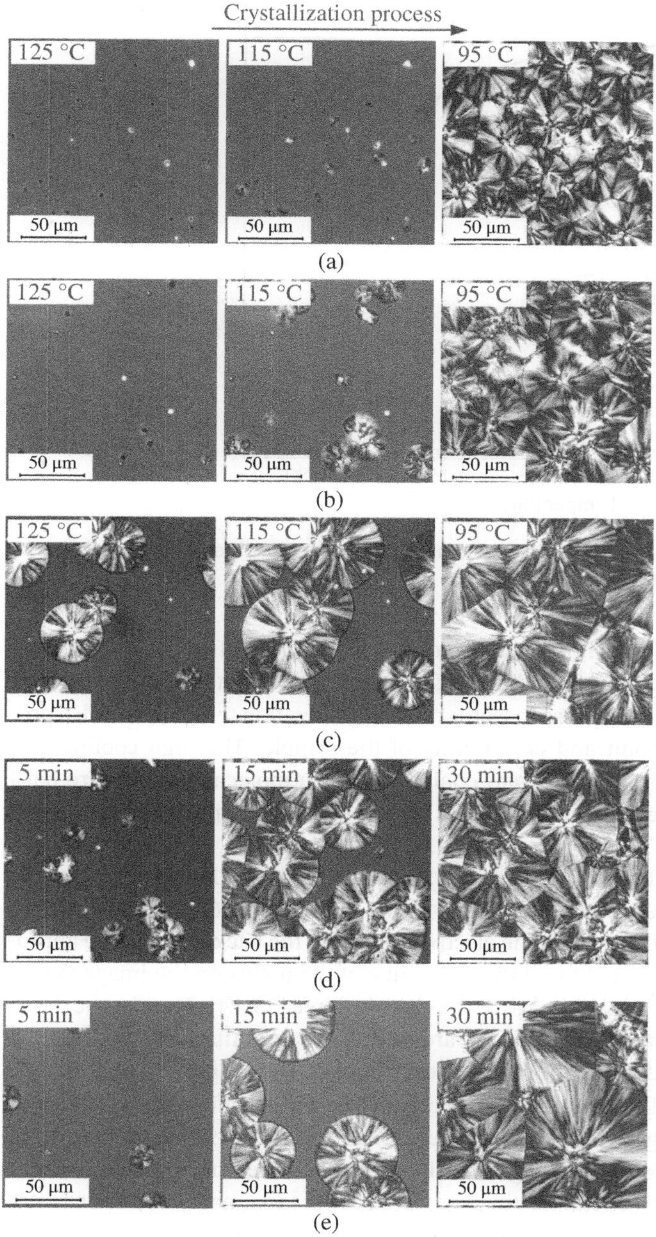

Figure 6.14 Crystalline morphologies of iPP with different cooling methods: (a) sample A, (b) sample B, (c) sample C, (d) sample D, and (e) sample E. *Source:* Li et al. [28]/with permission from IEEE.

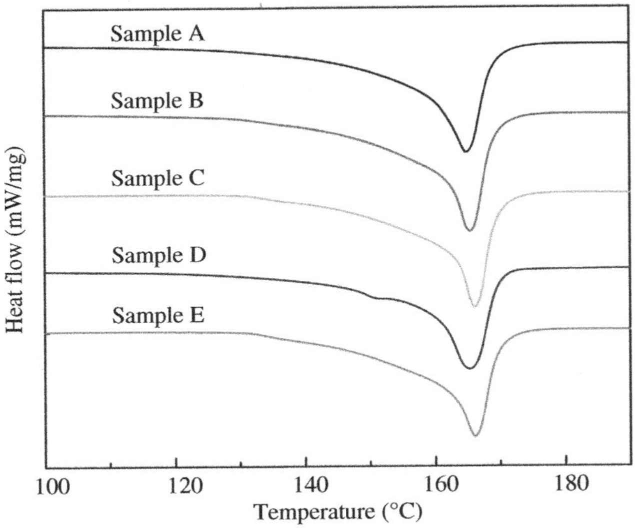

Figure 6.15 Melting curves of iPP samples with different cooling methods. *Source:* Li et al. [28]/with permission from IEEE.

decreases. Sample C which is crystallized at a cooling rate of 2.5 °C/min has the highest melting point and crystallinity. Compared with the non-isothermal crystallization process, the increase in isothermal crystallization temperature also promotes the melting point and crystallinity of the sample. The high cooling rate limits the mobility of iPP molecular chains, and then molecular chains are hard to fold to form lamellar and spherulites which makes the spherulite size small and the crystallinity low.

The mesoscopic electrical tree morphologies under the POM camera and the binary images of the electrical tree are shown in Figure 6.16. The thresholding operation is used to distinguish the electrical tree and the background. The appropriate threshold is selected to set the gray value of the points on the image to 0 or 255 and to obtain the binary image of the electrical tree [23, 24]. The electrical tree length and accumulated damage area are selected to quantitatively describe the degree of electrical deterioration of the test samples. The length of a mesoscopic electrical tree is defined as the distance between the needle electrode tip and the end of the longest branch of the electrical tree pointed by the arrow in Figure 6.16. Accumulated damage area is calculated based on the deterioration area covered by the electrical tree in pixels from the binary image [24]. Spherulites and the spherulites boundary of samples C, D, and E are clear, as depicted in Figure 6.16c–e. The main branches of the electrical tree of samples C, D, and E grow quickly along the spherulites boundary by the force of the electrical field

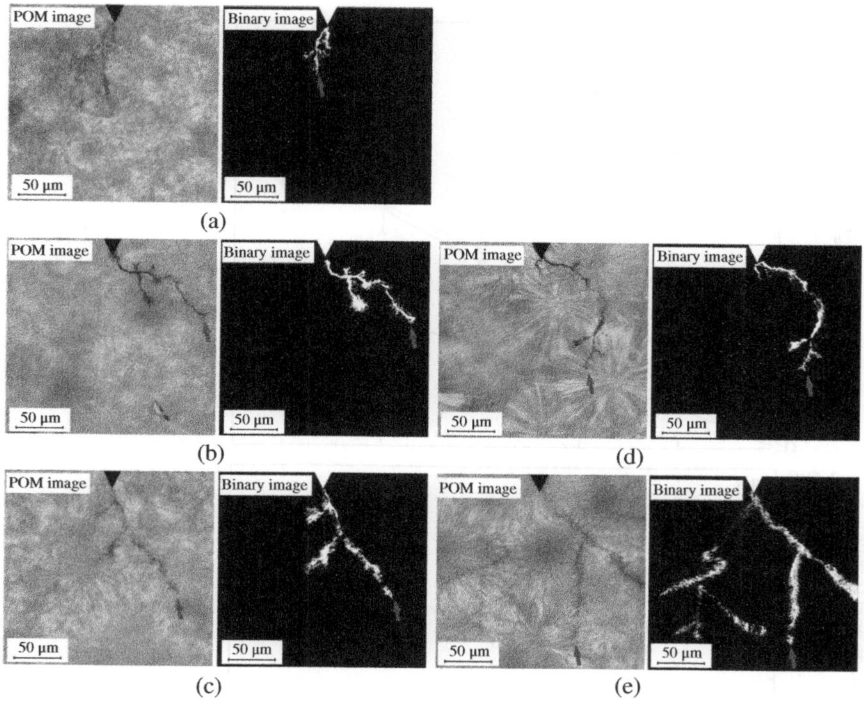

Figure 6.16 POM images and binarized images of mesoscopic electrical tree morphologies of iPP with different crystalline morphologies: (a) sample A, (b) sample B, (c) sample C, (d) sample D, and (e) sample E. *Source:* Li et al. [28]/with permission from IEEE.

while some branches grow slowly within the spherulites forming the bush tree. The electrical tree branches growing in the interior of spherulites are hard to grow forward and form large carbonation channels as illustrated in Figure 6.16c.

With the cooling rate increasing, the size of spherulite decreases significantly and the electrical tree length decreases which indicates that the growth of the electrical tree is hindered by the spherulites. The electrical tree length of sample A is the shortest and a number of tiny branches are initiated forming the pine tree as illustrated in Figure 6.16a. The electrical tree tends to form a bush tree in the sample with small spherulites. The amorphous region has lower resistance to electrical breakdown forming a weak point for the electrical tree. The electrical tree tends to grow in the amorphous region and form a long branch of the tree between the huge spherulites. The length and the accumulated damaged area of the mesoscopic electrical tree are shown in Figure 6.17.

The growth rate of the mesoscopic electrical tree gradually slows down over the treeing time. After applying AC voltage for 5 minutes, the length of sample A is the

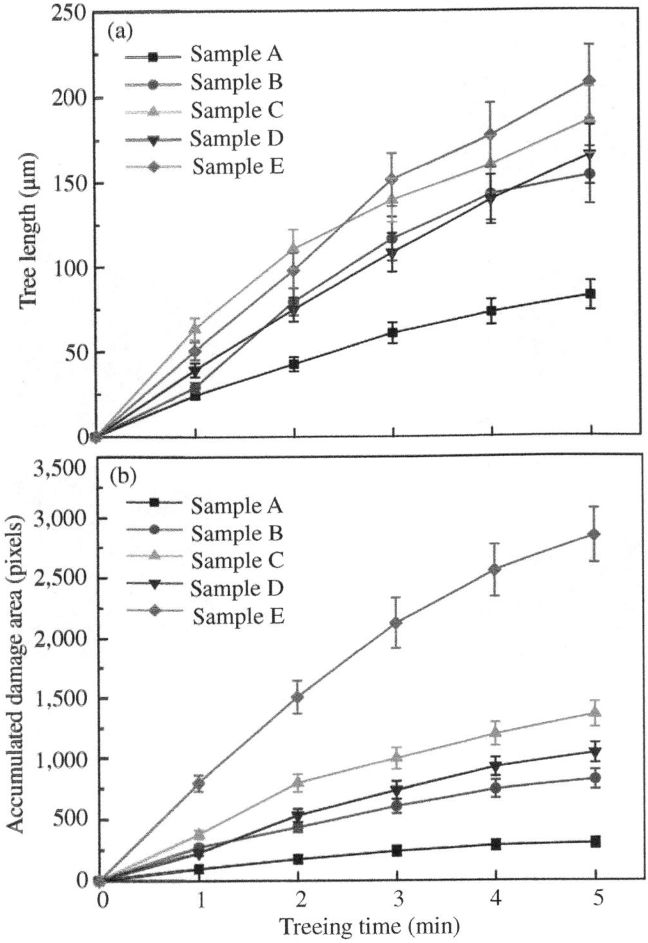

Figure 6.17 (a) Length and (b) accumulated damage area of mesoscopic electrical tree morphologies of iPP with different crystalline morphologies. *Source:* Li et al. [28]/with permission from IEEE.

shortest and the accumulated damage area is the smallest, indicating the electrical tree is hard to grow in the sample crystallized at the cooling rate of 10 °C/min. The length of the mesoscopic electrical tree of sample E is up to 207.4 μm which is 2.5 times that of sample A. The mesoscopic electrical tree is easy to grow in sample E which has the largest spherulite size.

The results of the mesoscopic electrical tree test reflect the growth characteristics of the mesoscopic tree in the crystal and amorphous regions, and the experimental phenomena of the macroscopic tree show the same growth characteristics. Therefore, there is a relationship between mesoscopic and macroscopic trees. The Pearson correlation coefficient is adopted to describe the relationship between the mesoscopic electrical tree and the macroscopic electrical tree which is calculated by Eq. (6.3):

$$\rho = \frac{\sum_{i=1}^{n} (X_i - \overline{X})(Y_i - \overline{Y})}{\sqrt{\sum_{i=1}^{n} (X_i - \overline{X})}\sqrt{\sum_{i=1}^{n} (Y_i - \overline{Y})^2}} \tag{6.3}$$

where ρ is the Pearson correlation coefficient. X_i represents the mesoscopic tree length or damaged area and Y_i represents the macroscopic tree length or damaged area. $\overline{X}$ is the mean value of X_i and $\overline{Y}$ is the mean value of Y_i [25].

The samples cooled in different methods have different mesoscopic spherulite morphologies, which leads to the morphologies of mesoscopic electrical trees being different. Molecular chains are damaged under the force of the electric field, and the carbonization channels appear with lots of molecular chains broken. The carbonization channels form the mesoscopic electrical tree. The mesoscopic electrical tree gradually develops into the macroscopic electrical tree with treeing time. Figure 6.18a indicates the fitting of the length of the mesoscopic and macroscopic electrical tree and Figure 6.18b indicates the fitting of the accumulated damaged area of the mesoscopic and the macroscopic electrical tree. The Pearson correlation coefficient of mesoscopic and macroscopic electrical trees length is 0.85 while that of mesoscopic and macroscopic electrical trees accumulated damaged area is 0.82. The Pearson correlation coefficient of mesoscopic and macroscopic electrical trees length and accumulated damage area are larger than 0.80 which represents that there is a close relation between mesoscopic electrical tree characteristics and macroscopic electrical tree. Sample A in the mesoscopic electrical tree test has the shortest length and smallest accumulated damage area while it shows the same growing characteristics in the macroscopic electrical tree test. The small and dense spherulites of sample A inhibit the tree growth resulting in the mesoscopic and macroscopic electrical tree branch being short and dense.

Previous studies have shown that the molecular chain in the amorphous region of iPP is relatively loosely packed while that in the crystal region, it is densely packed [26]. There are lots of defects in the amorphous region resulting in the local electrical field being higher than the crystal region. The distorted electrical field leads the partial discharge (PD) to occur easily in the amorphous region. The hot electrons and photons, resulting from PD, destroy the molecular chains and accelerate the formation of the low-density region. The PD and formation of the low-density region are iteratively continued, resulting in the electrical tree

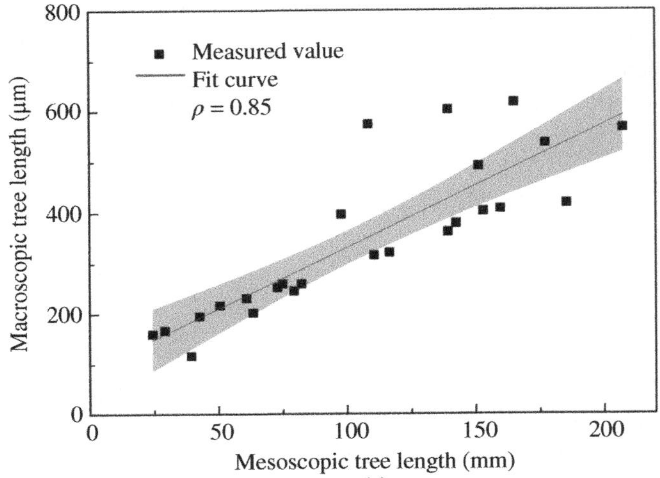

(a)

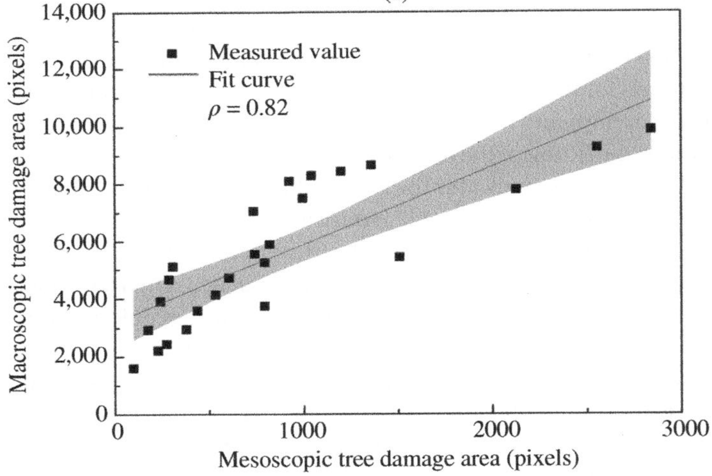

(b)

Figure 6.18 Pearson correlation coefficient between the mesoscopic and macroscopic tree growth characteristics: (a) tree length and (b) accumulated damage area. *Source:* Li et al. [28]/with permission from IEEE.

degradation along the amorphous region [27]. By comparison, the electrical field is relatively uniform and the PD is not easy to occur resulting in molecular destruction in the crystal region. Therefore, the electrical tree is easy to grow toward the ground electrode in the amorphous region while the growth of the tree is hindered by the crystal region.

The mobility of molecular chains of sample crystallized at a rapid cooling rate is limited, resulting in the orientation arrangement of the molecular chain being limited and the spherulite size being small. The length of the electrical tree is short and its morphology shows the bush-like tree in the sample with small spherulite contrary to the sample with small spherulites, the length of the electrical tree is long and the tree forms branch tree in the sample with large spherulites. Therefore, improving the cooling rate of the sample in the crystalline process can reduce the size of spherulites and enhance the resistance of the electrical tree.

6.3.2 Heterogeneous Crystalline Structure Modulation Enhancing Dielectric Strength

To further enhance the dielectric strength of PP insulation, some modulation experiments have been conducted on its crystal structure. It has been found that by increasing the cooling rate, the PP's spherocrystal size can be effectively reduced and its grain boundary structure improved, so that the growth shape of the electrical tree inside PP is shrubbier and the resistance to it is thus improved [28]. The addition of nucleating agents is another effective way to achieve crystal structure modulation. MMT, as a type of α-nucleating agent, plays a role in heterogeneous nucleation during the crystallization of PP, enhancing the spherocrystal density and crystallinity. The improved crystal structure also further inhibits the development of electrical tree at the initiation and growth stages [11, 29]. The added β-nucleating agent induces the production of β-crystal clusters with higher electron affinity and wider bending gap, which can effectively limit charge injection and migration, thus enhancing the breakdown strength and bulk resistivity of PP [30, 31]. Not only the long-chain branching increases the number of spherocrystals in PP by increasing the crystal nucleus density but the enhanced inter-chain entanglement also limits the excessive growth of spherocrystals. The thermal stability and dielectric properties of the material are thus synergistically improved [32–34].

In this subsection, an α-nucleating agent (TMP-1, sodium 2,2′-methylene bis-[4,6-di-*tert*-butylphenyl] phosphate) is employed to modulate the heterogeneous crystalline structure, thus further influencing the electrical tree characteristics. The compositions of the five samples are shown in Table 6.5.

Table 6.5 Index for PP samples with different nucleating agent.
Source: Wu et al. [69]/with permission from Elsevier.

Index	iPP (wt%)	TMP-1 (wt%)
PP-p	100	0
PP-α-0.025	99.975	0.025
PP-α-0.05	99.95	0.05
PP-α-0.1	99.9	0.1
PP-α-0.3	99.7	0.3

Figure 6.19 shows how the crystalline morphology changes during the three stages of isothermal crystallization: before, during, and after. Phase (a) corresponds to the initial moment of sample crystallization. It can be seen that the spherocrystal samples are in a completely molten state under the effect of high temperatures. There are no spherocrystals present at this point, as they have just undergone a cooling process from 200 to 130 °C. And, the bright dots in the visual field are the α-nucleating agent. As its own melting temperature is higher than 200 °C, it is stable in the form of solid particles inside the composite during this phase. At this point, by comparing Figure 6.19a1–a5, it can be found that the distribution density of the nuclei within the visual field gradually increases with the increase of the nucleating agent addition ratio. The crystal nucleus in PP-α-0.1 is small in size and uniformly dispersed, whereas those appearing in PP-α-0.3 are not only excessive in number but also have a large dispersion in size, ranging from 0.3 to 13.1 μm.

Phase (b) reflects the isothermal crystallization behavior of PP at 130 °C. As can be seen from Figure 6.19b1, when the crystallization temperature is high, the number of crystal nuclei formed within the PP material itself is extremely limited and the rate of homogeneous nucleation is low. The addition of nucleating agents, by contrast, provides a greater number of crystal nuclei for PP. By promoting the process of heterogeneous nucleation, not only does the spherocrystal number increase but also the crystallization rate is significantly enhanced [35].

It can be seen from part (c) that when the whole crystallization process is finished, the spherocrystal size of PP-p is the largest, with an average diameter above 100 μm. When the nucleating agent was added at 0.025, 0.05 wt%, the number of crystal nuclei introduced was small and could not be uniformly dispersed within the polymer, limiting the effect of heterogeneous nucleation and resulting in large and small spherocrystal sizes. And when it reaches 0.3 wt%, the excess nucleating agent can lead to agglomeration problems [13]. Meanwhile, the excessive crystal nucleus density also leads to spatially restricted crystal growth and the formation of a large number of imperfectly structured fine grains [36]. In contrast, the addition of 0.1 wt% not only reduces the spherocrystal size of PP to a certain extent, resulting in a considerable reduction in dispersion, but also significantly increases the material's crystallization rate.

The X-ray diffraction results of the samples are shown in Figure 6.20. The test curves of the samples all contain five main diffraction peaks of the α-crystal system at diffraction angles 2θ of 14.0°, 16.7°, 18.5°, 21.0°, and 21.7°, corresponding to the characteristic diffraction peaks of the (110), (040), (130), (111), and (131) crystal planes. Since the introduction of the α-nucleating agent does not change the crystal lattice structure of PP itself, the diffraction peak positions of the corresponding crystal planes remain unchanged. It can be found that with increasing nucleating agent addition ratio, the diffraction peak intensities of (040) and (131) crystal

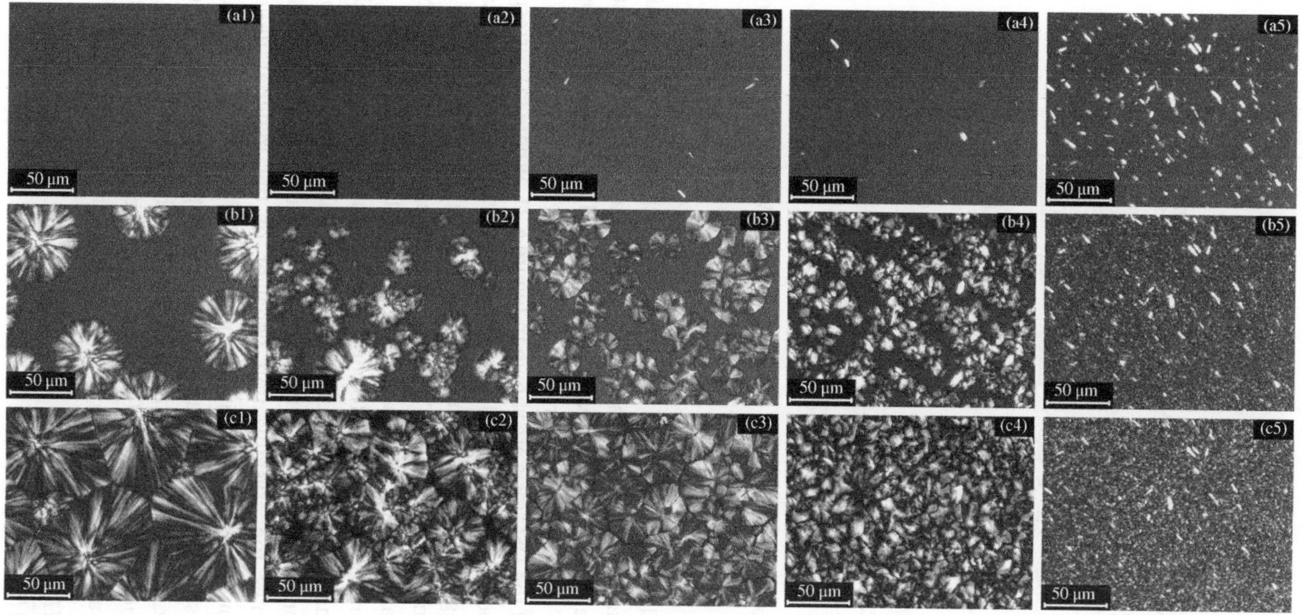

Figure 6.19 Heterogeneous structure of samples (a) before crystallization, (b) during crystallization, and (c) after crystallization varying with α-nucleating agent content. *Source:* Wu et al. [69]/with permission from Elsevier.

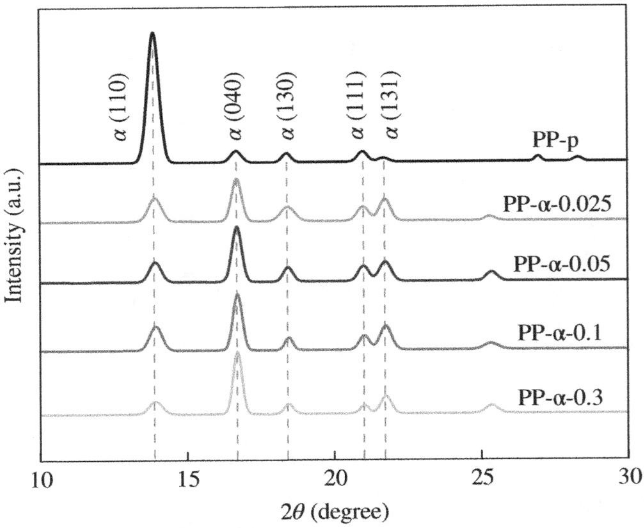

Figure 6.20 X-ray diffraction pattern of samples. *Source:* Wu et al. [69]/with permission from Elsevier.

planes gradually increase, while those of (110) crystal planes significantly decrease, compared to PP-p. This indicates that the change in the crystal structure of PP under the effect of the nucleating agent has led to a transformation in the arrangement of atoms in the original crystal cell.

The results of the DSC tests are shown in Figure 6.21. Where (a) corresponds to the melting process of the material during the temperature rise phase. As can be seen from the graph, the addition of the nucleating agent resulted in a small increase in the melt temperature (T_m) of the composites, where PP-α-0.1 increased by 4.5 °C compared to PP-p. This indicates that it effectively enhances the thermal stability of the material. Figure 6.21b corresponds to the isothermal crystallization process of the material during the temperature drop phase. The comparison shows that the addition of nucleating agents effectively enhances the nucleation rate of PP and promotes the crystallization behavior of the material under high-temperature conditions. At the high temperature of 130 °C, the crystallization phase of PP-α starts earlier and takes less time compared to PP-p. To quantify the effect of nucleating agents on the phase change behavior of PP during isothermal crystallization. This study used Eqs. (6.4)–(6.6) to calculate the crystallinity X_c of the samples and the semi-crystallization time $t_{1/2}$, respectively [16, 37].

$$X_c = \frac{\Delta H_m}{\Delta H_m^0} \tag{6.4}$$

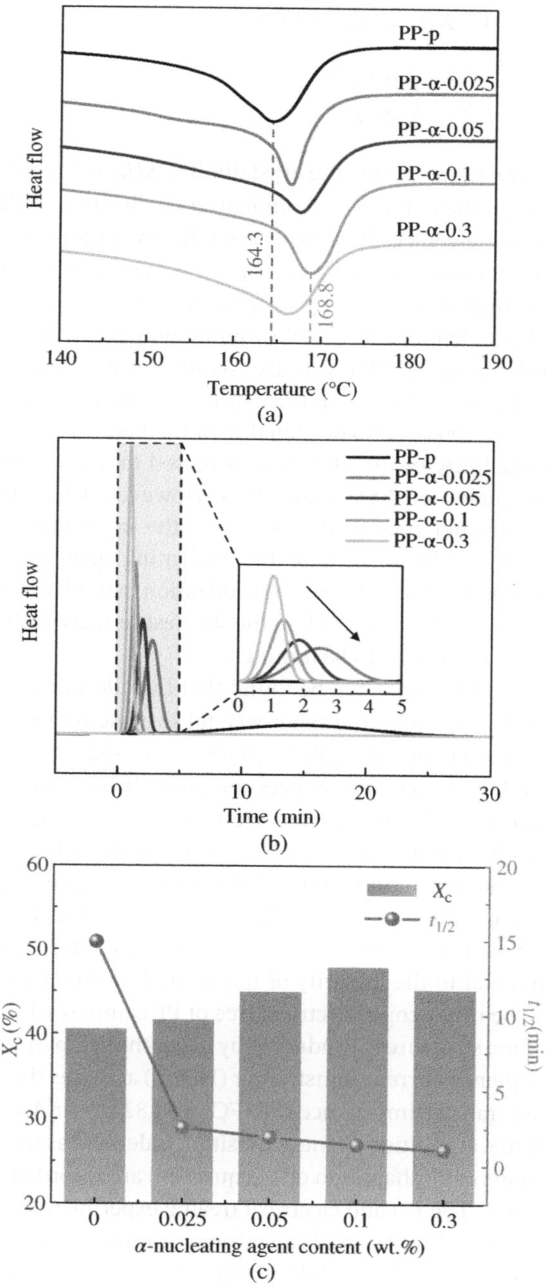

Figure 6.21 DSC curves of (a) heating melting, (b) isothermal crystallization, and (c) X_c and $t_{1/2}$ varying with α-nucleating agent content. *Source:* Wu et al. [69]/with permission from Elsevier.

$$1 - X(t) = \exp(-Kt^n) \tag{6.5}$$

$$t_{1/2} = \left(\frac{\ln 2}{K}\right)^{1/n} \tag{6.6}$$

where X_c represents the crystallinity, ΔH_m is the melting enthalpy of the sample, ΔH_m^0 is the enthalpy of complete crystallization of PP, $\Delta H_m^0 = 209$ J/g, $X(t)$ represents the change in relative crystallinity with isothermal crystallization, K is the crystallization rate constant, n is the Avrami index, and $t_{1/2}$ is the half crystallization time corresponding to when $X(t) = 50\%$.

The calculated data is shown in Figure 6.21c, which shows that the crystallinity of the modified PP-α was also significantly improved compared to that before the modification. Where in the 0–0.1 wt% addition range, the crystallinity of the samples was positively correlated with the percentage of nucleating agent added. The crystallinity of PP-α-0.1 was increased to 47.89% compared to the crystal region percentage of 40.51% for PP-p. However, when the nucleating agent content was further increased to 0.3 wt%, the crystallinity of the sample decreased to 45.22%. At the same time, the nucleating agent also greatly increased the crystallization rate of PP. The crystallization period was thus significantly shortened. Among them, PP-α-0.3 had the shortest semicrystallization period of 1.08 minutes, while PP-p took 14.88 minutes.

In summary, it can be found that the addition of nucleating agents can, on the one hand, reduce the spherocrystal size and refine the crystalline structure so that the percentage of crystal regions in the spatial range is increased. On the other hand, it also greatly reduces the crystallization period and increases the crystallization rate by introducing the process of heterogeneous nucleation. However, it should be noted that the above conclusions only apply to nucleating agents when the addition ratio is controlled within a suitable range. When too much is added, too high a crystal nucleus density in a limited spatial range will restrict the growth behavior of spherocrystals, leading to crystallinity reduction instead, which is detrimental to the integrity of the crystalline structure.

The mesoscopic electrical tree of PP samples with different heterogeneous crystalline structure is conducted by using the setup in Figure 6.13. In addition, a high-frequency current transformer (HFCT) acquires the PD signal of XLPE insulation. The transfer impedance of HFCT is 4.82 mV/mA (at 1 MHz). The maximum discharge amplitude in one acquisition is defined as the maximum PD amplitude. The counts of discharges in one acquisition are recorded and defined as the average PD counts. The PD and electrical treeing experimental results collected based on electrical tree D&PD simultaneous measurement are shown in Figure 6.22. Where the phase-resolved partial discharge (PRPD) spectrums are plotted in Figure 6.22a for the test data collected at 15 minutes. Among them, the horizontal coordinates

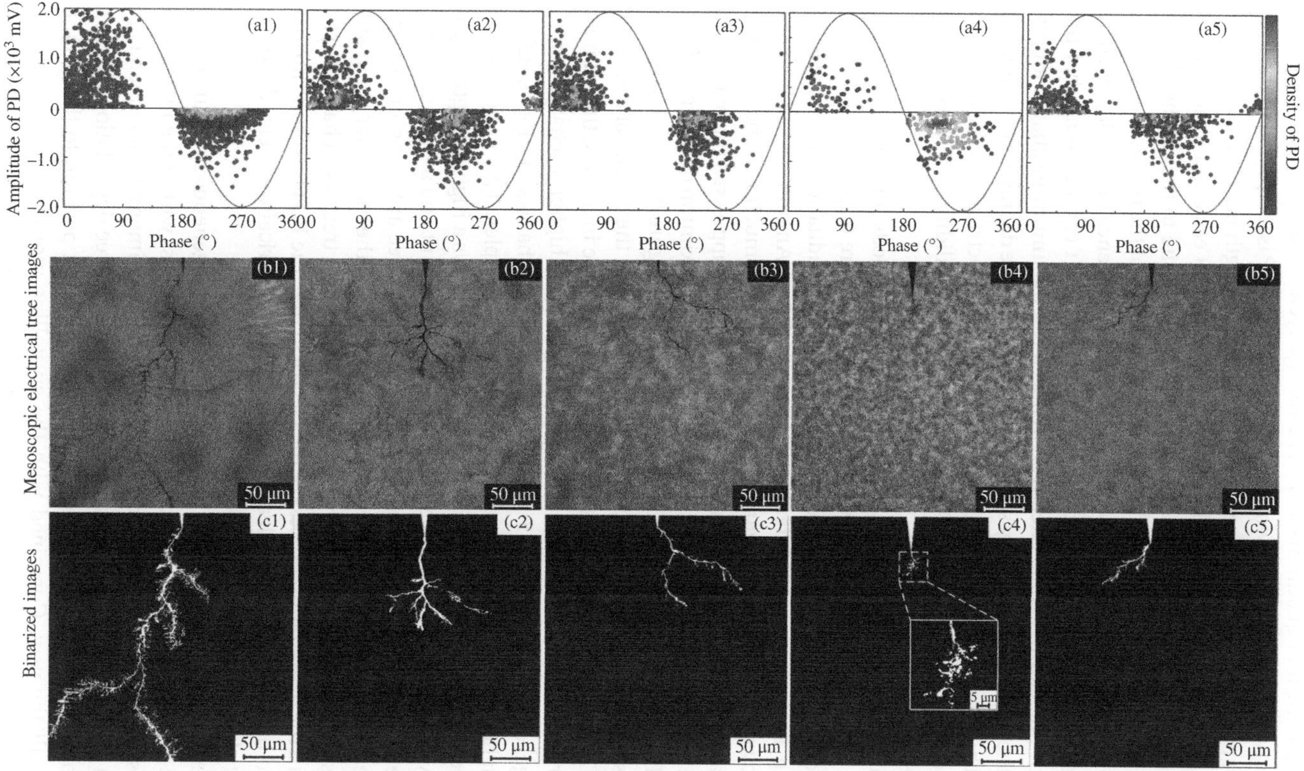

Figure 6.22 (a) PRPD spectra during electrical treeing. (b) POM images and their (c) binarized images of the mesoscopic electrical tree varying with α-nucleating agent content. *Source:* Wu et al. [69]/with permission from Elsevier.

correspond to the four quadrants of the AC voltage cycle transformation. The comparison shows that the PD behavior occurring inside the PP is all located in the first and third quadrants. This is because the electric field strength near the needle electrode gradually increases during the voltage rise, making it easier for charge injection to occur. The polarity of the charge injected is opposite to that of the charge injected in the previous quadrant, and the recombination of the two occurs accompanied by the release of energy, which leads to the occurrence of PD [38, 39]. The vertical coordinate corresponds to the amplitude of PD. It can be found that the addition of nucleating agents effectively reduces the maximum discharge amplitude and the number of discharges during the electrical tree degradation compared to PP-p. This phenomenon is also consistent with the electrical tree degradation within the samples in Figure 6.22b, which verifies the validity of the test results to some extent.

The parameters of PD data with 50 sinusoidal periods in one second are extracted by data-processing software and plotted in Figure 6.23a. As can be seen that the discharge times of PP-α samples decreased gradually with increasing nucleating agent addition content compared to PP-p. Among them, PP-α-0.1 has the lowest number of discharges, only 280, which is much less than 2414 of PP-p. Meanwhile, the addition of nucleating agents also suppressed the maximum discharge amplitude during the discharge process. The maximum discharge amplitude of PP-α-0.1 is reduced by 58.3% compared to PP-p. Figure 6.23b characterizes the correspondence between the discharge amplitude and the counts of discharge during PD. It can be found that as the discharge amplitude increases, the number of discharges within PP all show a decreasing trend. In the higher discharge amplitude range, especially above 1000 mV, the number of discharges in PP-p is significantly higher than in the nucleating agent-modified samples. This indicates that during the degradation of the electrical tree, there is always a high-strength, high-frequency PD inside the PP-p. The magnitude of the amount and amplitude of the discharge also reflects the growth state of the electrical tree at this point to some extent, i.e. the larger the value, the stronger the damage to the molecular chain and the correspondingly faster the growth rate of the electrical tree. As a characteristic quantity in the degradation process of the electrical tree, the PD test results further confirmed the inhibition of nucleating agents on the electrical tree degradation in PP under electrical stress.

To investigate the mechanism of the effect of the heterogeneous structure on the dielectric properties of PP, the electrical tree growth process within the material at the mesoscopic scale was recorded using POM. The experimental phenomena are shown in Figure 6.22b. It can be seen that the electrical tree degradation channel grows along the electric field direction from the pinpoint position to the ground electrode all the way. Combined with the crystalline morphology of PP at the mesoscopic scale, it can be seen that the electrical tree's growth path is not

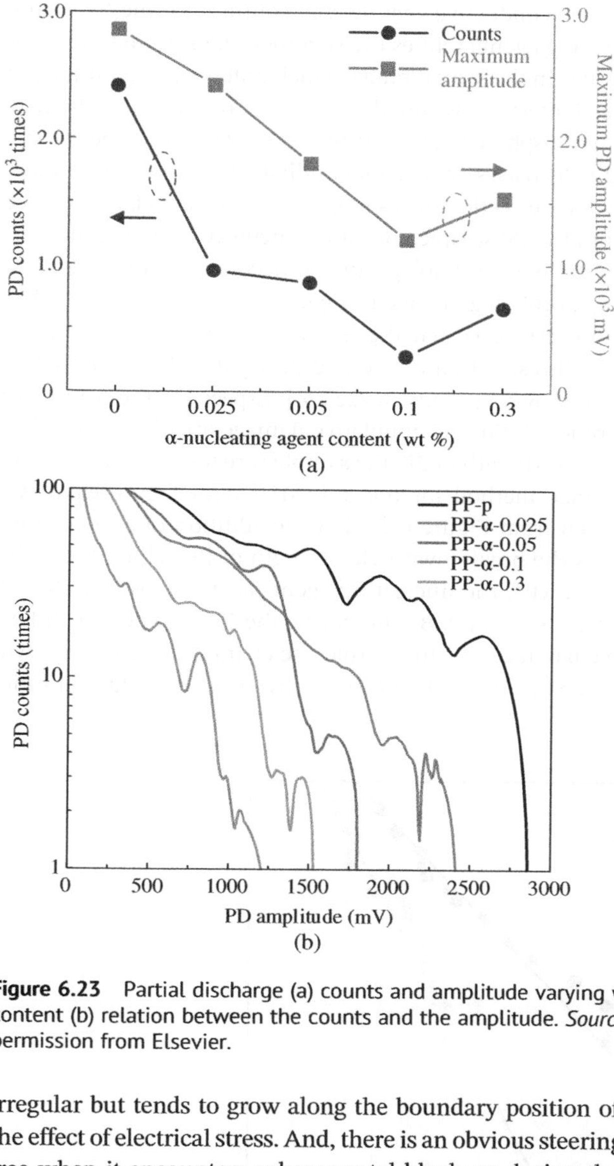

Figure 6.23 Partial discharge (a) counts and amplitude varying with α-nucleating agent content (b) relation between the counts and the amplitude. *Source:* Wu et al. [69]/with permission from Elsevier.

irregular but tends to grow along the boundary position of spherocrystals under the effect of electrical stress. And, there is an obvious steering effect of the electrical tree when it encounters spherocrystal blockage during the longitudinal growth. The tendency that the electrical tree grows along the spherocrystal interface is extremely evident within the PP-p in Figure 6.22b1. Due to the larger size of the PP-p crystal type, the continuity of the long-sized spherocrystal boundary is higher. The electrical tree degradation channels keep consistency with it and develop into continuous long conductive channels. In contrast, for the PP-α sample, the addition of nucleating agents refines the spherocrystal size and increases the number of spherocrystals in the spatial range, which reduces the continuity of the long-sized spherocrystal interfaces within the material. Consequently, the electrical tree is blocked by more spherocrystals during its growth along the electric field direction. Its length decreases in the longitudinal direction, while more branches appear in the lateral direction. So as to quantify the degradation degree of the electrical tree within the PP sample and simultaneously avoid the interference of the material crystalline morphology to the accuracy of the statistical results, the electrical tree morphology image in Figure 6.22b was binarized. The principle of this image processing is that the interference distortion is filtered out using a set grey-scale threshold parameter, resulting in a black-and-white image of the electrical tree shown in Figure 6.22c [24]. Especially, the parameter difference of the aspect ratio within the cumulative damage area is compared.

For testing the breakdown strength of PP, the samples are tested at 30 °C using a linearly increasing voltage method for the industrial frequency breakdown experiments. It can be seen from Figure 6.24 that the addition of a nucleating agent effectively enhances the breakdown field strength of PP. When the addition proportion reaches 0.1 wt%, the nucleating agent has the most significant effect on the electrical resistance of PP. By comparing the E_0 values of the different groups of PP in Table 6.6, it can be found that the characteristic breakdown field strength increases by 8%, 11%, and 20%, respectively, as the addition ratio

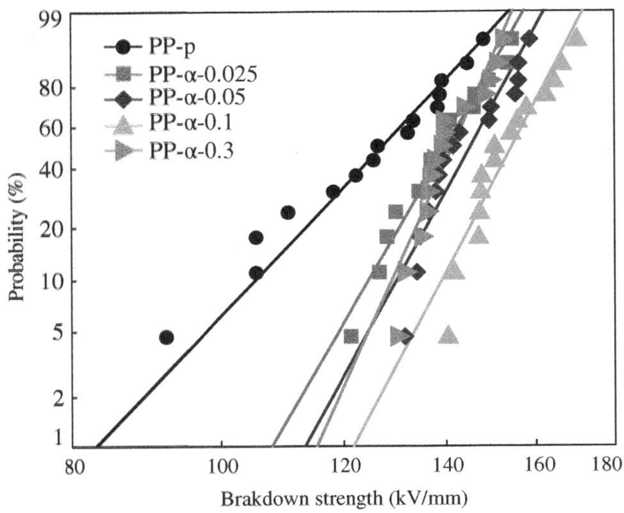

Figure 6.24 Effect of heterogeneous mesoscopic structure on breakdown strength. *Source:* Wu et al. [69]/with permission from Elsevier.

Table 6.6 Weibull distribution parameters of samples. *Source:* Wu et al. [69]/with permission from Elsevier.

Sample	E_0 (kV/mm)	β
PP-p	132.1	9.88
PP-α-0.025	143.4	16.18
PP-α-0.05	148.5	17.24
PP-α-0.1	158.0	17.94
PP-α-0.3	144.1	20.90

rises from 0.025 wt% to 0.1 wt% compared to PP-p. Although the breakdown properties of PP-α-0.3 are slightly reduced compared to PP-α-0.1, they are also higher than the breakdown strength of 132.1 kV/mm for the PP-p sample before modification. Meanwhile, the magnitude of the β value correlates with the dispersion degree of the data and also reflects the insulation reliability of the material to some extent from the side [40]. Compared to PP-p, the β values of the modified PP-α samples are significantly higher. The above-mentioned experimental phenomena show that the nucleating agent effectively enhances the breakdown strength of PP while synergistically improving the insulation reliability of the material and reducing its probability of weak point breakdown.

6.3.3 Electric Field Simulation of Heterogeneous Crystalline Structure

6.3.3.1 Heterogeneous Mesoscopic Structure Simulation

To address the heterogeneous structural characteristics of PP, this study uses the molecular dynamics simulation (MDS) method to investigate the dielectric property differences between the crystal and amorphous regions within the polymer caused by differences in the molecular chain structure [41, 42]. First of all, the α-iPP cell model and the single-chain PP molecular model were constructed using the Visualizar model of Materials Studio software. Then, the models were geometrically optimized based on the Forcite module of the Molecular Force Field Analysis method until energy convergence, obtaining the molecular structure model shown in Figure 6.25. Where the lattice size parameter for the crystal region is $22.1640 \times 24.6450 \times 22.8510$ Å^3 and the amorphous region is $22.1640 \times 24.6450 \times 22.8510$ Å^3.

Further, the constructed models are periodically annealed using the annealing tractor (NPT), and MDSs are carried out using the regular tractor (NVT). The MDS solution in this chapter is the dipole autocorrelation function (DACF). This function describes the change of dipole moments by comparing the positional transformations from group cluster to group cluster in the composite model. It is defined as

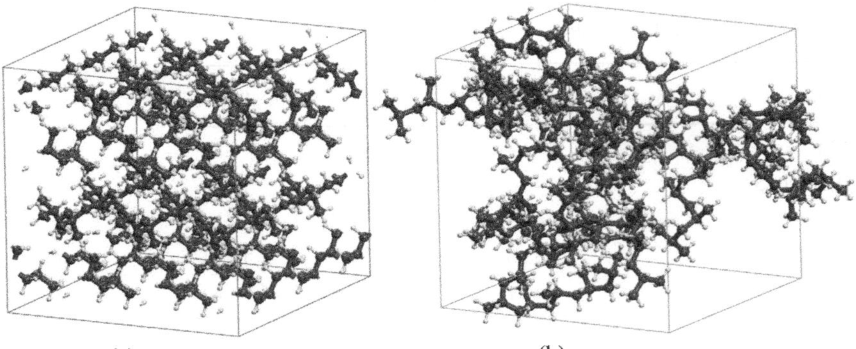

Figure 6.25 Heterogeneous mesoscopic structure of (a) crystal region and (b) amorphous region. *Source:* Wu et al. [69]/with permission from Elsevier.

$$\Phi(t) = \langle M(0) \cdot M(t) \rangle \tag{6.7}$$

where $M(t)$ is the dipole moment of the molecular chain at moment of t, and its value is the quantity product of the charge of the molecular chain and its position vector at that moment. $\langle M(0) \cdot M(t) \rangle$ is the average of the sum of the number of products of all molecular chains $M(0)$ and $M(t)$ at moment t in the system. The results of the calculations are shown in Figure 6.25a.

DACF reveals the contribution of atomic and dipole polarization to the macroscopic dielectric properties of the system but is unable to characterize the effect of electronic polarization on the dielectric properties. As the frequency of the applied electric field rises from 0 to infinity, the dipole-oriented polarization cannot keep up with the periodical change of the electric field, relaxation losses are generated, and the dielectric constant tends to decrease with the increase of the electric field frequency. The electron polarization does not relax with the change in frequency of the applied electric field, and thus only the electron polarization exists in the system at infinite frequency. Thus, the relationship between the frequency-varying dielectric constant $\varepsilon(\omega)$ and the DACF is satisfied as follows:

$$\frac{\varepsilon(\omega) - \varepsilon_{\infty}}{\varepsilon_{\mathrm{r}} - \varepsilon_{\infty}} = \int -\left(\frac{\mathrm{d}\Phi(t)}{\mathrm{d}t}\right) \cos(\omega t)\mathrm{d}t \tag{6.8}$$

where ε_{r} is the static dielectric constant, representing the value of the dielectric constant of the system when the frequency tends to 0 at infinity. $\varepsilon(\omega)$ is the value of the dielectric constant under the change of frequency. ε_{∞} is the optical frequency dielectric constant, representing the value of the dielectric constant of the system when the frequency is infinity. Since only electronic polarization exists in the system at frequency infinity, the optical frequency dielectric constant is equivalent to the contribution of electronic polarization to the dielectric constant. For PP, the value of the optical frequency permittivity is approximately 1.9.

Define the relaxation strength $\Delta\varepsilon = \varepsilon_r - \varepsilon_\infty$, which characterizes the contribution of atomic and dipole polarization to the permittivity, and is thus only related to the dipole moment

$$\Delta\varepsilon = \frac{4\pi}{3}\frac{\langle M(0)^2\rangle - \langle M(0)\rangle^2}{V\varepsilon_0 k_B T} \tag{6.9}$$

where V is the volume of the system, ε_0 is the vacuum dielectric constant, k_B is the Boltzmann constant, T is the temperature of the system, $\langle M(0)^2\rangle$ is the mean-square dipole moment of the dipoles in the system, and $\langle M(0)\rangle^2$ is the squared value of the average dipole moment of the dipoles in the system. Since PP is a non-polar medium, its mean dipole moment is zero, so $\langle M(0)\rangle^2 = 0$.

Based on the above two equations, the DACF curves of PP were brought to the rescue to obtain the dielectric constants of the crystalline and amorphous regions, and the results are shown in Figure 6.26. The time-dependent curve of the DACF for the PP crystal region and the amorphous region is presented in Figure 6.26a. It can be seen that as far as the initial value of the function is concerned, it is 2.9×10^{26} Debye^2 m^{-3} in the crystal region, which is higher than the 1.6×10^{26} Debye^2 m^{-3} in the amorphous region. Simultaneously, a comparison of the decay rate of the functions shows that the fluctuations of the DACF in the amorphous region are more lagging, while the decay of the function curve

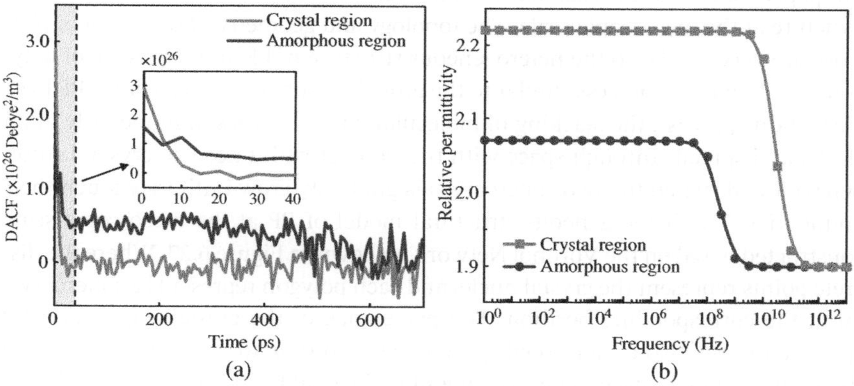

Figure 6.26 (a) Time-dependent curve of the DACF and (b) relative permittivity varying with frequency. *Source:* Wu et al. [69]/with permission from Elsevier.

in the crystal region is more rapid. The initial value of the function determines the static dielectric constant of the dielectric.

As can be seen from Figure 6.26b, the relative permittivity of the amorphous and crystal regions of PP is 2.06 and 2.22, respectively. The magnitude of the initial value of the function corresponds to the strength of the dipole provided by the model molecule per unit volume. Larger values imply a higher dipole polarization strength provided by the molecular chain. Compared to the amorphous region, the molecular chains in the crystal region are more regular, with a higher density of molecular chains per unit volume, and thus the dipole polarization strength provided by the molecular chains is higher. When a constant electric field is applied, the intrinsic dipole moments of the constituent dielectric molecules will align along the electric field direction, resulting in a higher static dielectric permittivity in the crystal region than in the amorphous region. The decay tendency of the function with time determines the variation of the dielectric permittivity with frequency. The DACF in the amorphous region decays more slowly and thus has a higher sensitivity to frequency. When the applied electric field frequency reaches 10^7 Hz, the dielectric permittivity of the amorphous region decreases first. This is because the molecular chains in the amorphous region are more randomly arranged than in the crystal region, and the intermolecular dipole moments are inhomogeneously distributed so that the DACF decays more rapidly. When the frequency of the electric field rises, the inhomogeneously distributed dipole moment is more likely to lag behind the change in the electric field. This explains why the relaxation in the amorphous region precedes that in the crystal region.

6.3.3.2 Electric Field Simulation in Mesoscopic Structure

Based on the growth and distribution properties of crystallization in PP materials, this paper uses a two-dimensional Voronoi Network to construct its heterogeneous structure at the mesoscopic scale. The topology and geometry of the Voronoi Network are very similar to the heterogeneous structure of PP and the Voronoi polygons are drawn in a process similar to the growth of spherocrystals within PP [43]. During this process, the stacking of molecular chains starts simultaneously at all nuclei and spreads through space with the same growth rate. The crystallization behavior ends when the two spherocrystals gradually grow until they touch their boundaries. The heterogeneous structural model of PP at the mesoscopic scale constructed based on the Voronoi Network is shown in Figure 6.27. Where the discrete points represent the crystal nuclei and each polygon represents a spherocrystal at the corresponding position. The percentage of the crystal region of a PP specimen in the two-dimensional space is set according to its crystallinity in the three-dimensional space. The AC module of COMSOL Multiphysics software can be used to simulate the electric field distribution inside the heterogeneous structure model of PP at the mesoscopic scale in order to further analyze the mechanism of PD and electrical tree degradation inside the polymer under the action of

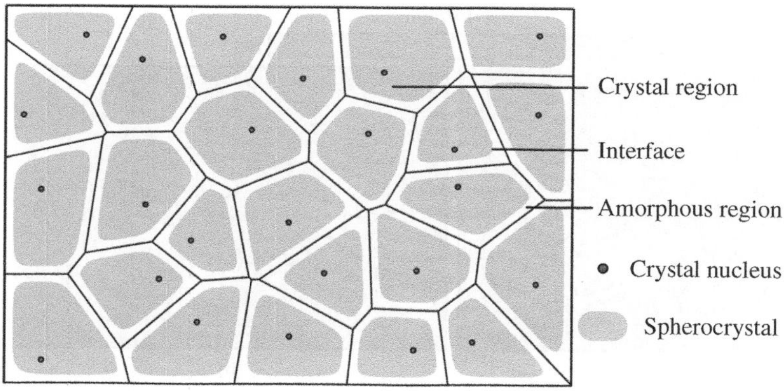

Figure 6.27 Voronoi Network-based PP heterogeneous mesoscopic structural model. *Source:* Wu et al. [69]/with permission from Elsevier.

electrical stress. The upper side of the model is the high-voltage electrode and the lower side is the ground electrode. The average strength of the applied AC field is 150 kV/mm. The dielectric parameters obtained from the MDS above are carried over into the crystal and amorphous regions, respectively. The electric field distribution within the model is calculated after the mesh has been dissected.

The structural models of PP-p and PP-α-0.1 were established based on the Voronoi Network, corresponding to the spherocrystal size and the proportion of crystal regions. The spherocrystals of PP-p are smaller in number and larger in size, with a larger size dispersion and a lower percentage of crystal regions in the spatial range, and the spherocrystals are more loosely arranged. The spherocrystals of PP-α-0.1 are more numerous and smaller in size. The spherocrystals are less dispersed in size and are more densely packed with each other, with a relatively higher percentage of area in the spatial range. The dielectric parameters of the crystal and amorphous regions obtained from the simulation above are brought into the model separately and an AC voltage with an average electric field strength of 150 kV/mm is applied. The electric field distribution inside PP is obtained from the simulation and the results are shown in Figure 6.28. It can be seen that there is a significant difference in the magnitude of the electric field between the crystal and amorphous regions within PP, with the electric field strength within the former being significantly smaller than that of the latter. This is due to the fact that the electric field distribution in the heterogeneous material is inversely proportional to its relative dielectric permittivity under AC voltage. The dielectric permittivity in the crystal region is higher and, therefore, the electric field strength inside it is lower. Specifically, strong electric field distortion occurs at the interface between the crystal and amorphous regions. This is extremely evident in the model of PP-p in Figure 6.28a. The maximum electric field strength

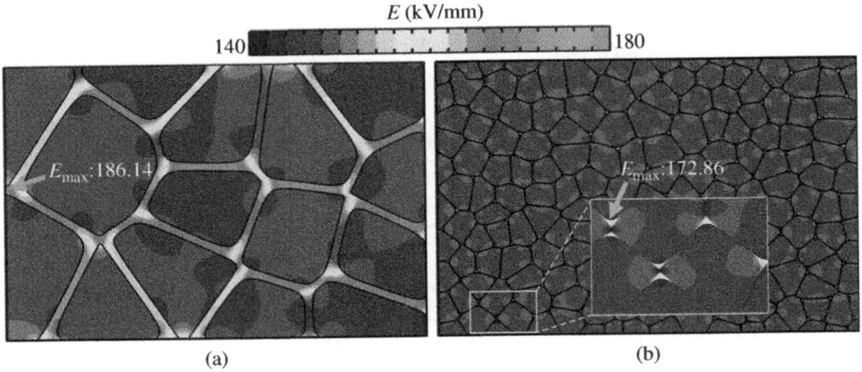

Figure 6.28 Electric field distribution in mesoscopic structure of (a) PP-p and (b) PP-α-0.1. *Source:* Wu et al. [69]/with permission from Elsevier.

at the interface reaches 186.14 kV, which is much higher than the average electric field strength of 150 kV/mm. The electric field distortion rate reaches 24%. In contrast, the electric field distortion inside PP-α-0.1 is weaker and more homogeneous. At the same time, the small size and the dense arrangement of the spherocrystals prevent the appearance of large areas of continuous electric field distortion caused by the long, highly continuous spherocrystal boundaries in PP-p.

The severe electric field distortion at the interface between the PP crystal and amorphous regions gives the electrons a large enough energy to be injected into the polymer interior. Under the effect of a periodically varying sinusoidal voltage, the continuous injection and extraction of high-energy charges promote the formation and expansion of the low-density region at the PP interface position. During the gradual expansion of the low-density region defects, the recombination of positive and negative polar charges gradually increases, which leads to the occurrence of PD phenomena. In particular, the PD process is accompanied by the release of energy in various forms such as thermal, optical, and mechanical. Under the continuous action of high-intensity, high-frequency PD, obvious conductive tree degradation channels are formed at the interface between the crystal and amorphous regions. The addition of nucleating agents effectively refines the size of the spherocrystals and increases the number of spherocrystals in the spatial range, resulting in a denser arrangement of spherocrystals. The improved crystalline structure, on the one hand, makes the electric field distribution more uniform and reduces the probability of strong local electric fields, effectively reducing the frequency and intensity of PD and thus suppressing the electrical tree degradation at the interface. On the other hand, the higher crystallinity enhances the proportion of crystal regions in the spatial range and reduces the inter-spacing between spherocrystals. The appearance of longer spherocrystal boundaries with higher continuity is prevented and the formation of large electrical tree degradation channels is somewhat suppressed. At the same time, the molecular chain arrangement in the crystal region is more regular than in the amorphous region, and the density of molecular chains per unit volume is higher. The probability of forming low-density regions within the crystal region is correspondingly lower. As a result, under the action of the electric field, the densely arranged spherocrystals are more effective in blocking the growth of electrical tree, which promotes the dispersed growth of electrical tree and reduces the probability of penetration channels occurring along the electric field direction. The electrical tree resistance and breakdown strength of PP have thus been improved to varying degrees.

6.4 Methods for Suppressing Electrical Treeing

6.4.1 Effect of Nucleating Agent and Cooling Rate on Dielectric Property of PP/POE

Based on the above research, it can be concluded that PP cable breakdown accidents are closely related to the difference in dielectric strength of its insulation layer in the radial direction. In particular, the differences in the crystalline structure of the inner, middle, and outer layers of the insulation caused by the cooling rate during cable processing are an important cause of cable insulation breakdown accidents. Based on the actual cable production conditions, this section proposes an electric tree inhibition method based on nucleating-agent-modified PP crystals, with the aim of avoiding insulation breakdown accidents caused by the differences in dielectric properties of the insulating layer in the radial direction, and realizing the improvement of the overall insulation reliability of cables.

It has been found that the addition of nucleating agents can alter the aggregation structure of PP and thereby enhance its macroscopic properties [12, 44, 45]. By adding nucleating agents, nucleation sites can be provided for the PP crystallization process, reducing the spherulite size and forming more complete spherulites. The addition of α-nucleating agents can increase the crystallization temperature of PP, thereby improving the efficiency of PP crystallization [46]. The effect of nucleating agents on the properties of PP insulation is often related to the addition content. Within a certain range, increasing the content of nucleating agent can enhance the insulation properties of PP. However, if the content of nucleating agents is too high, PP will form fragmented crystals, which can lead to a decrease in its insulation properties [31, 47, 48]. By adding nucleating agents, the mechanical and electrical properties of PP can be significantly improved. The type and the structure of nucleating agents have a significant effect on the crystallization process of PP. Among them, the sorbitol nucleating agent can effectively control the crystallization behavior of PP due to its structure bearing resemblance to the size of propylene [11]. The samples in this subsection are the PP/elastomer blends with different contents of nucleating agent, prepared by the melt blending method, and 0, 0.1, 0.2, 0.4, and 0.6 wt% of nucleating agent are added separately. The experimental samples are prepared by hot-pressing method at a temperature of 200 °C and a pressure of 15 MPa and cooled down at a rate of 2.5 and 10 °C/min, respectively, in order to simulate the difference in cooling rate between the outer and middle regions of the cable insulation layer. The abbreviations for samples are indicated in Table 6.7.

The crystalline morphologies of PP/elastomer blends with different nucleating agent contents are shown in Figure 6.29. Compared to rapid cooling, PP/elastomer blends under slow cooling will form larger spherulites. With the increase of the content of the nucleating agent, the number of crystal nuclei in PP/elastomer blends increases and the spherulite size of PP/elastomer blends

Table 6.7 Index for PP samples investigated in this subsection. *Source:* Du et al. [70]/with permission from IEEE.

Index	Nucleating agent (wt%)	Cooling rate (°C/min)	Index	Nucleating agent (wt%)	Cooling rate (°C/min)
S0	0	2.5	R0	0	10
S1	0.1		R1	0.1	
S2	0.2		R2	0.2	
S3	0.4		R3	0.4	
S4	0.6		R4	0.6	

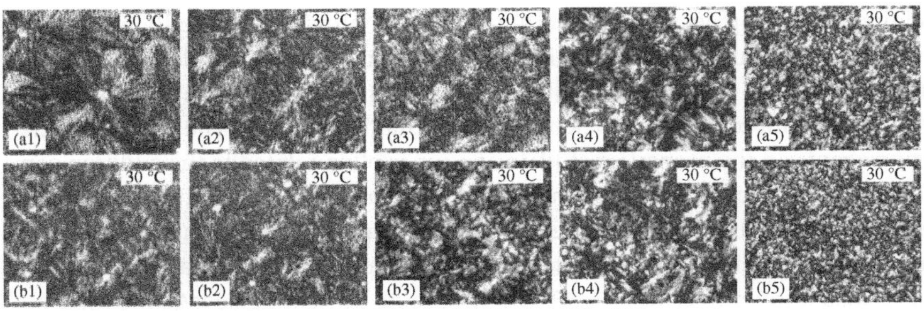

Figure 6.29 Crystalline morphologies of PP/elastomer blends with different nucleating agent contents at (a) slow and (b) rapid cooling rates. *Source:* Du et al. [70]/with permission from IEEE.

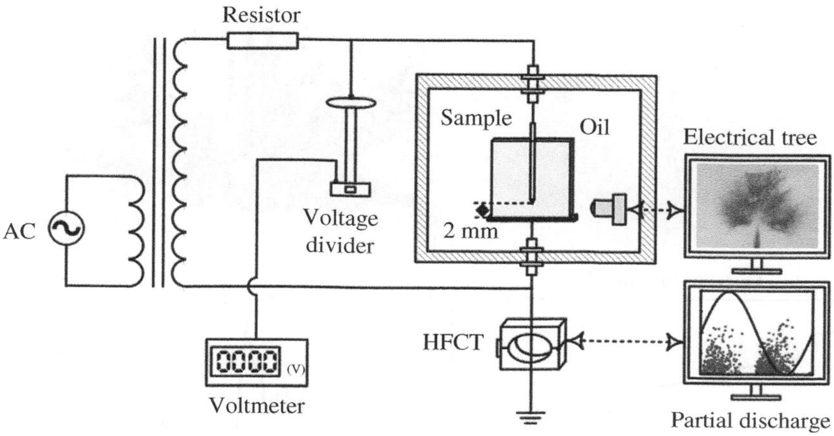

Figure 6.30 Schematic diagram of experiment system. *Source:* Du et al. [70]/with permission from IEEE.

decreases. This indicates that the addition of nucleating agent leads to a significant reduction in the spherulite size of PP/elastomer blends and results in a more compact spherulite stacking. When adding 0.4 wt% nucleating agent, PP/elastomer blends have a smaller spherulite size while maintaining a complete spherulite structure. When the content of the nucleating agent increases to 0.6 wt%, the spherulite size of PP/elastomer blends will be further reduced, but substantial fragmented crystals will form. Furthermore, with the addition of nucleating agent, the disparity in spherulite size between the PP/elastomer blends prepared with slow and rapid cooling rates is observed to decrease.

The schematic diagram of the PP/elastomer samples' electrical treeing characteristics experiment system is shown in Figure 6.30. To record PD signals from PP/elastomer samples, a HFCT with a transmission impedance of 4.82 mV/mA at 1 MHz is utilized. The maximum PD amplitude recorded during a single collection is defined as the maximum PD amplitude, while the average number of discharges recorded during a single collection is defined as the average PD counts. PP/elastomer samples prepared with different cooling rates are subjected to AC electrical treeing with a voltage of 7 kV for 30 minutes at 30 and 90 °C. To better compare the differences in electrical treeing characteristics of PP/elastomer samples modified with sorbitol nucleating agent, the electrical tree images are processed using the method shown in Figure 6.31. The electrical tree images are converted to binary images through image recognition, and the binary image is divided into $m \times n$ pixels based on its size as shown in Figure 6.31b, and the number of black pixels in each row and column is calculated using the following formula and plot the distribution of black pixels as

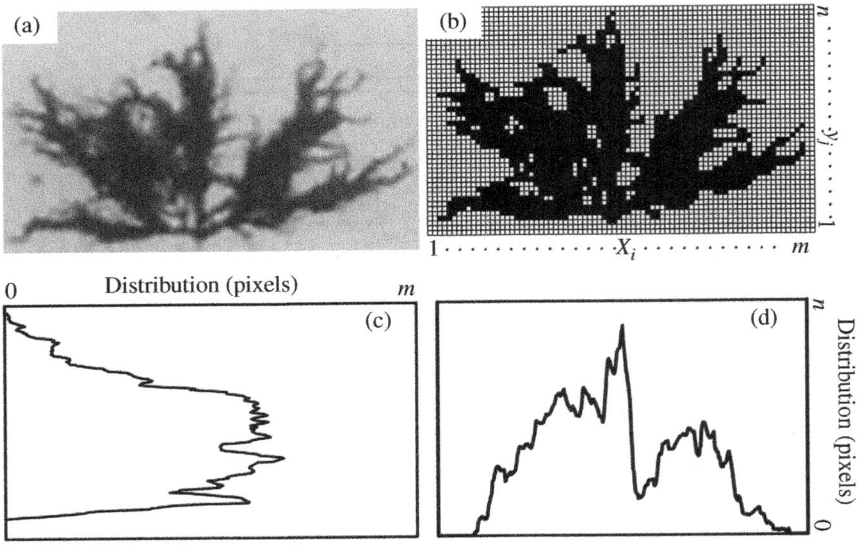

Figure 6.31 Schematic diagram of electrical tree morphology characterization: (a) electrical tree, (b) pixel recognition, and (c) and (d) pixel distribution in horizontal and vertical direction, respectively. *Source:* Du et al. [70]/with permission from IEEE.

shown in Figure 6.31c and d. The calculation method for pixel distribution in the horizontal direction is as follows:

$$C_j = \sum_{1}^{m} r_{(i,j)}, \quad \left[r_{(i,j)} = \begin{cases} 0, RGB, (255, 255, 255) \\ 1, RGB, (0, 0, 0) \end{cases} \right] \tag{6.10}$$

where C_j is the number of black pixels in the jth row and $r_{(i,j)}$ is the pixel value of the ith row and jth column, where black is represented by 1 and white by 0. The calculation method for pixel distribution in the vertical direction is the same.

The electrical tree morphologies at 30 minutes of unmodified PP/elastomer samples prepared with slow and rapid cooling rates at 30 and 90 °C are shown in Figure 6.32. The variation of electrical tree degradation area and PD characteristics over time are shown in Figure 6.33. PP/elastomer samples prepared with slow cooling rate exhibit fewer tree branching and tend to favor forward propagation. In contrast, PP/elastomer samples prepared by rapid cooling have more branches, presenting a jungle-like morphology. This is because the electrical tree tends to propagate along the boundaries of the spherulites [5]. The PP/elastomer blends prepared with slow cooling rate exhibited more severe deterioration in the electrical tree experiments at 30 and 90 °C. After applying AC voltage for 30 minutes, the

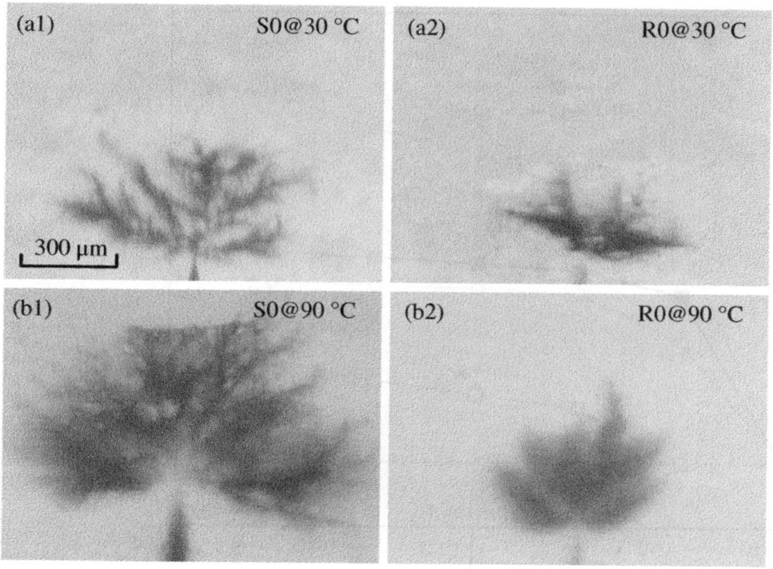

Figure 6.32 Electrical tree morphologies of unmodified PP/elastomer samples: (a) 30 °C and (b) 90 °C. *Source:* Du et al. [70]/with permission from IEEE.

electrical tree degradation area of the PP/elastomer blends prepared with rapid cooling rate measures 0.12 and 0.21 mm^2 at 30 and 90 °C, respectively. In comparison, the electrical tree degradation area of the PP/elastomer blends prepared with slow cooling rate measures 0.24 and 0.43 mm^2 at 30 and 90 °C, respectively. At 30 and 90 °C, PP/elastomer samples prepared with slow cooling rate exhibit higher maximum PD amplitudes of 1.82 and 2.79 V, respectively, while samples prepared with rapid cooling rate show values of 1.32 and 2.70 V, respectively. Additionally, PP/elastomer samples prepared with rapid cooling rate demonstrate lower average PD counts. The PD characteristics of PP/elastomer samples are consistent with the growth of electrical trees, demonstrating a correlation between the development of PD and the growth of electrical trees.

The electrical tree morphologies of PP/elastomer samples modified with different contents of the nucleating agent at 30 and 90 °C after applying voltage for 30 minutes are shown in Figure 6.34a–d show the experimental results at 30 °C, while Figure 6.34e–h shows the experimental results at 90 °C. As the content of the nucleating agent increases from 0 to 0.4 wt%, the forward propagation of electrical treeing in PP/elastomer samples prepared with slow cooling rate is significantly inhibited. Additionally, there is an increase in branching, leading to the formation of shrubby electrical trees.

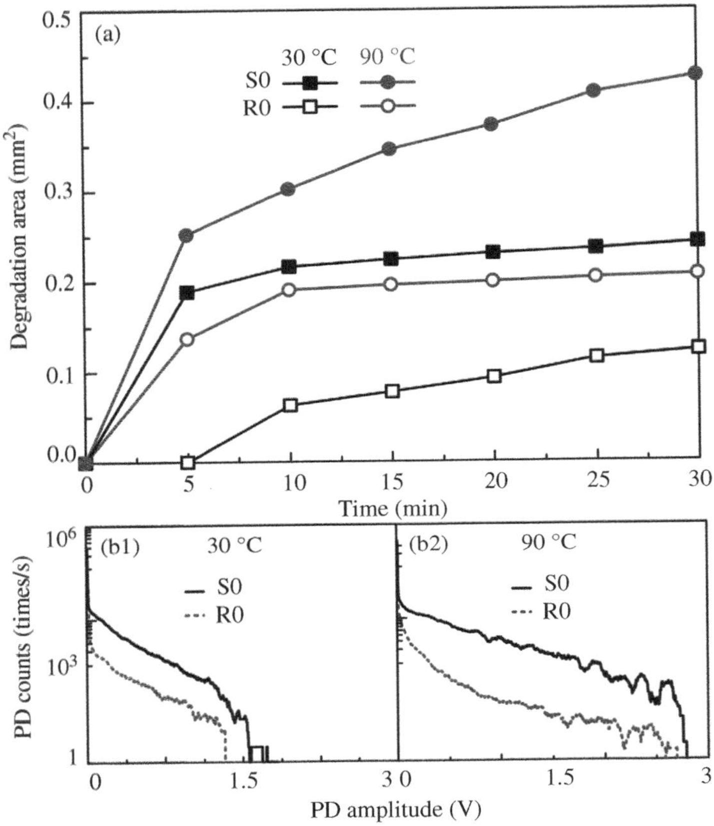

Figure 6.33 Degradation area and PD characteristics of electrical treeing: (a) degradation area and (b) PD characteristics. *Source:* Du et al. [70]/with permission from IEEE.

The discharge intensity distribution during 30 minutes of electrical tree growth is shown in Figure 6.35a and b. As the content of the nucleating agent increased from 0 to 0.4 wt%, both the maximum PD amplitude and the average PD counts of the PP/elastomer samples prepared with slow and rapid cooling rates show a significant reduction. With the addition of 0.4 wt% of nucleating agent, the maximum PD amplitude of the PP/elastomer samples prepared with slow and rapid cooling rates decreased by about 37% and 18%, respectively, compared to the unmodified samples at 30 °C. At 90 °C, the maximum PD amplitude of the samples, prepared with slow and rapid cooling rates, decreased by about 32% and 37%, respectively. Similarly, after the addition of the nucleating agent, there is a significant decrease in the average PD counts for the PP/elastomer samples. When the content of the

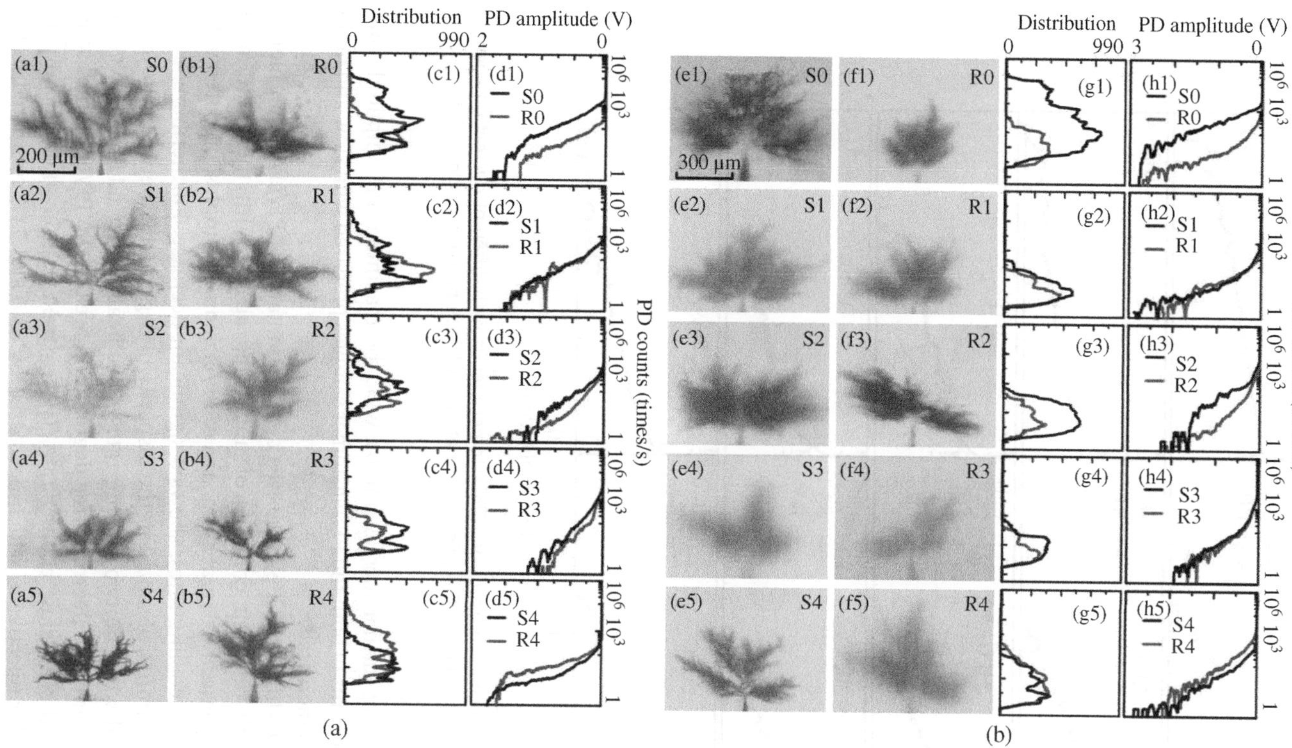

Figure 6.34 Electrical tree morphologies of PP/elastomer blends with different nucleating agent contents: (a) 30 °C and (b) 90 °C. *Source:* Du et al. [70]/with permission from IEEE.

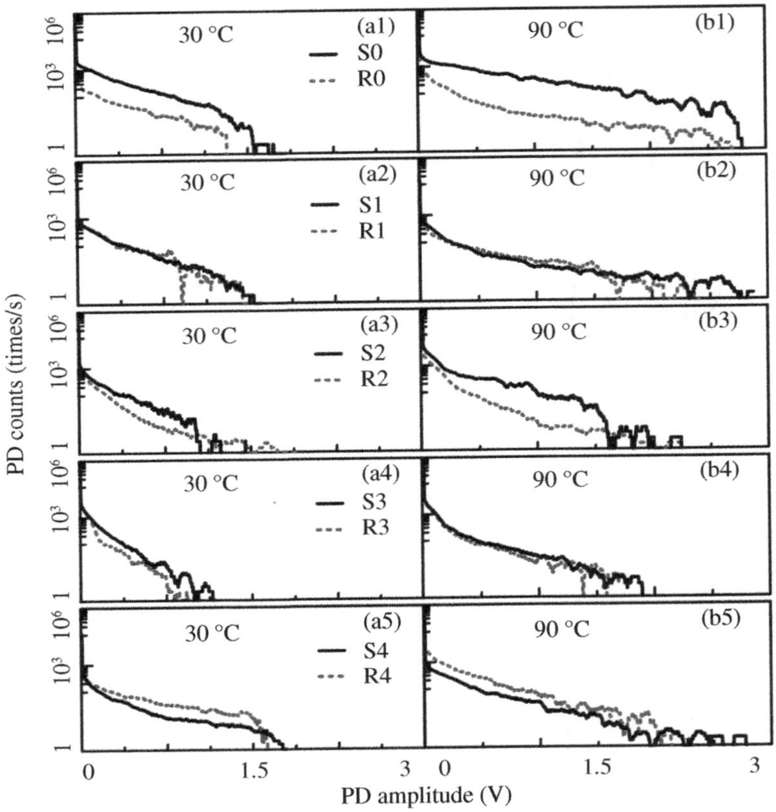

Figure 6.35 PD characteristics of PP/elastomer blends with different nucleating agent contents: (a) 30 °C and (b) 90 °C. *Source:* Du et al. [70]/with permission from IEEE.

nucleating agent further increases to 0.6 wt%, the maximum PD amplitude and average PD counts of both the PP/elastomer samples prepared with slow and rapid cooling rates increased.

Figure 6.36 shows the electrical tree degradation area of PP/elastomer blends modified with different contents of nucleating agent. With the addition of nucleating agents, the degradation area of PP/elastomer samples, prepared with slow and rapid cooling rates, shows a significant decrease. Furthermore, a significant reduction is observed in the disparity of degradation between the samples prepared with the two cooling rates. Specifically, after adding 0.4 wt% of nucleating agent, the electrical tree degradation area of the PP/elastomer samples, prepared with slow cooling rate, is measured as 0.11 and 0.15 mm^2 at 30 and 90 °C, respectively, which decreased by 63% and 74% compared to the

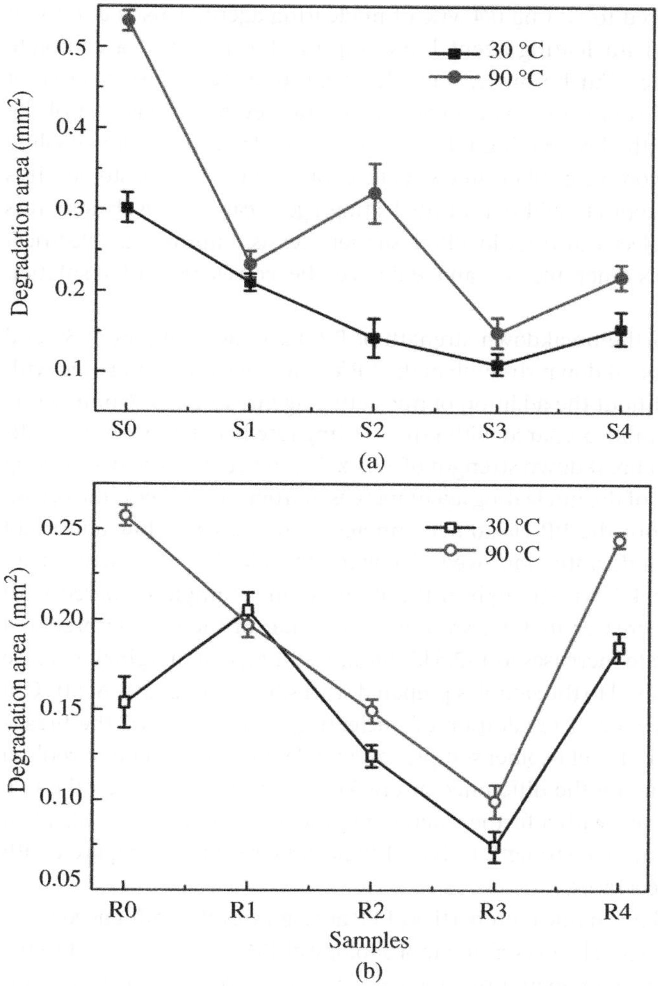

Figure 6.36 Degradation area of PP/elastomer blends with different nucleating agent contents: (a) slow cooling and (b) rapid cooling. *Source:* Du et al. [70]/with permission from IEEE.

unmodified PP/elastomer samples. For the PP/elastomer samples prepared with rapid cooling rate, the electrical tree degradation area is measured as 0.07 and 0.10 mm^2 at 30 and 90 °C, respectively, which decreased by 53% and 42% compared to the unmodified PP/elastomer samples. However, when the content of the nucleating agent is increased to 0.6 wt%, the electrical tree degradation of PP/elastomer samples prepared with both slow and rapid cooling rates becomes more severe compared to adding 0.4 wt% of nucleating agent. This is due to an excessive amount of nucleating agent hindering the formation of a complete spherulitic structure. Furthermore, the electrical tree degradation area of PP/elastomer samples prepared with rapid cooling rate even exceeds that of the samples prepared with slow cooling rate. This may be attributed to slow cooling resulting in a more complete spherulitic structure formation in PP/elastomer. It is proved that the appropriate addition of nucleating agent can effectively suppress the propagation of electrical trees in PP/elastomer blends, improve the electrical properties of PP/elastomer blends, and enhance the consistency of insulation properties.

Figure 6.37 shows the breakdown strength of PP/elastomer samples at 30 and 90 °C. At 30 °C, the breakdown strength of the PP/elastomer sample prepared with slow cooling rate without the addition of nucleating agent is 144.7 kV/mm, while the PP/elastomer sample prepared with rapid cooling rate is 164.8 kV/mm, resulting in a difference in breakdown strength of 20.1 kV/mm between the two cooling rates. As the content of the nucleating agent increased from 0 to 0.4 wt%, the breakdown strength of both the PP/elastomer samples prepared with slow and rapid cooling rates was significantly improved. Following the addition of 0.4 wt% nucleating agent, the breakdown strength of the PP/elastomer sample prepared with slow cooling rate increases to 178.7 kV/mm, while that of the sample prepared with rapid cooling rate increases to 182.5 kV/mm, resulting in a marginal increase of 3.8 kV/mm compared to the samples prepared with slow cooling rate. At 90 °C, a similar trend is observed. The addition of nucleating agent increases the breakdown strength of both PP/elastomer samples prepared with slow and rapid cooling rates, gradually reducing the difference in breakdown strength between the two cooling rates. However, with a further increase in the content of nucleating agent to 0.6 wt%, the breakdown strength of both PP/elastomer samples prepared with two cooling rates decreases.

The electrical and mechanical properties of PP are significantly influenced by its crystalline morphology. The crystalline morphology of PP is closely related to the cooling temperature, as temperature changes can cause variations in the conformation and mobility of alkyl groups along the PP main chain. These variations ultimately impact the structure and arrangement of PP's crystalline phase, and subsequently, its overall properties [13, 49, 50].

Research has indicated that electrical trees have difficulty propagating in spherulite and tend to propagate along the spherulite. There are numerous defects in the amorphous region between spherulites, leading to partial electric field distortion and subsequently leading to PD. The PD results in the emission of thermal electrons and high-energy photons. These particles accelerate within the amorphous regions, leading to collisions that disrupt the molecular chains of the PP insulation. Consequently, low-density region is formed, which further

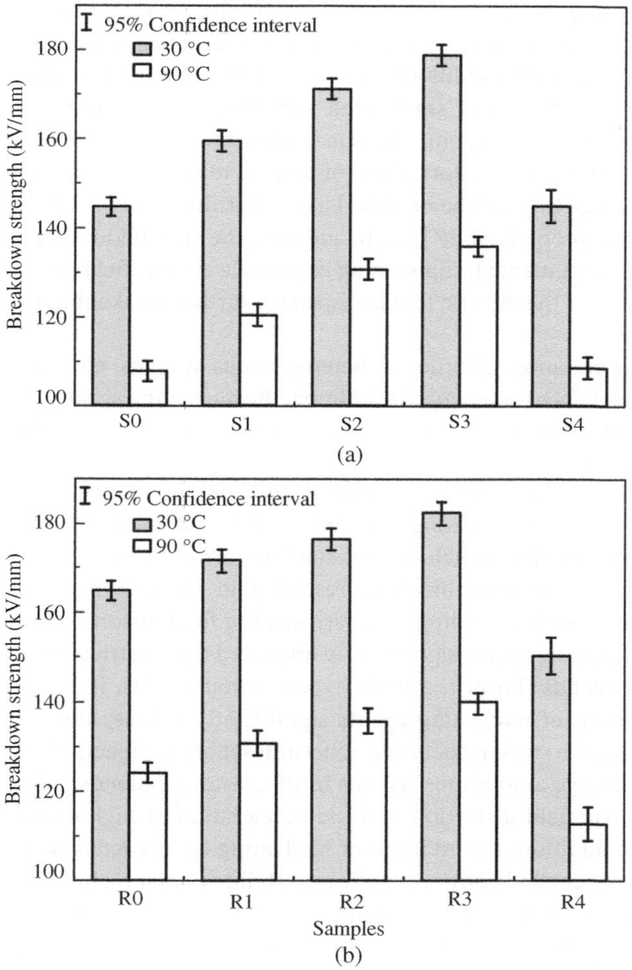

Figure 6.37 Breakdown strength of PP/elastomer blends with different nucleating agent contents: (a) slow cooling and (b) rapid cooling. *Source:* Du et al. [70]/with permission from IEEE.

increases the occurrence of localized discharge. Therefore, electric tree branches tend to propagate along noncrystalline regions [15].

Under slow cooling rate, PP molecule chains have enough time to orderly arrange and form crystalline nuclei, gradually forming large-sized spherulites. The increase in spherulite size leads to more amorphous regions appearing between the spherulites, which will lead to more severe electrical tree degradation and lower breakdown strength in PP/elastomer blends [28].

Under rapid cooling rate, PP molecular chains do not have enough time for regular arrangement and require a higher degree of undercooling to activate crystallization. Therefore, under such conditions, PP has a lower crystallization temperature, leading to the formation of small-sized spherulites. The amorphous region is generally considered a weak point of thermal breakdown [51]. The reduction of spherulite size can avoid the formation of a wide range of amorphous regions and reduce the edge traps of the original large spherulites in PP, thereby improving the breakdown property of PP [52]. In addition, the amorphous region between the spherulites is segmented, suppressing large-scale electric field distortion, and the propagation of the electric tree is dispersed, further weakening the electric field.

The nucleating agent enhances disorder in heterogeneous systems, elevating entropy (S), and lowering Gibbs free energy. In addition, the nucleating agent facilitates crystalline nucleus formation, reducing energy barriers and undercooling required for crystallization initiation.

In this chapter, without the addition of nucleating agent in PP/elastomer blends, PP molecular chains have strong flowability and tend to stack in an orderly manner to form lamellae, which further stack to form spherulite. Slow cooling allows more time for crystallization, resulting in larger spherulites and loose stacking of spherulites, leading to severe electric field distortion and PD. The addition of nucleating agents significantly improved the electrical properties of PP/elastomer blends. From the POM experimental results, it can be observed that the addition of nucleating agents significantly reduces the size of PP spherulites, making the stacking between spherulites more compact, which avoids the formation of large amorphous regions in PP/elastomer blends, effectively suppressing electric field distortion and the acceleration of high-energy particles during PD. In addition, the addition of nucleating agents reduces the undercooling required for crystallization, enabling crystallization at higher temperatures. This leads to the formation of a more complete spherulite structure, reducing defects in PP/elastomer blends. Moreover, the addition of nucleating agents significantly reduces the dielectric properties differences between PP/elastomer blends prepared with slow and rapid cooling, effectively improving the consistency of insulation properties. This can be attributed to the addition of nucleating agents, which reduces the difference in spherulite size of PP/elastomer blends prepared at different cooling rates.

The above method is further applied to enhance the resistance to electrical treeing of the 35 kV PP-insulated cables. Figure 6.38 shows the cross-sectional images of 35 kV PP-insulated cable (a) before and (b) after modulation. It can be seen that the bright-colored rings present in the middle position of the insulating layer before conditioning disappeared after conditioning. This phenomenon also proves that the modulation used in this chapter effectively avoids the differences in the

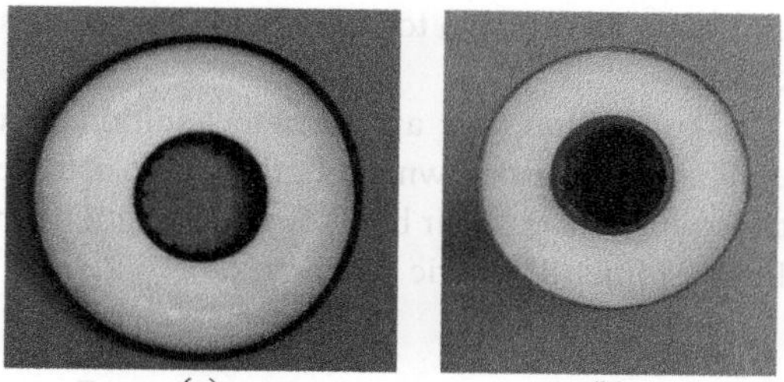

Figure 6.38 Cross-sectional images of 35 kV PP-insulated cable (a) before and (b) after modulation.

crystalline structure of the insulating layer at different locations in the radial direction caused by the differences in cooling rates.

Further, we also rolled and cut the insulation of the regulated cable to obtain slices of the inner and outer layers of the insulation with a thickness of 100 μm. The sliced samples were subjected to AC breakdown tests at different temperatures, and the test results were obtained as shown in Table 6.8. By comparing the test data in the table, it can be seen that the breakdown strengths of the insulating materials are significantly improved after the modulation compared with those before the modulation. At the same time, due to the nucleating agent regulation to reduce the cooling rate caused by the cooling rate of the ball crystal size difference, so that the insulation in the middle layer and the outer layer of the dielectric strength of the difference is significantly reduced. For example, at 30 °C, the breakdown strength of the outer layer of the insulation was 13% higher than that of the inner layer before the modulation, and this parameter decreased to 2% after the modulation. This demonstrates that the nucleating agents used in this chapter to modulate the crystalline behavior of PP cable insulation to achieve an overall improvement in the insulation performance of the cable are significantly effective

Table 6.8 Weibull distribution parameters of samples.

		Breakdown strength (kV/mm)	
Temperature (°C)	**Polypropylene insulation**	**Pre-modification**	**Post-modification**
30	Middle layer	145.0	178.7
	Outer layer	164.8	182.5
60	Middle layer	123.8	153.3
	Outer layer	142.1	159.2
90	Middle layer	107.8	135.9
	Outer layer	124.1	140.1

and have been shown to be usable in the processing and manufacture of practical cables.

In conclusion, the appropriate addition of nucleating agent can effectively enhance the breakdown strength of PP/elastomer samples and mitigate the effect of cooling rate on their breakdown strength, thereby improving the consistency of PP/elastomer dielectric properties.

6.4.2 Enhanced Dielectric Breakdown Property of Polypropylene Based on Mesoscopic Structure Modulation by Crystal Phase Transformation

As mentioned above, as a semicrystalline polymer material, the heterogeneous structure of PP at the mesoscopic scale is closely related to its dielectric properties. In this chapter, a method based on the mesoscopic structure modulation by crystal phase transformation was proposed, which effectively realized the synergistic enhancement of the breakdown properties and mechanical toughness of the material.

Some studies have shown that the materials obtained by the blend-modification method will appear agglomeration caused by phase separation [53], as well as increased loss due to interfacial polarization under the effect of long-term electro-thermal stress [54], which will further result in the material's insulation degradation. In contrast, intrinsic modification emphasizes the improvement of properties by modulating the polymer matrix material itself, which makes the modified material have more stable physical and chemical properties even under complex working conditions [55]. Therefore, how to achieve the goal of synergistic enhancement of mechanical and dielectric properties by means of material intrinsic modulation rather than inhomogeneous material introduction is now an imperative scientific issue.

In the heterogeneous structure of PP, spherulite is the main form of crystal region, and the spatial stacking mode of molecular chains in the lattice determines its crystal phase structure [56]. The common crystal phases include α, β, and γ, and their mechanical, thermal, and electrical properties are significantly different, and they can also be transformed into each other under certain conditions [57]. Therefore, an intrinsic modulation method for PP cable insulation based on mesoscopic structure modulation by crystal phase transformation is proposed in this study, which not only effectively improved the dielectric properties but also enhanced the mechanical toughness synergistically. In the experiments, a test system for insulation degradation at the mesoscopic scale is constructed and used to investigate the effect of heterogeneous structures on insulation.

PP/β-NA composites were prepared by the solution blending method. First, the PP granular material was placed in xylene solution and heated at 130 °C for

Table 6.9 Composition of the samples. *Source:* Wu et al. [71]/ with permission from American Chemical Society.

	Component (wt%)	
Sample	**iPP (T30)**	**β-NA (WBG-II)**
PP-p	100	0
PP-β-0.05	99.95	0.05
PP-β-0.1	99.9	0.1
PP-β-0.2	99.8	0.2
PP-β-0.4	99.6	0.4

3 hours until it was completely dissolved. Subsequently, a mixed suspension solution of xylene/β-NA, which was pre-sonicated, was slowly poured into it, and then methanol was added and stirred to cause flocculation. Finally, the flocculated material was vacuum dried at 80 °C for 12 hours to obtain PP/β-NA composites with different mass fractions (0, 0.05, 0.1, 0.2, and 0.4 wt%) (Table 6.9).

To investigate the effect of mesoscopic structure, especially the crystal phase transformation, on PP dielectric properties, the structure and properties of the modified samples were characterized in this chapter. First, the mesoscopic crystal structure of the samples before and after the modification is characterized to ensure the successful transformation from α-crystal to β-crystal. Second, the partial discharge and electrical tree properties of the material are tested at the same mesoscopic scale. Ultimately, the validity of the insulation enhancement is verified by testing its dielectric breakdown strength at the macroscopic scale. Regarding the logic between the contents of this chapter, the experimental arrangement meets the hierarchical progression from structure to performance and from mesoscopic to macroscopic.

Figure 6.39 shows the mesoscopic crystal morphology of the samples during the various stages of isothermal crystallization. Among them, in the initial stage of crystallization shown in Figure 6.39a, there is no obvious spherocrystal in the sample at this time, and only the crystal nucleus exists within the visual field in the form of bright spots. Compared with PP-p, the addition of β-NA makes the density of the crystal nucleus increase significantly. Moreover, with the content increasing to a certain proportion, it also self-assembles to form a dendritic crystal nucleus structure.

In the middle stage in Figure 6.39b, the spherical crystalline structure centered on the crystal nucleus can be observed clearly inside the samples. Different from PP-p, which only consists of α-crystal spherulites with radially stacked wafer structures, PP-β samples contain β-crystals in which the wafer is arranged parallelly in

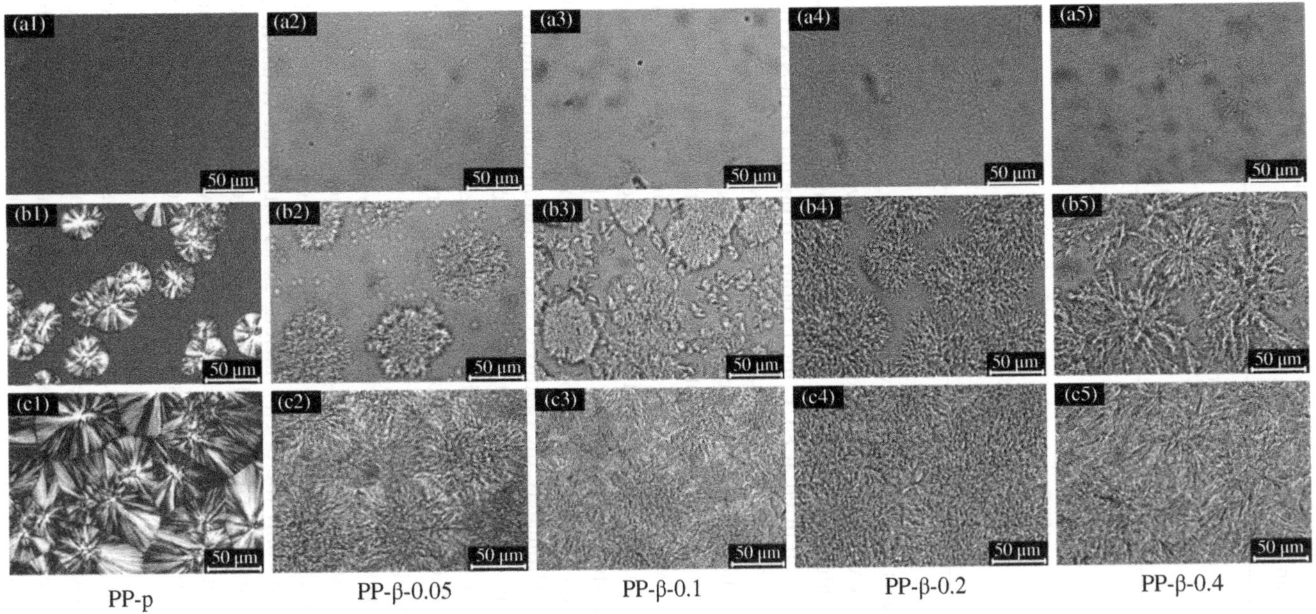

Figure 6.39 Mesoscopic crystal morphology of the samples during the isothermal crystallization (a) initial stage, (b) middle stage, and (c) final stage varying with β-NA content. *Source:* Wu et al. [71]/with permission from American Chemical Society.

the form of bundles. And, as the NA content increases, both the density and the size of β-crystal gradually increase.

At the final stage of crystallization in Figure 6.39c, the spherocrystal has basically filled the entire field. At this point, it can be noticed by comparison that, unlike PP-p, which has darker α-crystals and clear spherocrystal boundaries, the β-crystals inside the PP-β samples are much brighter and no obvious spherocrystal boundaries can be observed.

By comparing the characteristic diffraction peaks at the corresponding angles in the XRD curves of the samples from Figure 6.40a, it can be found that the five diffraction peaks α(110), α(040), α(130), α(111), and α(131) existing in the PP-p, which are all characteristic peaks of α-crystal, correspond to the angles of 13.9°, 16.7°, 18.4°, 20.9°, and 21.7°, respectively [58]. Meanwhile, the characteristic diffraction peak of β-crystal appeared in the PP-β sample at $2\theta = 15.8°$ due to the addition of β-NA, and its peak value rose continuously with the increase of the nucleating agent content. The above experimental results also further proved that the addition of NA successfully realized the crystal phase transformation from α-crystal to β-crystal within PP.

Figure 6.40b and c, respectively, corresponds to the melting and crystallization curves during the heating and cooling processes of the samples obtained from the DSC tests. Wherein, as can be seen from Figure 6.40b, unlike PP-p, where there is only a single α-crystal melting peak corresponding to $T_c = 166.86$ °C, the addition of β-NA resulted in the appearance of a second distinctive melting peak in the PP-β samples at the temperature rise to 155.33 °C. It is known from the related literature that the formation of this melting peak corresponds to the melting behavior of β-crystal during the heating process [59]. Simultaneously, by comparing the crystallization curves of different samples in Figure 6.40c, it can be found that the addition of β-NA not only lowered the nucleation barrier through heterogeneous nucleation [60], which enabled the crystallization behavior to start earlier, but also greatly improved the crystallization efficiency, and the crystallization period was thus greatly shortened. Compared with the 19.8 minutes crystallization period of PP-p, PP-β-0.4 took only 2.1 minutes from the beginning to the end.

Furthermore, separating and fitting the DSC heating curves in Figure 6.40b, and the crystallinity X_m and relative crystallization percentage K of the samples were calculated as shown in Figure 6.40d. By comparing the crystallinity of the samples shown in the upper part of Figure 6.40d, it can be concluded that within a certain addition ratio, the NA as the crystal nucleus effectively promoted the heterogeneous nucleation process within the dielectric. In this process, the molecular chains are more stacked in a regular arrangement to form the spherocrystal structure, rather than existing as amorphous regions arranged disorderly, and the overall crystallinity of the samples is thus remarkably raised. But, when the addition ratio is further increased to 0.4 wt%, the β-NA content is higher than the PP itself

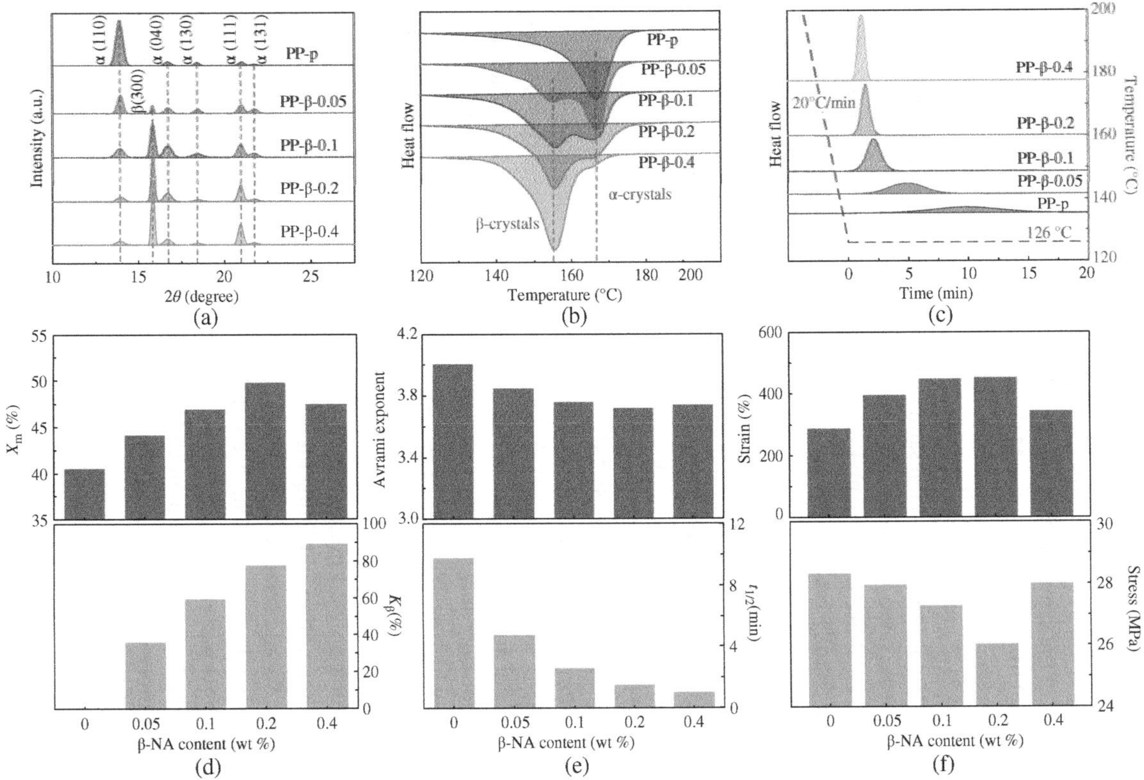

Figure 6.40 (a) X-ray diffraction pattern, (b) DSC curves within the heating melting stage, (c) DSC curves within the isothermal crystallization stage, (d) crystallinity X_m and β-crystal relative crystallization percentage $K_β$, (e) Avram exponent n and crystallization half-time $t_{1/2}$, and (f) strain and stress in tensile tests varying with β-NA content. *Source:* Wu et al. [71]/with permission from American Chemical Society.

solubility, and the excess NA will be agglomerated and precipitated as an impurity, which will also restrict the improvement of the crystal structure, resulting in a lower percentage of the crystal region within the spatial range.

There is competition between the two crystal phase structures of spherocrystals in the growth process within a limited spatial range. Therefore, the addition of β-NA promotes the formation of β-crystal while effectively reducing the percentage of α-crystal inside the sample. It is also proved by comparing the changes in the K values of the β-crystal shown in the lower part of Figure 6.40d. In particular, this promotion effect was especially significant when the β-NA addition ratio was controlled within 0.1 wt%.

For analyzing the influence mechanism of NA on the crystallization behavior within the materials, this study extracts the parameters of Avrami exponent n and the half-crystallization time $t_{1/2}$ from the isothermal crystallization curves in Figure 6.40c based on the Avrami equation, and the results are shown in Figure 6.40e. It can be seen that under the condition of isothermal crystallization, the average value of the n of PP-p is 4.01, which indicates its internal nucleation mode is homogeneous nucleation. With the addition of β-NA, the values of n in PP-β are all reduced to below 4. It means that the nucleation model within modified samples includes both homogeneous and heterogeneous nucleation modes [61], which also further proved that the addition of the NA promoted the heterogeneous nucleation inside the samples.

Compared with the homogeneous nucleation mode, heterogeneous nucleation provides more nuclei numbers for the crystallization within the samples, which greatly increases the crystallization rate. In comparison with the $t_{1/2}$ of about 9.8 minutes for PP-p, the whole process from the beginning of the crystallization to the relative crystallinity of 50% only takes about 1 minute in PP-β-0.4.

The strain and stress in tensile tests were measured by tensile test, and the results are shown in Figure 6.40f. As can be seen from it, with the addition of β-NA, the toughness of the samples was enhanced to different degrees. Among them, the toughening effect on the matrix material is most prominent when the addition ratio is 0.2 wt%. At this time, the elongation at break reached a maximum of 451%, which is increased by 164% in value compared to PP-p. Correspondingly, the tensile strength of the samples also decreased at the same time.

Based on previous studies, it has been known that the net-like wafer structure gives α-crystal greater rigidity, the parallel wafer structure makes the β-crystal lower in rigidity modulus [62]. Therefore, within a certain range, this range is 0.05–0.2 wt% in this paper. The higher the percentage of β-crystal, the more obvious the enhancement of material toughness. But, when the content of β-NA is further increased to 0.4 wt%, it exceeds the solubility of material. Excess NA will agglomerate and precipitate as impurities. As an inorganic material, NA has poor toughness compared with organic materials. Therefore, when its addition ratio is too high, it will lead to the decline of the mechanical properties of composites. Meanwhile, the agglomeration caused by too high content of NA will also reduce the number of crystal nuclei, leading to the excessive growth of single spherulite, and the large-sized spherulite effect will further reduce the overall mechanical toughness [63].

As one major form of insulation degradation in cables during operation, the process of start and growth during electric treeing is simultaneously accompanied by partial discharges. Figure 6.41a and b, respectively, shows the partial discharge signal and mesoscopic tree images within the samples collected by the measurement system when the AC voltage is applied for 20 minutes.

Among them, Figure 6.41a shows the PRPD spectrums obtained after processing, which reflects the partial discharge information inside the samples under the AC voltage signals varying in the form of sinusoidal waves. From it, it is not difficult to find that, with the sinusoidal form of the AC voltage signal changes periodically, a significant discharge behavior occurs within the sample and mainly occurs in the 0–90° and 180–270° two voltage rise stage. Considering that the electric field intensity is significantly enhanced in this stage, a large amount of energy is released by charge injection and recombination, which leads to the destruction of molecular chain structure, and the local defects formed further trigger the occurrence of partial discharge. Moreover, comparing different samples, it can be observed that the addition of β-NA effectively suppresses the frequency and intensity of partial discharges, and both the number and the amplitude of discharges in the samples are obviously reduced as a result.

Unlike the irregularity in the growth of the macroscopic electrical tree, the degradation channel of the mesoscopic electrical tree is closely related to its crystal structure, which has been proved by the mesoscopic electrical tree image shown in Figure 6.41b. It can be found that during the insulation degradation process, the electrical tree tends to grow along the amorphous region between the spherocrystal boundaries, which is especially reflected in the fact that the electrical tree turns significantly when it encounters the spherocrystals during its forward development along the direction of the electric field. This experimental phenomenon has also appeared in related studies. The addition of β-NA weakens the prominence of the boundary between the spherocrystals to some extent, and the electrical tree degradation channels become finer and shorter. Furthermore, the suppression effect of samples with different additive ratios on the electrical tree degradation is also consistent with the experimental results of partial discharges mentioned above, which verifies the validity of the test data to some extent.

Furthermore, extracting the parameters in the partial discharge signals, the data obtained are shown in Figure 6.42a and b. Therein, Figure 6.42a statistics the counts and amplitudes of the partial discharges of various samples during the

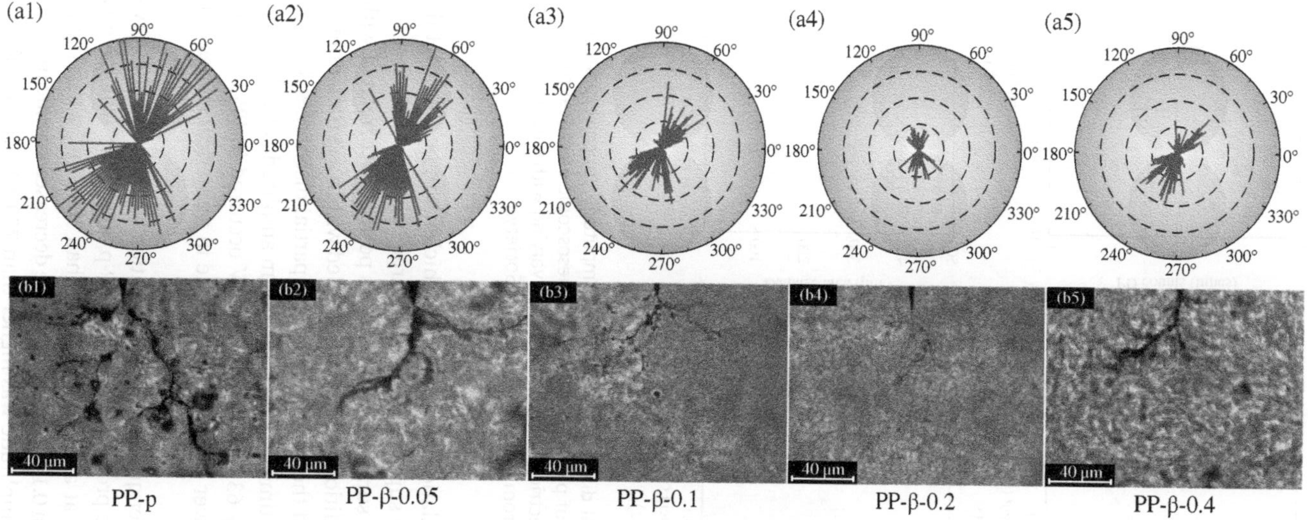

Figure 6.41 (a) Phase-resolved partial discharge spectrums and (b) mesoscopic electrical tree images, varying with β-NA content under 8 kV AC voltage at ambient temperature. *Source:* Wu et al. [71]/with permission from American Chemical Society.

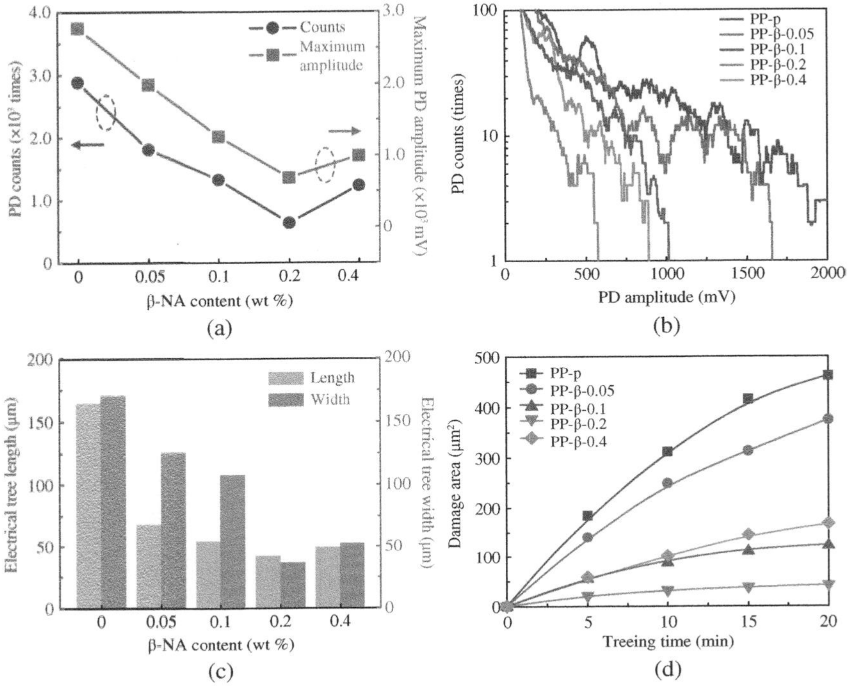

Figure 6.42 (a) Partial discharge counts and amplitude, (b) correspondence between counts and amplitude of partial discharge, (c) mesoscopic electrical tree length and width, and (d) mesoscopic electrical tree damage area varying with treeing time. *Source:* Wu et al. [71]/with permission from American Chemical Society.

electrical tree degradation process, from which it can be seen that the addition of the β-NA effectively suppressed the discharge behavior.

Among them, the suppression effect was positively correlated with the content of β-NA when its addition ratio was controlled within a certain range, and the sample with 0.2 wt% had the best resistance to partial discharges. Compared with the total counts of 2890 times and the maximum amplitude of 2779 mV of partial discharge in PP-p, only 632 times and 688 mV occurred in PP-β-0.2 samples, which are 78% and 75% lower than the former one in numerical values of the two parameters, respectively.

Figure 6.42b further quantifies the correlation between the above two parameters during the discharge process. Unlike the PP-p samples in which a great number of discharges occurred at amplitudes higher than 1000 mv, when the β-NA addition ratio was higher than 0.1 wt%, the amplitude decreased significantly, where the total number when the amplitude was higher than 750 mV was less than 10.

To quantify the degree of insulation degradation within the samples, while avoiding the unnecessary errors caused by the crystalline morphology to the statistical results, the mesoscopic electrical tree image shown in Figure 6.42b is further binarized in this study. The MATLAB software was used to process the image, counting the number of black pixels corresponding to the electrical tree channels, and referring to the scale bar to determine the size range of images. Finally, the length, width, and cumulative damage area of the electrical tree branches were calculated [64].

Among them, Figure 6.42c specifically quantifies the length and width of the electrical tree, where the former is parallel to the electric field direction, while the latter is perpendicular. It can be seen that with the increase of β-NA content, the values of both decrease monotonically in the range of 0–0.2 wt%, but when the content further increases to 0.4 wt%, there occurs a small increase on the contrary.

Moreover, Figure 6.42d compares the cumulative damage areas of the electrical tree degradation channels within different samples. Analyzing together with Figure 6.42b, it is not hard to find that the addition of β-NA refines the boundaries between the spherocrystals, thus avoiding the formation of degradation channels with thicker branches, and the cumulative damage area is also significantly reduced as a result. The cumulative damage area within the PP-β-0.02 sample was only 42 μm^2, which is about 90% less compared to 462 μm^2 before modification.

In order to reflect the breakdown strength of the material, the breakdown data of the samples under AC voltage were summarized and fitted, the obtained Weibull distribution is shown in Figure 6.43. The scale parameters and shape parameters

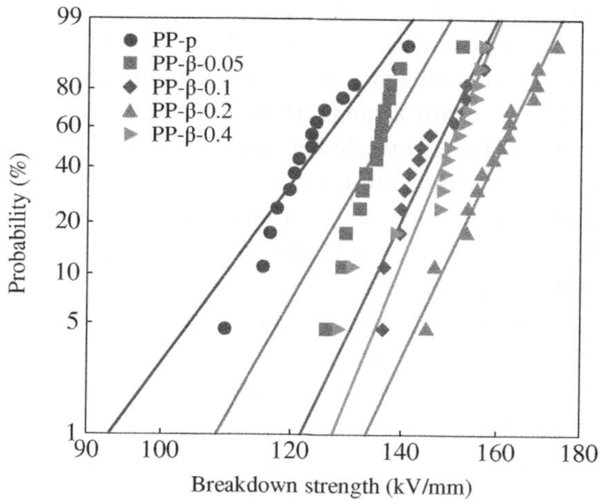

Figure 6.43 Weibull distribution of dielectric breakdown strength. *Source:* Wu et al. [71]/with permission from American Chemical Society.

Table 6.10 Weibull distribution parameters of samples.
Source: Wu et al. [71]/with permission from American Chemical Society.

Sample	Scale parameter (kV/mm)	Shape parameter
PP-p	127.6	14.53
PP-β-0.05	138.0	18.92
PP-β-0.1	149.3	22.56
PP-β-0.2	163.6	22.52
PP-β-0.4	151.5	26.52

are detailed in Table 6.10. Among them, the scale parameter corresponds to the characteristic breakdown strength of the samples. Comparing the values of this parameter between PP-p and PP-β, it is found that the addition of β-NA effectively improves the breakdown strength with a maximum enhancement of 28%.

At the same time, the shape parameter is also raised in value and can reach 26.52 at the maximum. From the mathematical point of view, the numerical value of shape parameters is negatively related to the dispersion degree of breakdown data. And, from the physical point of view, the weak point breakdown theory shows that the smaller the dispersion degree of breakdown data, the fewer the defects in the material. Therefore, the increase of shape parameters means that the insulation reliability of the modified material has been effectively improved in a certain range.

For investigating the mechanism of crystal phase transformation on the dielectric properties' enhancement of PP, the differences in intrinsic dielectric properties among α-crystal, β-crystal, and amorphous regions are characterized in this chapter based on the MDS method. The three different structures at the mesoscopic scale mentioned above were modeled separately based on the differences in their molecular chain arrangements, and their respective dielectric parameters were further obtained by calculations. On this basis, the obtained dielectric parameters are brought into the corresponding structures in the PP mesoscopic structure model constructed on the basis of the Voronoi diagram distribution principle. Finally, the electric field distribution at the mesoscopic scale is calculated using COMSOL physical field simulation software.

According to the relevant crystal parameters in Table 6.11 [65], two crystal phase models of α-crystal and β-crystal were constructed, as shown in Figure 6.44a1 and a2. The amorphous model in Figure 6.44a3 was obtained by geometrical optimization of the molecular model of single-chain PP. By comparison, it can be seen that, unlike the amorphous region model where the molecular chains are irregularly and loosely arranged, the molecular chains in the crystalline region structure

Table 6.11 Lattice parameter of crystal. *Source:* Wu et al. [71]/with permission from American Chemical Society.

Sample	Crystal system	Space group	Lattice length parameter (Å)	Lattice angle parameter (°)
α-Crystal	Monoclinic form	C2/C	a = 6.630 b = 20.78 c = 6.50	α = 90 β = 90.50 γ = 90
β-Crystal	Trigonal form	P31	a = 11.01 b = 11.01 c = 6.50	α = 90 β = 90 γ = 120

are periodically ordered and have a higher density per unit volume. Moreover, comparing the differences between the two crystal phases of α-crystal, and β-crystal, it can be seen that even though they are both in the same crystal structure, the former molecular chains are narrower and thus have a higher filling density in the same spatial range.

The models constructed above were sequentially annealed, simulated by molecular dynamics, and then obtained their DACF, as shown in Figure 6.44b. Previous studies have shown that a series of physical information related to the intrinsic properties of the material, including the relaxation strength and the permittivity, can be obtained by parametric extraction from the DACF of the model. Among them, the magnitude of the initial value of this function determines the intensity of the relaxation strength. By comparison, it can be found that in terms of the initial value magnitude, the amorphous region is 1.6×10^{26} Debye^2·m^{-3}, which is apparently lower than that of the crystal region. Furthermore, by comparing the two crystal phases, α-crystal and β-crystal, it can be found that the latter is lower than the former in the functional initial value of 0.8×10^{26} Debye^2·m^{-3}.

These differences are mainly related to the arrangement of the molecular chains per unit volume within the various structures. More regular and densely arranged molecular chains produce higher polarization strengths under the effect of an electric field, and the model with the above structures, therefore, has larger relative permittivity. Eventually, the relative permittivity of the α-crystal, β-crystal, and amorphous region in PP can be calculated to be 2.22, 2.13, and 2.06, respectively.

There are three different structural models of PP at the mesoscopic scale as shown in Figure 6.45. Among them, Figure 6.45a1 shows the sample of the PP-p before modification, at this time, the spherocrystals are all α-crystal. In comparison, Figure 6.45a2 corresponds to the PP-β-0.1 sample, which contains spherocrystals of both α- and β-crystal phases, and the proportion of the two in the overall crystal phase structure differs slightly. The model shown in Figure 6.45a3 represents the PP-β-0.2 sample in this paper, in which the crystal

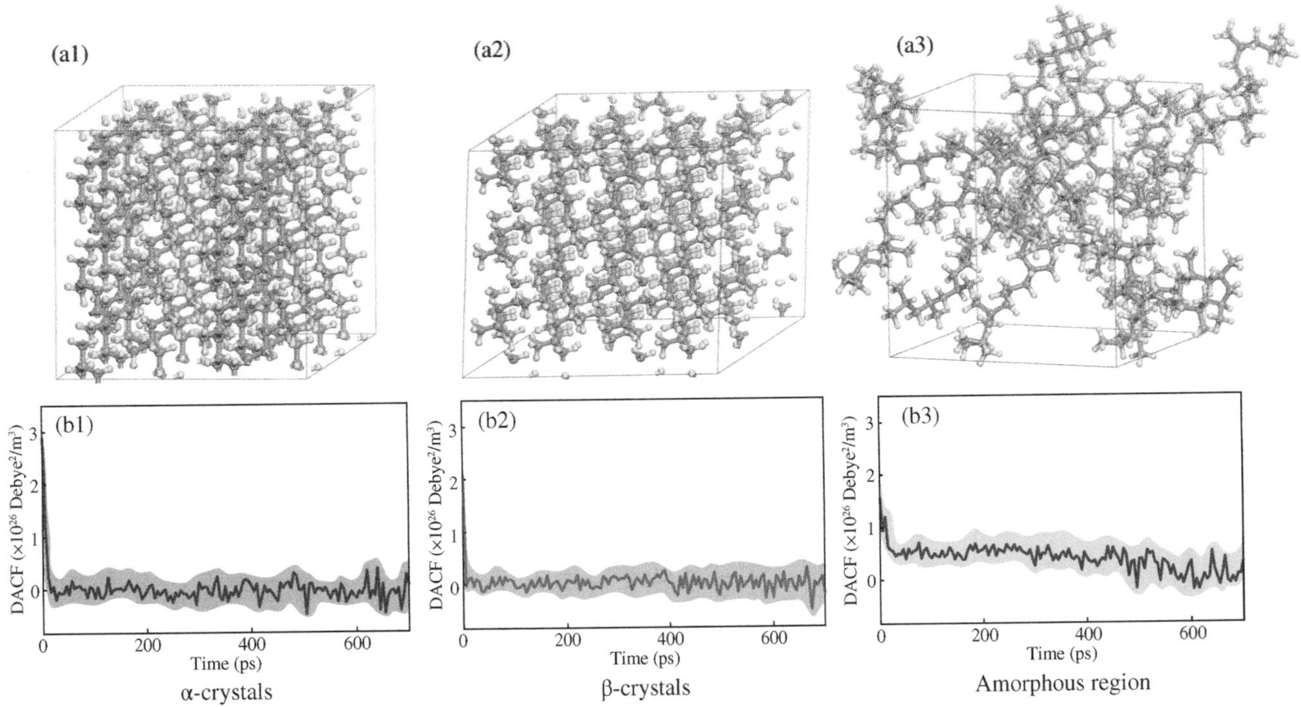

Figure 6.44 Heterogeneous mesoscopic structure: α-crystal, β-crystal, and amorphous regions (a) molecular configuration model. (b) Time-dependent curve of the dipole autocorrelation function. *Source:* Wu et al. [71]/with permission from American Chemical Society.

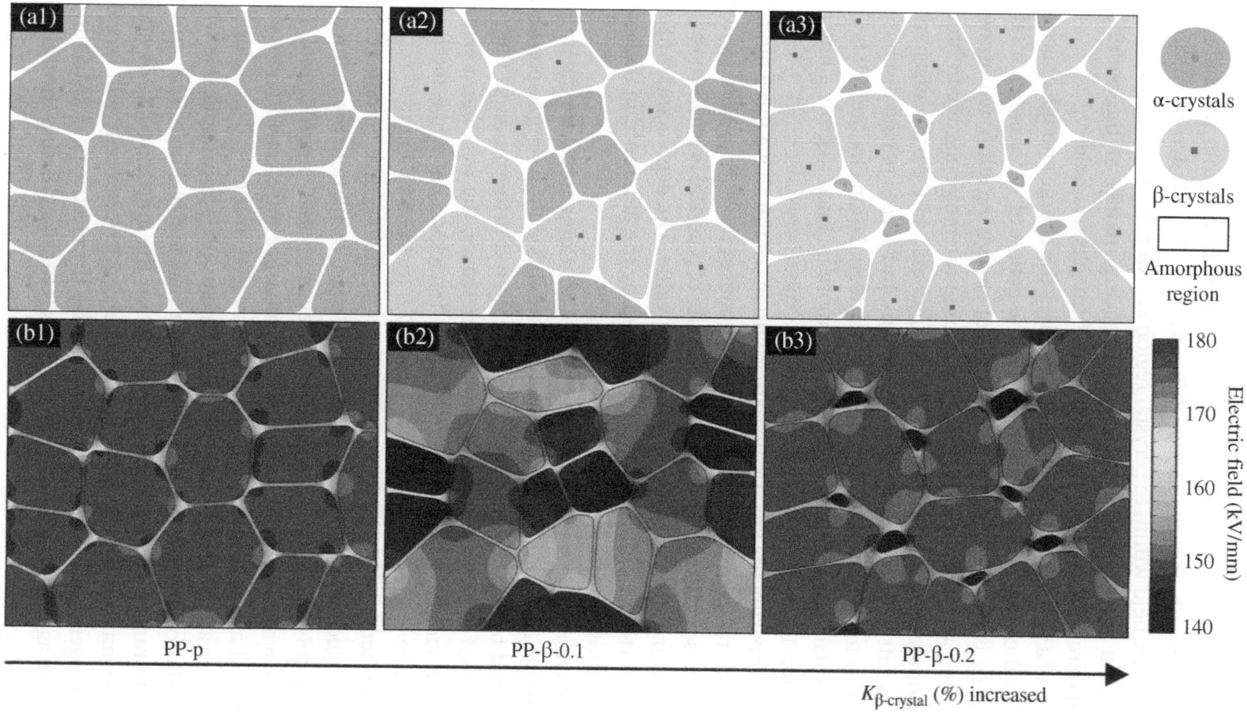

Figure 6.45 Mesoscopic heterogeneous structure of PP with various crystal phase structures: (a) models constructed based on the Voronoi diagram distribution principle and (b) electric field distribution obtained based on simulation under AC voltage. *Source:* Wu et al. [71]/with permission from American Chemical Society.

phase within the material has basically all been transformed from α-crystal to β-crystal.

The electric field distribution inside the dielectric at the mesoscopic scale is shown in Figure 6.45b. It can be noted that the electric field within the dielectric is not uniformly distributed due to the differences in the relative permittivity. Therein, the electric field within the crystal region is lower compared to the amorphous region, and the electric field distortion occurs mainly at the boundary between the spherocrystals along the direction of the electric field.

The presence of a partially strong electric field gives the charge enough energy to be injected into the material. Especially under the action of AC voltage, a large amount of energy is released during the process of charge injection, withdrawal, and compounding at a high frequency, which caused the molecular chain breakage and development of local defects in the polymers, and partial discharges occurred as a result [66]. Moreover, the continuous and severe partial discharges further contribute to the formation of electrical tree degradation channels. It is also reasonable to explain why the electrical tree was more inclined to grow along the amorphous region between the boundaries of the spherocrystals.

Further, by comparing the electric field distributions of the above three models, it can be seen that the introduction of β-crystal enhances the overall homogeneity of the dielectric to some extent since the permittivity of β-crystal is between that of α-crystal and the amorphous region in numerical value. The degree and range of electric field distortion within the dielectric caused by the difference in dielectric parameters are thus significantly reduced.

The above phenomenon is quantified in Figure 6.46. Compared with the unmodified PP-p sample containing only α-crystal, the modified PP-β sample possesses β-crystal obtained by the crystal phase transformation. Furthermore, as the proportion of β-crystal in the overall crystal phase increases, the maximum electric field E_{max} in the sample decreased from 183.6 to 166.8 kV/mm, while the corresponding breakdown strength increased from 127.6 to 163.6 kV/mm.

In view of the fact that electric field distortion is the main source of induced partial discharges, electrical tree degradation, and even insulation breakdown in dielectrics [67], this problem was effectively alleviated after modification. This is also the reason why both resistance to partial discharges and electrical trees were enhanced, as well as the breakdown strength increased in the PP-β samples.

It is necessary to add the point that the above conclusion is only applicable when the addition ratio of β-NA is controlled within a reasonable range, and, in this paper, this ratio ranges from 0.05 to 0.2 wt%. When the amount of β-NA added exceeds the maximum solubility of PP itself, the excess NA not only will not play the role of heterogeneous nucleation to homogenize the electric field distribution, but it will also be agglomerated to form impurities [68], exacerbating the intensity of partial discharges and the degree of electrical tree degradation. This explains

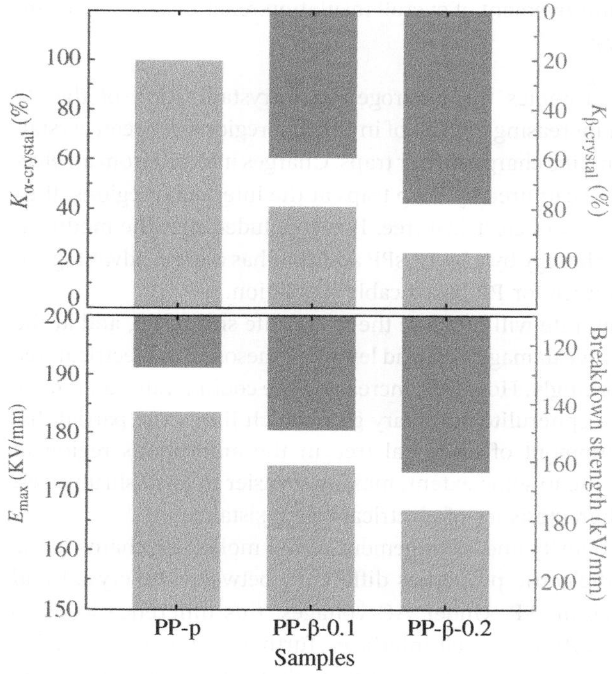

Figure 6.46 Effect of crystal phase structures on E_{max} and breakdown strength. *Source:* Wu et al. [71]/with permission from American Chemical Society.

why compared to PP-β-0.2, PP-β-0.4 samples showed a decrease in insulation properties despite a higher percentage of β-NA addition.

6.5 Conclusions

This chapter focuses on the electrical tree deterioration and insulation breakdown of PP cable insulation under AC voltage and studies the influence of material crystal structure on its electrical tree resistance characteristics from macro- and meso-scales, respectively. The differences in dielectric properties of PP insulation samples with different crystal structures were studied experimentally, and the molecular model of the corresponding structure was established to analyze the correlation mechanism between mesoscopic structure and macroscopic dielectric properties. On this basis, the directional control of PP cable insulation is carried out to reduce the difference of dielectric properties inside the insulation layer and realize the effective improvement of overall insulation reliability. The relevant conclusions are as follows:

1) The addition of sPP promotes the heterogeneous crystallization of the PP matrix, resulting in an increasing density of interfacial regions between crystalline regions, which contains charge carrier traps. Charges injected from an electrode into a polymer are captured by deep traps at the interfacial regions, thus inhibiting the propagation of electrical tree. It is concluded that the modification of crystalline morphology by 15 wt% sPP addition has a great advantage in electrical treeing resistance for PP-based cable insulation.
2) The decrease of cooling rate will increase the spherulite size of PP, and at the same time, the cumulative damage area and length of mesoscopic electrical tree will also increase accordingly. However, increasing the cooling rate can reduce the spherulite size and spherulite boundary gap, which limits the partial discharge and the development of electrical tree in the amorphous region at the interface of spherulite to some extent, making it easier to form shrub trees, thus improving the characteristics of electrical tree resistance.
3) The difference in the density and arrangement of the molecular chains is the main reason for the dielectric properties difference between the crystal and amorphous regions within PP. Under AC voltages, this difference leads to strong electric field distortions at their interfaces, further leading to partial discharge. The electrical trees thus tend to grow more along the amorphous region at the spherocrystal boundary.
4) The addition of an appropriate amount of nucleating agent acts as a heterogeneous nucleating agent, reducing the spherulite size in PP/POE blends and promoting dense spherulite stacking. Consequently, it significantly inhibits electrical treeing and partial discharge, reduces conductivity, and improves breakdown strength in PP/POE blends. Furthermore, the addition of the nucleating agent effectively reduces the differences in spherulite size and electrical properties caused by different cooling rates, enhancing the consistency of insulation properties.
5) The addition of β-NA-enabled heterogeneous nucleation replacing homogeneous nucleation as the dominant nucleation mode inside the dielectric. With the increase of β-NA content, β-crystal gradually dominated in the process of growth competition with α-crystal, and the proportion of it in the overall crystal system gradually increased, so the elongation at the break of the material was greatly improved. The density and the regularity of the molecular chains arranged within the β-crystal structure are between the α-crystal and amorphous region, so the introduction of β-crystal can effectively alleviate the problem of electric field distortion caused by the excessive dielectric difference, so that the insulation of the material can be improved.

In this chapter, based on the way of adding nucleating agent, the polypropylene cable insulation material is directionally controlled, which reduces the difference of crystal structure of materials at different radial positions of the insulation layer, improves the overall dielectric strength inside the insulation layer, reduces the difference of insulation performance at different positions, avoids the cable from electrical tree deterioration or even insulation breakdown, and ensures the safe and reliable operation of the cable.

References

1 Uehara, H. and Kudo, K. (2011). Directional properties of positive impulse tree propagation in oriented PET and PP barrier films molded in EVA. *IEEE Transactions on Dielectrics and Electrical Insulation* 18 (1): 162–167.

2 Zha, J., Wu, Y.H., Wang, S.J. et al. (2016). Improvement of space charge suppression of polypropylene for potential application in HVDC cables. *IEEE Transactions on Dielectrics and Electrical Insulation* 23 (4): 2337–2343.

3 Doedens, E.H. and Jarvid, E.M. (2020). Considerations on the impact of material mesostructure on charge injection at cable interfaces. *IEEE Electrical Insulation Magazine* 36 (5): 43–51.

4 Zhang, P., Zhang, Y., Wang, X. et al. (2021). Effect of acetylated SEBS/PP for potential HVDC cable insulation. *Materials* 14 (7).

5 Meng, P.F., Zhou, Y., Yuan, C. et al. (2019). Comparisons of different polypropylene copolymers as potential recyclable HVDC cable insulation materials. *IEEE Transactions on Dielectrics and Electrical Insulation* 26 (3): 674–680.

6 Zhang, C., Zha, J.W., Yan, H.D. et al. (2018). High improvement in trap level density and direct current breakdown strength of block polypropylene by doping with a beta-nucleating agent. *Applied Physics Letters* 112 (9).

7 Hosier, L., Vaughan, A.S., and Swingler, S.G. (2011). An investigation of the potential of polypropylene and its blends for use in recyclable high voltage cable insulation systems. *Journal of Materials Science* 46 (11): 4058–4070.

8 Zhao, Y., Vaughan, A.S., Sutton, S.J., and Swingler, S.G. (2001). On the crystallization, morphology and physical properties of a clarified propylene/ethylene copolymer. *Polymer* 42 (15): 6587–6597.

9 Rytoluoto, L., Gitsas, A., Pasanen, S., and Lahti, K. Effect of film structure and morphology on the dielectric breakdown characteristics of cast and biaxially oriented polypropylene films. *European Polymer Journal* 95: 606–624.

10 Yang, K., Liu, Y., Yan, Z. et al. (2020). Enhanced morphology-dependent tensile property and breakdown strength of impact polypropylene copolymer for cable insulation. *Materials* 13 (18): 3935.

11 Su, J.G., Du, B., Li, J., and Li, Z. (2020). Electrical tree degradation in high-voltage cable insulation: progress and challenges. *High Voltage* 5 (4): 353–364.
12 Ran, Z.Y., Du, B.X., Xiao, M., and Li, J. (2021). Crystallization morphology-dependent breakdown strength of polypropylene films for converter valve capacitor. *IEEE Transactions on Dielectrics and Electrical Insulation* 28 (3): 964–971.
13 Ran, Z.Y., Du, B.X., Xiao, M. et al. (2021). Effect of crystallization regulation on the breakdown strength of metallized polypropylene film capacitors. *IEEE Transactions on Dielectrics and Electrical Insulation* 28 (1): 175–182.
14 Zhou, S.F., Fan, M., Li, Z., et al. (2021). Effect of β-spherulite on electrical tree characteristics of isotactic polypropylene insulation. *3rd International Conference on Electrical Materials and Power Equipment (ICEMPE)*, 18 August 2021, Chongqing, China: IEEE, 1–4.
15 Du, B.X., Zhu, L.W., and Han, T. (2017). Effect of ambient temperature on electrical treeing and breakdown phenomenon of polypropylene with repetitive pulse voltage. *IEEE Transactions on Dielectrics and Electrical Insulation* 24: 2216–2224.
16 Jeziorny, A. (1978). Parameters characterizing the kinetics of the non-isothermal crystallization of poly(ethyleneterephthalate) determined by D.S.C. *Polymer* 19: 1142–1144.
17 Shi, Z., Zhang, X., Wang, K. et al. (2017). Investigation on electrical tree propagation in polyethylene based on etching method. *AIP Advances* 7: 115012.
18 Teyssedre, G. and Laurent, C. (2005). Charge transport modeling in insulating polymers: from molecular to macroscopicscale. *IEEE Transactions on Dielectrics and Electrical Insulation* 12: 857–875.
19 Vaughan, A.S., Hosier, I.L., Dodd, S.J., and Sutton, S.J. (2006). On the structure and chemistry of electrical trees in polyethylene. *Journal of Physics D: Applied Physics* 39: 962–978.
20 Kao, K.C. (1984). New theory of electrical discharge and breakdown in low-mobility condensed insulators. *Journal of Applied Physics* 55: 752–755.
21 Alapati, S., Meledath, J.T., and Karmarkar, A. (2014). Effect of morphology on electrical treeing in low density polyethylene nanocomposites. *IET Science, Measurement and Technology* 8: 60–68.
22 Yan, Z., Yang, K., Wang, S. et al. (2020). Electrical tree in silicone rubber: roles of silicone grease and switching impulses. *IEEE Transactions on Dielectrics and Electrical Insulation* 27: 829–836.
23 Chen, G. and Tham, C.H. (2009). Electrical treeing characteristics in XLPE power cable insulation in frequency range between 20 and 500 Hz. *IEEE Transactions on Dielectrics and Electrical Insulation* 16 (1): 179–188.
24 Du, B.X., Li, Y.M., Han, C.L. et al. (2021). Graphene nanoplatelets and temperature gradient affecting electrical tree in graphene/SiR nanocomposites. *IEEE Transactions on Dielectrics and Electrical Insulation* 28 (4): 1255–1263.

25 Zhang, P.Z., Wang, H., Li, Z. et al. (2021). Molecule diffusion behaviours of waterproof sealants into silicone rubber insulation for submarine cable joint based on molecular dynamics simulations. *High Voltage* 6 (2): 230–241.
26 Xu, J.F., Yan, P., Yoshimura, N., and Nakagawa, Z. Morphology of polypropylene and tree initiation. *Japanese Journal of Applied Physics* 37 (7): 4021–4026.
27 Zhou, S.F., Yu, F., Yang, W. et al. (2021). Effect of sPP content on electrical tree growth characteristics in PP-blended cable insulation. *Materials* 13 (23): 5360.
28 Li, Z.L., Dai, F., Wu, Y., and Du, B. (2023). Effect of crystalline morphology on electrical tree morphology and growth characteristics of PP insulation: from mesoscopic to macroscopic. *IEEE Transactions on Dielectrics and Electrical Insulation* 30 (3): 989–996.
29 Ma, J.S., Zhang, S., Qi, Z. et al. (2002). Crystallization behaviors of polypropylene/montmorillonite nanocomposites. *Journal of Applied Polymer Science* 83 (9): 1978–1985.
30 Yun-Hui Wu, J.W., Zha, W.K.L. et al. (2015). A remarkable suppression on space charge in isotactic polypropylene by inducing the β-crystal formation. *Applied Physics Letters* 107 (11): 112901.
31 Zhang, W., Xu, M., Huang, K. et al. (2019). Effect of β-crystals on the mechanical and electrical properties of β-nucleated isotactic polypropylene. *IEEE Transactions on Dielectrics and Electrical Insulation* 26 (3): 714–721.
32 Xiao, M., Zhang, M., and Du, B. (2023). Crystallisation regulation of long-chain branched polypropylene on dielectric performance and energy density for metallised film capacitors. *High Voltage* 8 (5): 921–930.
33 Ran, Z., Du, B., Xiao, M., and Liu, H. (2022). Breakdown performance of polypropylene and its long-chain branched blending for metalized film capacitors. *IEEE Transactions on Dielectrics and Electrical Insulation* 29 (1): 145–152.
34 Xiao, M., Zhang, M., Du, B. et al. (2022). Breakdown strength improvement of polypropylene by introducing long-chain branched structures for film capacitors. *Journal of Physics D: Applied Physics* 55 (24): 245501.
35 Romankiewicz, A., Sterzynski, T., and Brostow, W. (2004). Structural characterization of α- and β-nucleated isotactic polypropylene. *Polymer International* 53 (12): 2086–2091.
36 Gui, Q., Xin, Z., Zhu, W., and Dai, G. (2003). Effects of an organic phosphorus nucleating agent on crystallization behaviors and mechanical properties of poly(propylene). *Journal of Applied Polymer Science* 88 (2): 297–301.
37 Furushima, Y., Toda, A., Nakada, M. et al. (2022). Enthalpy relaxation of unconstrained and constrained amorphous phase for low isotacticity polypropylene. *Polymer* 253: 124991.
38 Tanaka, T. (1992). Charge transfer and tree initiation in polyethylene subjected to AC voltage stress. *IEEE Transactions on Electrical Insulation* 27 (3): 424–431.

39 Zhu, X., Wu, J., Wang, Y., and Yin, Y. (2020). Characteristics of partial discharge and AC electrical tree in XLPE and MgO/XLPE nanocomposites. *IEEE Transactions on Dielectrics and Electrical Insulation* 27 (2): 450–458.

40 Laghari, J.R., Cygan, P., and Khechen, W. (1990). A short method of estimating lifetime of polypropylene film using step-stress tests. *IEEE Transactions on Electrical Insulation* 25 (6): 1180–1182.

41 Lao, S., He, H., Wang, L., and Yin, T. (2021). Simulation and analysis of polypropylene dielectric constant based on molecular simulation method. *22nd International Symposium on High Voltage Engineering*, 26 November 2021, Xi'an, China: IET, 1831–1834.

42 Nagarajan, K. and Myerson, A.S. (2001). Molecular dynamics of nucleation and crystallization of polymers. *Crystal Growth & Design* 1 (2): 131–142.

43 Cheng, Y.H., Zhang, K., Xie, X., and Yu, S. (2011). Research on dielectric property of polyethylene with defects based on molecular dynamics. *IEEE International Symposium on Applications of Ferroelectrics (ISAF)*, 08 September 2011. Vancouver, BC: IEEE, 1–4.

44 Li, J.Y., Yang, K., Wu, K.N. et al. (2023). Eco-friendly polypropylene power cable insulation: present status and perspective. *IET Nanodielectrics* https://doi.org/10.1049/nde2.12048.

45 Ran, Z.Y., Du, B.X., and Xiao, M. (2021). High-temperature breakdown performance improvement of polypropylene films based on micromorphology control. *IEEE Transactions on Dielectrics and Electrical Insulation* 28 (5): 1547–1554. https://doi.org/10.1109/TDEI.2021.009583.

46 Supaphol, P., Dangseeyun, N., and Srimoaon, P. (2004). Non-isothermal melt crystallization kinetics for poly(trimethylene terephthalate)/poly(-butylene terephthalate) blends. *Polymer Testing* 23 (2): 175–185. https://doi.org/10.1016/j.polymertesting.2015.04.002.

47 Ketsamee, P., Andritsch, T., and Vaughan, A. (2023). The effects of humidity on dielectric permittivity of surface-modified TiO_2- and MgO-based polypropylene nanocomposites. *IEEE Transactions on Dielectrics and Electrical Insulation* 30 (1): 82–89. https://doi.org/10.1109/TDEI.2022.3209640.

48 Zhou, S.F., Fan, M.S., Li, Z.L. et al. (2021). Effect of β-spherulite on electrical tree characteristics of isotactic polypropylene insulation. *3rd International Conference on Electrical Materials and Power Equipment (ICEMPE)*, 18 August 2021. Chongqing, China: IEEE, 1–4. https://doi.org/10.1109/ICEMPE51623.2021.9509207.

49 An, Y., Wang, S., Li, R. et al. (2019). Effect of different nucleating agent on crystallization kinetics and morphology of polypropylene. *e-Polymers* 19 (1): 32–39. https://doi.org/10.1515/epoly-2019-0005.

50 Kim, D.W. and Yoshino, K. (2000). Morphological characteristics and electrical conduction in syndiotactic polypropylene. *Journal of Physics D: Applied Physics* 33 (4): 464–471. https://doi.org/10.1088/0022-3727/33/4/321.

51 Cho, K., Li, F.K., and Choi, J. (1999). Crystallization and melting behavior of polypropylene and maleated polypropylene blends. *Polymer* 40 (7): 1719–1729. https://doi.org/10.1016/S0032-3861(98)00404-2.

52 Du, B.X., Hou, Z.H., Li, Z.L., and Li, J. (2019). Temperature dependent space charge and breakdown strength of PP/ULDPE/graphene nanocomposites for HVDC extruded cable insulation. *IEEE Transactions on Dielectrics and Electrical Insulation* 26 (3): 876–884. https://doi.org/10.1109/TDEI.2019.007858.

53 Du, B.X., Hou, Z.H., Li, J., and Li, Z.L. (2018). Effect of graphene nanoplatelets on space charge and breakdown strength of PP/ULDPE blends for HVDC cable insulation. *IEEE Transactions on Dielectrics and Electrical Insulation* 25 (6): 2405–2412. https://doi.org/10.1109/TDEI.2018.007271.

54 Kim, D.K., Lee, S.H., Kwon, T.H. et al. (2023). Study on high-temperature and high-voltage insulation characteristics of polypropylene blend with highly packed elastomeric domains for power cable applications. *Polymer Testing* 120: 107942.

55 Li, Z., Zhong, Z., and Du, B. (2019). Dielectric relaxation and trap-modulated DC breakdown of polypropylene blend insulation. *Polymer* 185: 121935.

56 Vallentyne, P. (1997). Intrinsic properties defined. *Philosophical Studies: An International Journal for Philosophy in the Analytic Tradition* 88 (2): 209–219.

57 Binsbergen, F.L. and De Lange, B.G.M. (1968). Morphology of polypropylene crystallized from the melt. *Polymer* 9: 23–40.

58 Wang, Z.G., Hsiao, B.S., Srinivas, S. et al. (2001). Phase transformation in quenched mesomorphic isotactic polypropylene. *Polymer* 42 (18): 7561–7566.

59 Liu, F., Guo, C., Wu, X. et al. (2012). Morphological comparison of isotactic polypropylene parts prepared by micro-injection molding and conventional injection molding. *Polymers for Advanced Technologies* 23 (3): 686–694.

60 Li, J.X. and Cheung, W.L. (1997). Effect of mould temperature on the formation of α/β polypropylene blends in injection moulding. *Journal of Materials Processing Technology* 63 (1–3): 472–475.

61 Nagarajan, K., Levon, K., and Myerson, A. (2000). Nucleating agents in polypropylene. *Journal of Thermal Analysis and Calorimetry* 59 (1–2): 497–508.

62 Yazdi, M., Asl, V.H., Pourmohammadi, M., and Roghani-Mamaqani, H. (2019). Mechanical properties, crystallinity, and self-nucleation of carbon nanotube-polyurethane nanocomposites. *Polymer Testing* 79: 106011.

63 Chen, H.B., Karger-Kocsis, J., Wu, J.S., and Varga, J. (2002). Fracture toughness of α-and β-phase polypropylene homopolymers and random-and block-copolymers. *Polymer* 43 (24): 6505–6514.

64 Huan, Q., Zhu, S., Ma, Y. et al. (2013). Markedly improving mechanical properties for isotactic polypropylene with large-size spherulites by pressure-induced flow processing. *Polymer* 54 (3): 1177–1183.

65 Ferro, D.R., Meille, S.V., and Brückner, S. (1998). Energy calculations for isotactic polypropylene: a contribution to clarify the β crystalline structure. *Macromolecules* 31 (20): 6926–6934.

66 Tanaka, T. and Greenwood, A. (1978). Effects of charge injection and extraction on tree initiation in polyethylene. *IEEE Transactions on Power Apparatus and Systems* 5: 1749–1759.

67 Li, Z., Wu, Y., Zheng, Z., and Du, B. (2024). Temperature-gradient-dependent space charge and electric field evolutions of ± 500 kV HVDC cables with different thicknesses. *IEEE Transactions on Dielectrics and Electrical Insulation* 31 (3): 1620–1627.

68 Langhe, D.S., Keum, J.K., Hiltner, A., and Baer, E. (2011). Fractionated crystallization of α-and β-nucleated polypropylene droplets. *Journal of Polymer Science Part B: Polymer Physics* 49 (2): 159–171.

69 Wu, Y., Du, B., Li, Z., Wang, H., & Zheng, Z. (2023). Heterogeneous mesoscopic structure modulation enhancing dielectric strength of polypropylene insulation. *Polymer*, 280, 126072. https://doi.org/10.1016/j.polymer.2023.126072.

70 Du, B., Liu, H., & Li, Z. (2024). Effect of Nucleating Agent and Cooling Rate on Dielectric Property of PP/POE Cable Insulation. *IEEE Transactions on Dielectrics and Electrical Insulation*, 31 (2), 929–936. https://doi.org/10.1109/tdei.2023.3330694.

71 Wu, Y., Li, Z., Wang, H., Zheng, Z., & Du, B. (2024). Enhanced Dielectric Breakdown Property of Polypropylene Based on Mesoscopic Structure Modulation by Crystal Phase Transformation for High Voltage Power Cable Insulation. *ACS Applied Polymer Materials*, 6 (5), 3031–3041. https://doi.org/10.1021/acsapm.4c00251.

7

Electrothermal Aging and Lifetime Modeling

7.1 Introduction

In HVAC cable, the conductor temperature varies with time, but the electric field in insulation is constant [1]. In order to consider the above problems, a model of the influence of stress on the insulation lifetime of power cables is proposed, that is the "lifetime model" of power cables. The "lifetime model" of power cables is the mathematical relationship between the "life" of cable insulation and the level of stress applied. Generally speaking, the stress is inversely proportional to the life, i.e. the higher the stress, the shorter the lifetime. We usually consider that the stresses include: electrical stress E', mechanical stress M', thermal stress T, and environmental stress A'.

E' is generated by voltage, we usually use electric field strength instead of voltage in lifetime models.

M' is produced by various factors that can cause mechanical deformation such as tension and coiling and bending.

T' is generated by temperature, thermal stress is directly related to temperature, so the temperature is commonly used to characterize thermal stress.

A' is related to aging factors in the laying environment, any factor related to the environment will affect the environmental stress, we often consider corrosion, humidity, ultraviolet radiation, etc.

Lifetime models are usually concluded from the accelerated life test (ALT) under higher than rated stress, but it is difficult to evaluate the model parameters under high stress. Most HVDC cables have the following three-lifetime models:

- Single electric stress lifetime model, generated only under voltage or electric field.
- Single thermal stress lifetime model, determined by temperature.
- Comprehensive electrothermal lifetime model, voltage electric field, and temperature are both considered at the same time.

Polypropylene Cable Insulation, First Edition. Boxue Du and Zhonglei Li.

Published 2025 by John Wiley & Sons, Inc.

The lifetime model, also referred to as the reliability model, is a critical concept in understanding insulation failure. This failure can be attributed to various factors such as charge injection, transportation, storage, dielectric loss, local field strength concentration, overheating of the conductor or interface, partial discharges in air, Maxwell stress, water tree, and thermal mechanical stress. It is hypothesized that the time of failure is associated with the failure probability F (or reliability probability $R = 1 - F$). By presuming the probability distribution of failure time, lifetime models evolve into comprehensive probabilistic lifetime models (or reliability models). These models can predict the lifespan of the insulation and disclose the relationship between applied stress and cumulative failure probability or reliability.

Lifetime models can be divided into two categories: physical (microscopic) lifetime models and phenomenological lifetime models. To put it simply, the physical lifetime models refer to the microcosmic interpretation of aging, while the phenomenological lifetime models refer to the discovery of phenomena and the summary of rules through ALT tests. The biggest difference between the physical lifetime models and the phenomenological lifetime models is that the phenomenological lifetime models refine the life curve and have different models for different insulation materials [2].

7.2 Aging Mechanism and Lifetime Models

7.2.1 Physical Lifetime Models

The physical lifetime model focuses on the local micro-defects in insulation as the cause of aging. It attempts to describe the evolutions of damages from insulation defects, such as cavities, conductive particles, and protuberances and then provides expressions for damage growth breakdown or occurrence of different mechanisms [3]. The general physical lifetime model is derived from the aging model, which is used to describe the chemical–physical process of local degradation in order to predict the time to reach the critical level of aging as the lifetime. The definition of lifetime in the physical model is the time from the start of operation to the generation of electrical branches. We can note that the lifetime ignores the time from the electrical branch to the breakdown, because the transfer of energy in the electrical branch and, therefore, the growth of the damaged area does not accelerate in time. Figure 7.1 shows the differences between electrical branch development and breakdown:

The followings will introduce the three main types of physical models: thermodynamic models, space-charge-based models, and models based on PD-induced damage.

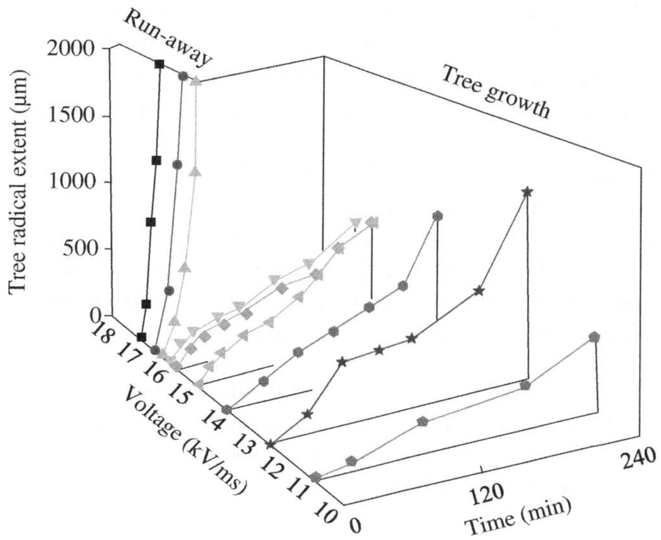

Figure 7.1 The difference between electrical branch growth and breakdown. *Source:* Dissado [4]/with permission from IEEE.

7.2.1.1 Thermodynamic Models

According to the thermodynamic model, aging under the dual action of electric field E and temperature T is thermal activation [5].

The specific explanation is as follows: in the curve shown in Figure 7.2, the solid line shows that no electric field is added, and the dashed line shows that the electric field is added.

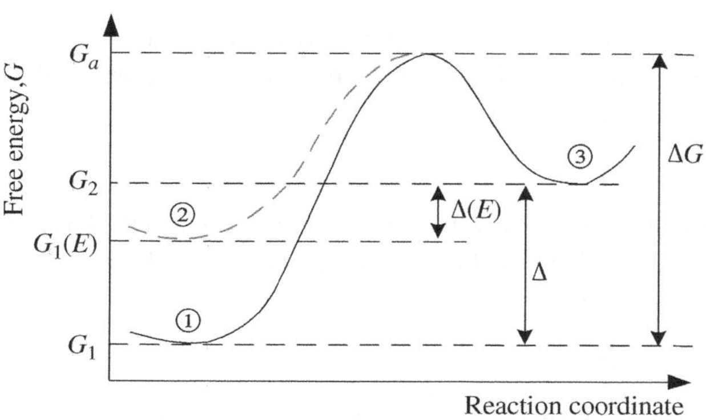

Figure 7.2 Free energy diagram: the solid line is in no electric field and the dashed line is in the electric field. *Source:* Dissado [6]/with permission from IEEE.

1) Regardless of whether there is an electric field, the activation-free energy barrier ΔG must be overcome to change the insulation from the reactant state (unaged state, free energy G_1, and reaction coordinate c_1) to the product state (aged state, free energy G_2, and reaction coordinate c_2). And, the free energy at the top of the energy barrier is G_a.
2) Adding electric field results in locally stored electromechanical energy, which increases the free energy of the original unaged state (G_1 to $G_1(E)$). The aging state free energy G_2 does not change, because we assume that the locally stored electromechanical energy disappears once the transition state G_a is entered. Therefore, we can find that the addition of an electric field leads to a decrease of ΔG, so that the process from unaged to aged is easier.

The main thermodynamic models for HVDC cable insulation are as follows. Among them, the Arrhenius model and the Eyring model can be considered as both physical models and phenomenological models (as phenomenological models, they will be explained in Section 7.2.2.2).

The Eyring Model

The main aging mechanism of dielectrics is considered to be chemical. The Eyring model holds that the aging rate is exponentially related to energy activation value ΔG and temperature T. By quantum mechanics, Eyring theoretically deduces a correct equation in which the deterioration rate of electronic components (including power cables, etc.) is proportional to the rate of chemical reactions that control the main failure mechanisms. The expression is as follows [7]:

$$L(T) = \frac{h_P}{k_B T} e^{\frac{\Delta G}{k_B T}} \tag{7.1}$$

where h_P is the Planck constant, k_B is the Boltzmann constant, and ΔG is the activation free energy barrier.

Through the Eyring model of reaction rate, it is proved that the relationship ($S_T{}^n t$ is *constant*, S_T is the thermal stress, and t is lifetime), which is usually considered to be completely empirical, has a theoretical basis.

The Endicott Model

The model is obtained from the single-stress thermal Eyring model. On the basis of the original, it introduced an ad hoc field-dependent free energy barrier. The expression is as follows (7.2) [2, 3]:

$$L(E,T) = C_E T^{-w} e^{\frac{\Delta G}{k_B T} - E\left(k_1 + \frac{k_2}{T}\right)} \tag{7.2}$$

where $L(E,T)$ is the "electric-thermal" lifetime of the insulation, which is the function of the electric field strength E and absolute temperature T; ΔG is the activation free energy barrier; k_B is the Boltzmann constant; C_E is a parameter that depends on the dielectric which usually depends on E and T; w, k_1, and k_2 are the quantities that primarily characterize the dielectric.

As we can see, the Endicott model can be obtained by letting $E = 0$, $w = 1$, and $C_E = h_P/k_B$ in the Endicott model.

The Lewis Model

T. J. Lewis believes that mechanical stress also plays an important role in insulation aging. In the case of the electrical failure in solid insulation, it can be considered that mechanical stress changes the shape of the dielectric, resulting in changes in electrical strength and breakdown activity. Therefore, it is necessary to consider the relationship between mechanical stress and electrical stress.

Using the theory of electro-induced mechanical forces, and inferring the significant mechanical stresses that can be expected to act in the plane perpendicular to the electric field line, the Lewis model is derived. It is based on the assumption that the reason why electromechanical stress causes insulation failure and electrical tree is the expansion of the existing small micro-cavity. The propagation of electric trees is described by the Griffith crack theory. Lifetime is the time from the start of operation to the beginning of the crack, assuming that it is the longest time interval of the total aging duration, and, therefore, a good approximation of the total failure time is [8]:

$$t_C = \int_{N_b}^{\eta_C} \left\{ \frac{k_B T}{h_P} \left[(N_b - \eta) e^{\frac{-G_b(E)}{kT}} - \eta e^{\frac{-G_f(E)}{k\Theta}} \right] \right\}^{-1} d\eta \tag{7.3}$$

where h_P is the Planck constant, T is the absolute temperature, k is the Boltzmann constant, and G_f and G_b are the energies for bond formation and bond breaking, respectively (related to G_1 and G_2 in Figure 7.2). N_b is the number of bonds that can be broken; η is the number of broken bonds whose critical value is η_C.

The Lewis model has a field threshold below which aging does not occur and the lifetime tends to infinity.

The Arrhenius Model

Arrhenius model is the most popular thermal lifetime model, which is suitable not only for HVDC cable insulation but also for electrical insulation. What is more, it is not only the physical lifetime model, but also the phenomenological lifetime model, because it fits well the results of ALTs with single thermal stress for lots of materials and at a wide range of temperatures. According to Dakin's theory [9], temperature accelerates the rate of thermally activated chemical degradation of dielectrics exponentially, with oxidation as the beginning, then depolymerization, then hydrolysis, and cross-linking [10]. Assuming that there is a main degradation reaction (activation energy ΔW) for dielectric in the temperature range considered, the expression of Arrhenius thermal lifetime model is [9]:

$$L(T) = R_T e^{\frac{\Delta W}{k_B T}} \tag{7.4}$$

where R_T is a material-dependent constant, which is inversely proportional to the reaction rate constant, ΔW is the activation energy, k_B is the Boltzmann constant, and T is the absolute temperature.

7.2.1.2 Space-Charge-Based Models

The understanding of the space charge effect depends on how to understand concurrent phenomena to a large extent, such as thermal and electrical aging [11]. At the same time, as ultra clean, dry materials have been used in high-voltage insulation systems, the level of defects that may lead to degradation is approaching the microscopic or submicroscopic scale [6]. At present, the mainstream space charge models are the DMM model and "trapping/de-trapping" space charge model.

DMM Model

DMM model is an aging lifetime model proposed by L. A. Dissado, G. Mazzanti, and G. C. Montanari at the micro level. The model mainly studies the aging process of polymer-insulating materials under thermal–electrical stress. Based on the method of thermodynamics, the effect of storage energy of space charge on accelerating the aging rate is considered. The derived model is the result of multi-stress lifetime test on polyethylene terephthalate (PET) film under DC voltage. The model can also fit the lifetime data obtained from AC aging very well. In these cases, the parameters of the model have a certain physical meaning [6].

The DMM model is based on the following assumptions:

I) Assume that the insulating weakness in the HVDC cable is a spherical micron or nanometer void with a radius of r_V.
II) The amount of charge that can be stored in the spherical void is $q_c = C_q E^{bq}$, where C_q and bq are parameters related to the space charge accumulation characteristics of the dielectric.
III) At the center of space charge, the effect of Poisson field on space charge is greater than that of applied field E on DC voltage.
IV) Chemical bonds and cross-linked bonds are considered to be degradation microstructural unit (the number of it is N_m) located in the shell with thickness λ on the void boundary.
V) Degradation proceeds as a competition between forward and reverse reactions, both of which are partially reversible.

VI) Once the value A of the degraded part exceeds a critical value A^* depending on the dielectric, an electrical branch is considered to be generated, that is the termination of life.

According to the above model assumption, the electrothermal lifetime expression of DMM model can be obtained (7.6) [12, 13]:

$$L(E,T) = \frac{h_P}{2k_B T} e^{\frac{\Delta H/h_B - C' E^{2bq}/2}{T} - \frac{\Delta S}{k_B}} \ln\left[\frac{A_{eq}(E)}{A_{eq}(E) - A^*}\right] \left[\cosh\left(\frac{\Delta/k_B - C' E^{2bq}}{2T}\right)\right]^{-1} \tag{7.5}$$

$$A_{eq}(E) = \frac{1}{1 + e^{\frac{\Delta/k_B - C' E^{2bq}}{T}}} \tag{7.6}$$

$$C' = \frac{B_q \alpha_E \delta_S \lambda C_q^2}{16\pi\varepsilon^2 r_V^2 N_m k_B} \tag{7.7}$$

where ΔH is the enthalpy of activation of each moiety, ΔS is the activation entropy of each moiety, Δ is the free energy change between the unaged and aged states (Figure 7.2), $A_{eq}(E)$ represents the value of A at equilibrium between the forward reaction and the reverse reaction, C' is the typical constant of the dielectric, the proportionality constant B_q is between the electromechanical energy stored by each moiety and its contribution to the degradation free energy barrier, α_E is the electrostriction coefficient, δ_S is the elemental strain, and k_B is the Boltzmann constant.

The DMM model also has an electrical threshold. L. A. Dissado et al. show us that DMM model fitted the ALT results for 0.15 mm XLPEs of high-voltage DC cable (DC-XLPE) under DC voltage at 20 °C, as shown in Figure 7.3 [12].

"Trapping/De-trapping" Space Charge Model

After DMM model, an updated lifetime model based on physical space charge is proposed, which focuses on polymer insulation and emphasizes the relationship between manufacturing features in polymers (e.g. free volume, sub-micropores, and low-density regions) and the process of trapping/de-trapping of space charge. The following are the assumptions and processes:

I) The polymer insulation of HVDC-extruded cable is nonuniform, which has a mixed crystalline/amorphous structure composed of many parts. It is believed that each part has different nature.
II) Amorphous phase has more free volume and space than the crystalline phase, so the injected electrons can travel more easily than in the crystalline phase. It is believed that the process of obtaining energy for electrons in the amorphous phase under electric field is similar to that in gas [14].

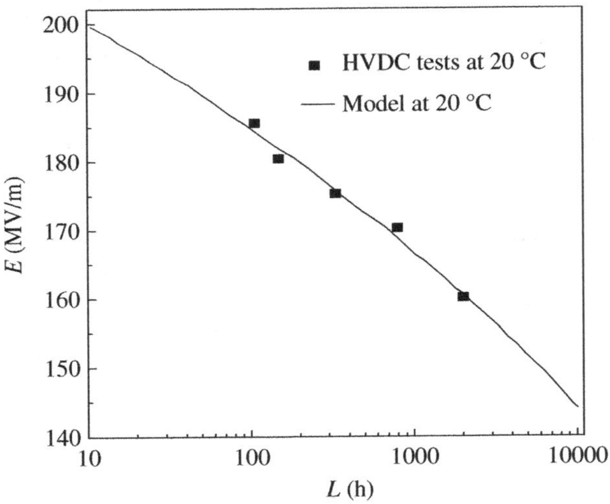

Figure 7.3 DMM model and HVDC lifetime test results. *Source:* Dissado [12]/with permission from IEEE.

III) Charge dynamics during electrothermal aging is defined by the trapping and de-trapping of electrons in the free volume according to a set of kinetic Eq. (7.9) [15]:

$$\begin{cases} \dfrac{\mathrm{d}n_{\mathrm{T}\rightarrow}(t)}{\mathrm{d}t} = K_{\mathrm{T}}[N_{\mathrm{T}} - n_{\mathrm{T}}(t)] \\ \dfrac{\mathrm{d}n_{\mathrm{T}\leftarrow}(t)}{\mathrm{d}t} = -K_{\mathrm{D}} n_{\mathrm{T}}(t) \end{cases} \tag{7.8}$$

where $n_{\mathrm{T}}(t)$ and N_T are trapped charges and total traps, respectively. The trapping and de-trapping coefficients K_{T} and K_{D} are expressed as [15]:

$$\begin{cases} K_{\mathrm{T}} = \dfrac{J_{\sigma\mathrm{T}}}{q} \\ K_{\mathrm{D}} = N_{\mathrm{c}} \nu_{\mathrm{th}} e^{-\frac{E_{\mathrm{T}} - W_{\mathrm{em}}}{kT}} \end{cases} \tag{7.9}$$

where k is the Boltzmann constant, T is the absolute temperature, J is the injected current density, which directly affects the trapping rate, σT is the capture cross section, q is the charge amount, N_{c} is the effective density of states in the conduction band, W_{em} is the electromechanical energy associated with the storage of space charges in a trap, ν_{th} is the thermal velocity of the charge, which mainly affects the de-trapping rate, and we often consider that thermal de-trapping occurs until the trapped charge carrier gains its energy for de-trapping from thermal lattice vibrations, and E_{T} is the trap depth.

IV) Low-density polyethylene (LDPE) film, which is the focus of the model.
V) We often adjust thermal de-trapping (ν_{th}) to include W_{em} which is related the electric field E.
VI) The lifetime $t_c(E,T)$ never ends until the value of the captured charge $X(t)$ (7.11) [15] reaches a critical value C^* such that W_{em} can break the chemical bonds and lead to the insulation fail:

$$X(t) = C\left[1 - e^{-(K_T + K_D)t}\right] + C_0 e^{-(K_T + K_D)t} \tag{7.10}$$

where C is the equilibrium value of X, C_0 is the initial value of C, K_T and K_D are expressed as (7.10), and t is the end of the life.

Hence, the lifetime model is obtained by setting $X(t_c) = C^*$ in the kinetic equations integrated over $t_c(E,T)$, which gives [15]:

$$t_C(E,T) = \left[\frac{\sigma_T}{q_e} A T^2 e^{-\frac{\Phi - \beta_s E^{1/2}}{k_B T}} + N_C \nu_{th} \sigma_T e^{-\frac{G_t - W_{em}}{k_B T}}\right]^{-1} + \ln\left[\frac{C - C_0}{C - C^*}\right] \tag{7.11}$$

$$\beta_s = \left(\frac{q_e}{4\pi\varepsilon_r\varepsilon_0}\right)^{1/2} \tag{7.12}$$

where σ_T is the capture cross-section, q_e is the electron charge, A is the Richardson–Dushman constant, T is the absolute temperature, Φ is the electrode insulation barrier height, E is the electric field, k_B is the Boltzmann constant, W_{em} is electromechanical energy associated with the storage of space charges in a trap, N_C is the effective density of states of conduction band, ν_{th} is the thermal velocity of charge, G_t is the trap depth, C is the equilibrium value of X, and C_0 is the initial value of C.

A good feature of this model is that except for C^*, its parameters can come from other studies other than ALT. Using this model, the results of electrothermal ALT of cable insulation at 80 °C DC voltage are well fitted, as shown in Figure 7.4.

7.2.1.3 PD-Induced Damage Model

G. Mazzanti proposed a physical electrothermal lifetime model, which solves the situation of PD in the microcavity of cable insulation since its production [14, 17]:

The model assumes that:

I) For the PE-insulated HVDC cable, the insulation is characterized by air-filled micro-voids. We assume that the micro-voids are cylindrical with radius R, circular base S, and height d along the field direction.
II) Assume that the electrons jump through the polymer after thermal activation to reach the PE gap interface, where the electronegativity of PE promotes

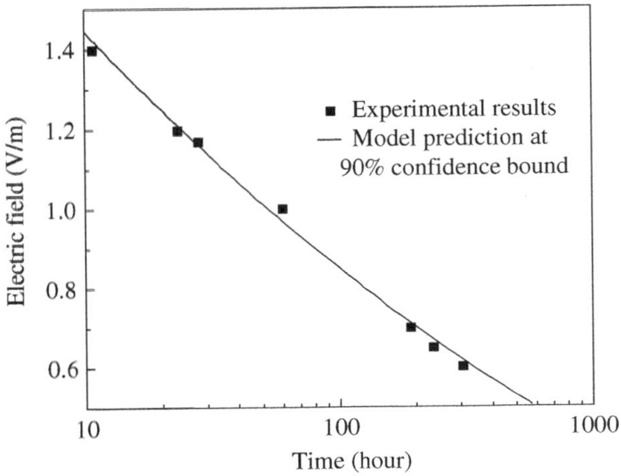

Figure 7.4 Results of electrothermal ALT at 80 °C DC voltage. *Source:* [16]/AIP Publishing LLC / CC BY 4.0.

charge capture. The captured electrons are then injected into the void by the Schottky effect.

III) The thermos-electron avalanche starts from the PE/ micro-void surface and hits the other PE/micro-void side.

IV) The impact causes the CH bond to break or even the PE body to be damaged. With the gradual increase of the crater after the impact, the electric tree is generated when the critical value is reached, i.e. the insulation lifetime is terminated. Based on the above assumptions, the following three steps explain the degradation process up to breakdown and give explanations of some parameters in the equation:

Step 1: Electrons are injected from the surface of the micro-void into the interior. The electron injection rate is R_{inj}.

Step 2: Electron avalanche occurs in the micro-void gradually. This step produces an electron energy distribution: the "hot" fraction F_{hot} of the electron that overcomes the bond breaking energy of the CH bond, equal to 8 eV. The effective fraction of F_{hot} is F_{eff}; $R_{el} = R_{inj}N_{el}$ is the number of electrons impacting the void-PP surface, where N_{el} the number of electrons generated in the cavity from 1 injected electron.

Step 3: Progress from degradation on the void/PE surface to degradation in the polymer lattice. This step produces: t_{dis} is the rupture time of the plate of thickness D_{dis}, where the number of CH bonds exposed to damage is N_{CH}; R_{dis} is the degradation growth rate in the polymer lattice; $L(E,T)$ is the electrothermal lifetime, indicating the time to reach the critical size d_c of the degraded region. So the expression $L(E,T)$ of the model is [14]:

$$L(E,T) = \frac{d_C}{R_{dis}} \tag{7.13}$$

where the R_{dis} is calculated as follows:

$$R_{inj} = \frac{J_C S}{q_e k_B T} \frac{2q_e^2}{4\pi\varepsilon_0 R} \left\{ \ln\left[1 + \frac{2q_e^2}{4\pi\varepsilon_0 R k_B T} e^{\frac{G_t - q_e\sqrt{\frac{q_e \varepsilon_r E}{4\pi\varepsilon_0}}}{k_B T}}\right]\right\}^{-1} \tag{7.14}$$

$$t_{dis} = \frac{N_{CH}}{2R_{el}F_{eff}F_{hot}} = \frac{N_{CH}}{2R_{inj}N_{el}F_{eff}F_{hot}} \tag{7.15}$$

$$R_{dis} = \frac{D_{dis}}{t_{dis}} \tag{7.16}$$

where J_C is the conduction current density.

7.2.2 Phenomenological Lifetime Models

The phenomenological lifetime model of DC cable insulation under constant stress looks like the lifetime model of AC cable insulation, and the main difference is the value of parameters of models. Its general form is:

$$L = L(S_1, ..., S_N) \tag{7.17}$$

where L is the lifetime and $S_1,...,S_N$ is a set of applied stresses (number is N), which is taken as a constant with time. Among them, electrical stress and thermal stress are the most important.

Due to the fact that phenomenological models primarily rely on ALT tests to summarize patterns based on experimental phenomena for lifetime prediction, let us first introduce the ALT tests.

ALT includes constant stress tests, step-stress tests, and progressive-stress tests, with variations in stress application patterns. Lifetime acceleration tests involve the superposition of electrical stress with thermal and mechanical stress, forming dual stress conditions. Currently, ALTs for insulation materials mainly focus on constant stress and step-stress tests at different temperatures.

7.2.2.1 Accelerated Life Tests Under Constant Stress

Due to the fact that phenomenological models primarily rely on ALT tests to summarize patterns based on experimental phenomena for lifetime prediction, let us first introduce the ALT tests under constant stress.

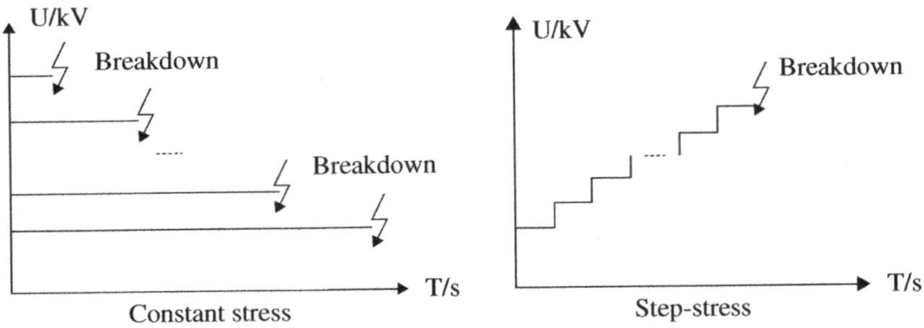

Figure 7.5 Different stress application methods.

Constant stress ALTs primarily target the test voltage, denoted as Si, which is selected based on the short-term breakdown voltage of the insulation material. Test voltages are incrementally applied from low to high, designated as $S1$, $S2$, $S3$, ..., Si ($i \geq 6$). High-voltage control gradually increases the voltage to reach the test voltage and maintains it while recording the duration until the material insulation fails (breakdown). The test is repeated under the same conditions, as shown in Figure 7.5.

Although accelerated aging under constant stress can closely simulate the degradation of samples under actual operating conditions and significantly reduce the testing time compared to natural aging conditions, conducting repetitive sample tests is cumbersome, and individual sample tests still take a long time. In comparison, step-stress tests accelerate the electrical aging testing, which is the more widely applied testing mode.

7.2.2.2 Accelerated Life Tests Under Step Stress

As shown in Figure 7.5, the parameters for step-stress electrical aging tests include the initial voltage (U_0), voltage increment (ΔU), and duration of each voltage level (Δt). In each test, timing starts from the voltage being raised to the initial voltage (U_0). After each duration (Δt), the voltage is increased by (ΔU). After increments (i), the sample breaks down, and the duration (t) is considered as the sample's lifetime. The breakdown voltage is (U_{i+1}), where ($U_{i+1} = (U_0 + i\Delta U)$). The duration of the last voltage level ($\Delta t'$) can be determined based on the number of increments (i), (Δt), and (t). Insulating materials do not possess self-recovery characteristics for insulation damage. Each insulation damage accumulates, and thus, the breakdown process can be regarded as the cumulative effect of electric stress applied to the material. The remaining lifetime of insulating material is determined by the currently applied voltage and the probability of breakdown at that voltage.

Compared to electrical aging tests under constant stress, the step-stress method results in a faster and more efficient failure process. However, this model is based on the principle of approximate equivalence. If the time increment (Δt) is set to infinity, it can be considered as constant stress. Therefore, constant stress electrical aging can be viewed as a special mode within the step-stress electrical aging method. Considering practical operating conditions, the impact of step-stress is more severe. There is a certain margin of error in parameters such as the endurance index (n) and cumulative damage (C) obtained, making the results less accurate than the constant stress method.

7.2.2.3 Single-Stress Electrical Lifetime Models

The simplest and the most popular phenomenological model of HVDC cable insulation is the electric inverse power model (IPM) (V–t characteristic) [18] which represents the insulation life t (or t_F or L) as the inverse power of the applied voltage U or electric field E, as follows:

$$U^n t = C \tag{7.18}$$

where n is the life index, also known as voltage endurance coefficient (VEC). The larger the n value, the flatter the curve, indicating better insulation withstand capability and resistance to electrical stress in cables. Therefore, the n value, which is the cable's life index, is a fundamental parameter of cable insulation characteristics, reflecting the rate of electrical aging. t is the lifetime at this voltage. C is a constant that depends on the insulation material of the cable and other influencing factors, determined through experimentation.

According to Eq. (7.19), the lifespan t of the cable under the current electric field strength E can be described as:

$$t = CE^{-n} \tag{7.19}$$

Taking the logarithm of Eq. (7.20) on both sides:

$$\mathrm{Ln}t = -n\mathrm{Ln}E + \mathrm{Ln}C \tag{7.20}$$

In a logarithmic coordinate system, a straight line is obtained. The cable V–t characteristic curve is shown in Figure 7.6, and it illustrates the relationship between voltage and the time to insulation breakdown. The curve is divided into three regions: I, II, and III.

In region I, which represents high electric field strength, the failure time t_1 is often less than one hour. Region II, on the other hand, represents insulation life t_2 which can span from several years to several decades. Region III demonstrates insulation failure occurring at lower field strengths, enduring pressure for a longer period. In this region, the insulation failure mechanism includes not only thermal breakdown but also electrochemical breakdown playing a dominant role.

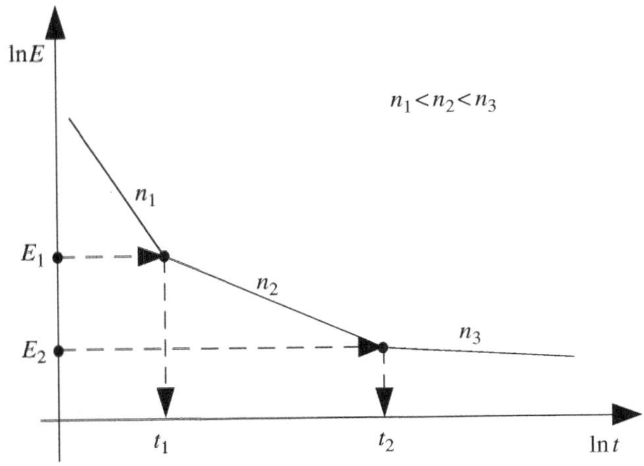

Figure 7.6 *E–t* characteristics of specimens.

Research indicates that regions II and III are closely related to the lifespan of cables and are in good agreement with the power-law model (7.21) that describes electrical aging.

From Figure 7.6, the n value obtained using the constant stress method is highly dependent on the applied field strength on the samples. If the test voltage is too high, the n value obtained in region I is too small. If the test voltage is too low, it leads to longer breakdown times, introducing other uncertainties, and a larger n value, which could result in a smaller safety margin in design. Therefore, the test generally uses region II as the basis for obtaining the n value. Based on the n value obtained from region II, it ensures a certain level of redundancy without making cable sample remaining life predictions overly conservative, and it is also more reasonable in terms of the time and cost of testing.

Equation (7.19) can use the rated voltage U_0 to design the life L_0 or to calculate the duration t_1 ($\ll L_0$) under the test voltage U_1 ($\gg U_0$), so there are:

$$t_1 = L_0 \left(\frac{U_1}{U_0} \right)^{-n} \tag{7.21}$$

Table 7.1 lists the PQT voltages of various cables. It can be seen that n depends on the waveform of voltages and insulation characteristics. At the same time, n also depends on temperature as shown in Figure 7.7.

The single stress electrical lifetime model is Eq. (7.18) in which $N = 1$, then S_I = electrical stress = E'. Where E' is given by U, but more often by local electric field intensity E, e.g. using E instead of U, assuming that the highest test electric

Table 7.1 Derivation of PQT voltage for EHVAC-extruded cable, HVDC-extruded cable, and HVDC lap cable.

Cable type	Standard	Rated life L_0	Duration t_1	n	PQT voltage U_1
AC extrude	IEC 62067	40	365	7	$\sqrt[7]{40 \times 365/365} = 1.7$
DC extrude	TB 852	40	360	10	$\sqrt[10]{40 \times 365/360} = 1.45$
DC lapped	TB 853	40	360	11	$\sqrt[11]{40 \times 365/360} = 1.4$

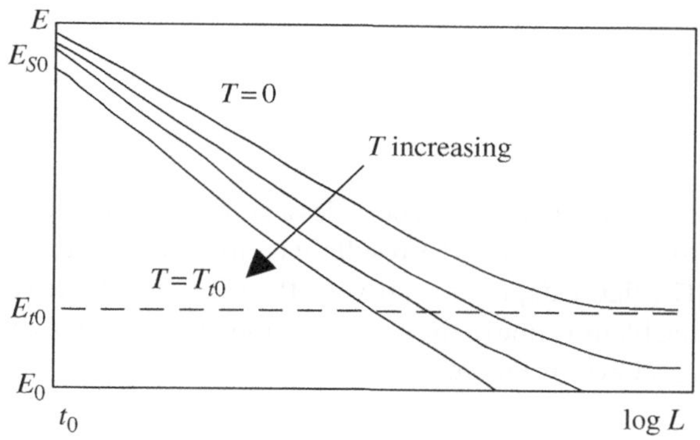

Figure 7.7 Electrical stress lifetime curve at various temperatures with constant temperature *T*. *Source:* Montanari et al. [18]/with permission from IEEE.

field E_H (usually E_H = breakdown strength) or the much lower rated electric field E_0 as the reference field, the following two equations can be obtained:

$$L(E) = L_H \left(\frac{E}{E_H}\right)^{-n} \tag{7.22}$$

$$L(E) = L_0 \left(\frac{E}{E_0}\right)^{-n} \tag{7.23}$$

where L_H is the lifetime under the field strength E_H and L_0 is the lifetime under E_0. The subscript corresponds to one by one, and this chapter will not repeat it later.

7.2.2.4 Single-Stress Thermal Lifetime Models

The phenomenological model of single thermal stress lifetime is Eq. (7.18) in which $N = 1$, S_1 = thermal stress = T.

Arrhenius Model

The physical expression of the Arrhenius model has been introduced in Eq. (7.4).

Similar to the IPM model, by taking the highest test temperature T_H or the much lower temperature T_0 (usually the rated temperature or room temperature of cable insulation) as reference, two more applicable expressions of Arrhenius model are obtained from Eq. (7.4):

$$L(T) = L_H e^{-B\left(\frac{1}{T_H}-\frac{1}{T}\right)} = L_H e^{-BT''} \tag{7.24}$$

$$L(T) = L_0 e^{-B\left(\frac{1}{T_0}-\frac{1}{T}\right)} = L_0 e^{-BT'} \tag{7.25}$$

$$B = \frac{W}{k_B} \tag{7.26}$$

$$T'' = \frac{1}{T_H} - \frac{1}{T} \tag{7.27}$$

$$T' = \frac{1}{T_0} - \frac{1}{T} \tag{7.28}$$

IEC standard 60 216 provides a graph for estimating the heat resistance of dielectrics based on the ALT results at temperature T based on the Arrhenius model [19]. The figure depicts the curve of the relationship between log (L) and $-1/T$, so that the lifetime line of the Arrhenius model becomes a straight line with a slope of $-B$, as shown in Figure 7.8.

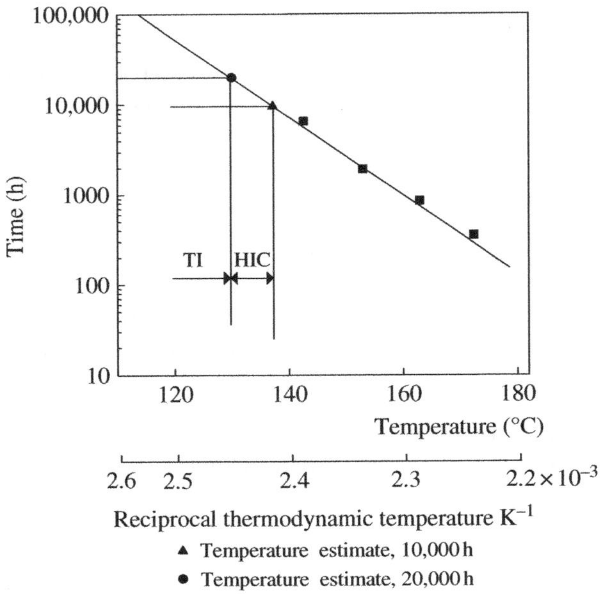

Figure 7.8 Thermal aging life curve by the Arrhenius model. *Source:* Hirose [19]/with permission from IEEE.

The significance of the curve lies in:

I) Get the temperature index (TI), which is the temperature that produces a life of 20,000 hours in Celsius.
II) Get the halving interval in Celsius (HIC), which is the temperature rise from the temperature of TI yielding a life of 10,000 hours, i.e. half of the life at the TI.

Eyring Model

The Eyring model shows that TI and HIC are similar to the Arrhenius model. There is only a slight difference between the TI and HIC values given by the Eyring model and the Arrhenius model, as shown in Figure 7.9. However, the latter is chosen by the IEC standard 60216 as it is simpler than the former.

7.2.2.5 Combined Electrothermal Lifetime Models

For practical cables, they are subjected to both electrical and thermal stresses during operation, and the magnitude of their lifespan results from the combined effect of these two stresses. Multiple aging factors interact simultaneously, and there is coupling between different forces during their application. The impact of aging is far greater than the results of individual stress applications. In the 1990s, Fallou and others introduced the Fallou model based on exponential relationships, Montanari and others proposed a probabilistic model, and Crine put forward the Crine model based on physical parameters. Additionally, by combining the inverse

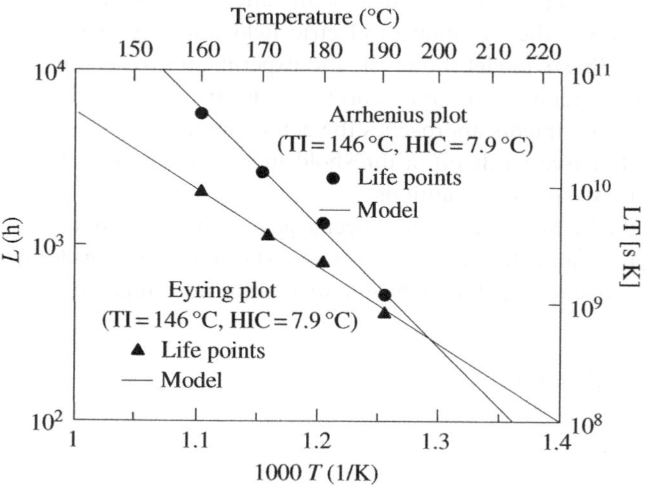

Figure 7.9 TI and HIC under the Eyring and Arrhenius models. *Source:* Adapted from Montanari and Mazzanti [20].

power model and threshold electric field, Simoni and Ruma, among others, introduced the Simoni and Ruma models [21–23].

Simoni considers a property (*p*) that depends on the aging process within the insulating material and varies with time. This property is used to indicate the aging state of the insulation. The model explains that aging of the insulating material under the combined action of electrical and thermal stresses is the result of an electric-thermal cumulative effect. Simoni assumes that the aging quantity (*F*) at the time of insulating failure serves as a characteristic parameter of the insulation system. Therefore, the insulation lifespan and aging rate become inversely proportional quantities. The aging rate (*R*) must be able to describe the process of combined thermal and electrical aging. Simoni represents the combined aging rate based on thermodynamic principles as follows:

$$R = A\exp\left(-\frac{B}{T}\right)\exp\left(\left(a+\frac{b}{T}\right)f(E)\right) \tag{7.29}$$

where $f(E)$ is an unspecified electric field function, T is the absolute temperature, and A, B, a, and b are constants. This rate equation represents the impact of the electric field on the thermal aging process, but the specific form of this impact is not described. Based on the aging rate model, the Simoni electrical–thermal combined aging lifetime model is derived as shown in Eq. (3.10).

$$L = \frac{1}{A}\exp\left(\frac{B}{T}\right)\exp\left(-\left(a+\frac{b}{T}\right)(E-E_0)\right), E \geq E_0 \tag{7.30}$$

The major advantage of this model lies in its ability to not only encompass the relationship between the electrical and thermal factors affecting aging and insulation life but also describe the variation of electric field intensity over time. The Simoni model is a linear model; however, the experimental results under low-voltage conditions sometimes do not conform to a linear model. Therefore, if the stress applied in experiments approaches the actual operational stress and the cable insulation life curve levels off, a threshold model is needed to better describe aging under these specific conditions.

The background of the RAMU model is the inverse power-law behavior of classical single-stress thermal aging. It interprets the synergistic effects by considering the constant of the inverse power-law behavior in relation to temperature, as shown in Eq. (7.32).

$$\begin{cases} L = K(T)\exp(-B \bullet DT)U^{-n(T)} \\ DT = \dfrac{1}{T_0} - \dfrac{1}{T} \\ K(T) = \exp(K_1 - K_2 \bullet DT) \\ n(T) = n_1 - n_2 DT \end{cases} \tag{7.31}$$

where L represents the lifespan of the insulation, U is the voltage in the ALT, T is the absolute temperature in the ALT, and B, K_1, K_2, n_1, and n_2 are constants determined through experimentation.

The Fallou model is derived from an exponential model, and its expression is as shown in Eq. (7.33).

$$L = \exp\left(A_1 + A_2U + \frac{B_1 + B_2U}{T}\right) \tag{7.32}$$

where A_1, A_2, B_1, and B_2 are constants determined through experimentation. The Fallou model embodies the interrelationship between electrical aging, thermal aging, and electrical–thermal aging.

The above three electrical–thermal combined aging lifetime models are mostly based on the experimental results. In contrast, the Crine model leans more toward being a physical model. The foundation of the Crine model is rate theory, which posits that factors affecting insulation aging such as electrical, thermal, and other environmental stresses can be considered as energy barriers to the aging process. Insulation life is the time, denoted as t, required to overcome these barriers. The expression of the Crine model is as shown in Eq. (7.34).

$$t = \frac{h}{KT} \exp \frac{\Delta G}{kT} \operatorname{csch}\left(\frac{e\lambda U}{kT}\right) \tag{7.33}$$

where h represents the Planck constant, k is the Boltzmann constant, ΔG stands for free energy, and λ denotes the average free path.

There are numerous models used to describe multifactor aging. However, it is essential to note that some of these models are similar to each other. For instance, the RAMU model and the Fallou model can be represented using the Simoni model. What sets the Simoni model apart from the other two models is its detailed description of how the model is constructed. Additionally, Simoni provides a graphical representation through a lifetime equation, which can help in a more intuitive understanding of the model.

From Eqs. (7.31) and (7.34), it is evident that both the Simoni model and the Crine model are based on thermodynamic relationships. Thermodynamics describe how thermal excitation processes are influenced by external electric fields. Thermal excitation processes could be a chemical reaction or the transport of charged particles. However, neither Simoni nor Crine describe which one they consider as the primary aging process. Crine might go a step further by assuming the involvement of charged particles in the aging process, as seen in Eq. (7.34). The most significant difference between the Simoni model and the Crine model lies in how they relate cable life to aging rate. In the Simoni model, lifetime and aging rate are inversely proportional, whereas Crine asserts that lifetime is equal to the reciprocal of the aging rate. Generally, thermodynamics is based on probabilistic considerations, where the reaction rate provides the average frequency of a reaction occurring. Crine employs thermodynamics to describe the crossing of an energy barrier as a single event, which is somewhat unconventional.

Although thermodynamics is the common foundation for both the Simoni and Crine models, when it comes to describing combined thermal and electrical aging, Simoni adjusts empirically and derives different conclusions from Crine. In the Crine model, the synergistic effect of thermal and electrical aging is manifested in the disruption of energy barriers, thus accelerating the aging process. The Simoni model also describes the impact of electric fields in a similar manner, but it includes an additional factor, a single-factor model that describes electrical aging. Consequently, Crine views combined thermal and electrical aging as an accelerated thermal excitation process, while Simoni describes it as a simultaneous occurrence of accelerated processes and electrical aging.

Representing the RAMU and Fallou models using the Simoni model can reduce the complexity of selecting aging models and facilitate the choice of the appropriate aging model for different insulation systems. Evaluating the effectiveness of any model involves assessing its ability to fit experimental data, which is of paramount importance.

7.3 Thermal Aging

7.3.1 Effect on Physical–Chemical Properties

When the cable is in operation, the current passing through the conductor will result in the heating of the cable conductor. This will cause the cable insulation to work at a higher temperature for a long time, leading to aging of the insulation layer and a decrease in insulation performance inevitably [24].

The unused PP cable is used for experimentation, and the PP insulation layer is subjected to roll-cutting treatment, with a thickness set at 200 μm. The obtained samples are cut into rectangular samples with dimensions of 8 cm in length and width. These samples are suspended in a high-temperature test chamber for the electrothermal ALT with clips on it, with a distance of more than 20 mm between each sample. The temperature of the high-temperature test chamber is set at 135 °C according to the IEC60216 standard, with a ventilation frequency of 160 times per hour. After aging for 7, 14, 30, and 60 days, several samples are taken out to test the changes in material properties after aging.

7.3.1.1 FT-IR Test

During the aging process, molecular structure and functional groups of PP undergo changes, which can be characterized by the Fourier transform infrared spectroscopy (FT-IR). Figure 7.10 shows the FT-IR spectra of PP at different aging

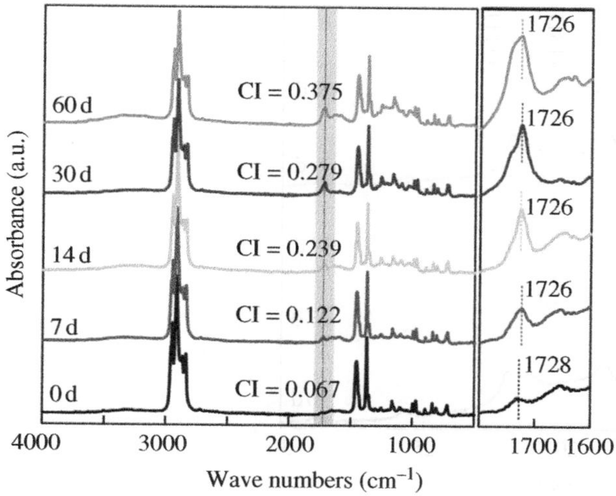

Figure 7.10 FT-IR spectra of PP samples with different aging times. *Source:* Li et al. [34]/ with permission from IEEE.

times. The selected spectral range is from 400 to 4000 cm^{-1}, with a resolution of 4 cm^{-1}. The peaks at 2962 cm^{-1}, 2872 cm^{-1}, and 1375 cm^{-1} correspond to the stretching vibrations of methyl groups ($-CH_3$). The peaks at 2926 cm^{-1}, 2853 cm^{-1}, and 1460 cm^{-1} correspond to the stretching vibrations of methylene groups ($-CH_2-$), A new absorption peak appears at 1720 cm^{-1} with increasing aging time, which corresponds to the carbonyl group (=C=O) absorption peak. This indicates that PP has been oxidized and the appearance of this absorption peak can be used as a characteristic peak for thermal aging.

During the process of thermal aging, the carbonyl index is calculated by maintaining the relatively stable methylene vibration peak at 1460 cm^{-1} as a reference value, this phenomenon is described by the following equation:

$$CI = \frac{D_1}{D_2} \tag{7.34}$$

where D_1 represents the strength of the carbonyl peak at 1720 cm^{-1}, while D_2 represents the strength of the methylene absorption peak at 1460 cm^{-1}. The changes in CI with aging time are shown in Figure 7.10. It can be observed that as the aging time increases, the carbonyl index of PP is increased, indicating a deeper degree of material aging. The carbonyl index of PP aging for 60 days increases by 460.3% compared to unaged PP samples.

7.3.1.2 XRD Test

Figure 7.11 shows the X-ray diffraction patterns (XRD) of PP at different aging times. Diffraction peaks of (110), (040), (130), (131), and (111) crystal planes are

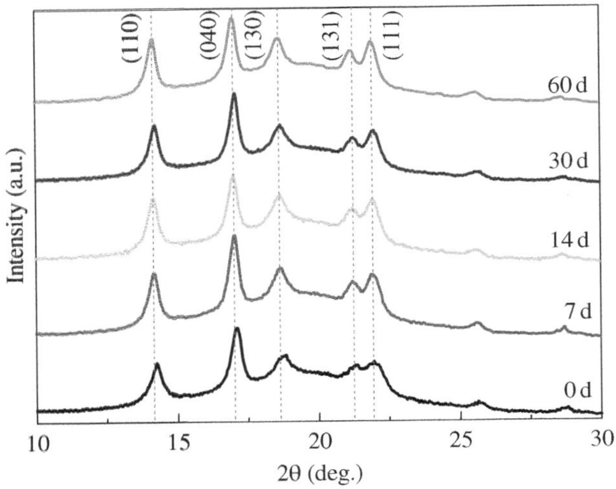

Figure 7.11 XRD spectra of PP samples with different aging times. *Source:* Li et al. [34]/ with permission from IEEE.

observed at 14.1°, 16.9°, 18.6°, 21.1°, and 21.9°, respectively. This is the α diffraction peak of crystals in PP. In the early stages of aging, these diffraction peaks become sharper. The half peak width decreases and the diffraction peak increases, indicating that the spherical crystal structure of PP becomes more perfect. At the same time, the diffraction peak tends to shift toward low angles, indicating a slight increase in the size of the spherical crystal. With the further lengthening of the aging time, a decrease in the amplitude of diffraction peaks on crystal planes such as (040) and (130) can be observed. Taking the (040) crystal plane with the strongest diffraction peak as an example, normalization treatment is performed on the unaged sample. The diffraction peak values are divided into 1, 1.15, 1.04, 1.03, and 0.94, and the area is divided into 1, 0.95, 0.91, 0.88, and 0.81 after aging for 0, 7, 14, 30, and 60 days. This indicates that as aging progresses, the spherical crystal structure of PP is destroyed. And, during the aging process, small-sized spherulites are more prone to be destroyed.

7.3.1.3 DSC Test

Differential scanning calorimetry (DSC) can characterize the changes in heat absorption of PP during melting. The PP sample is heated at a rate of 10 °C/min to 200 °C and held for 5 minutes, Subsequently, the temperature is reduced to 30 °C at a rate of 10 °C/min to remove any thermal history. This process is then repeated to obtain the melting curve. Figure 7.12 shows the DSC melting curves of PP at different aging times. By calculating the area between the solid line and the dashed line in Figure 7.12, the melting enthalpy (ΔH_m) of PP at different aging

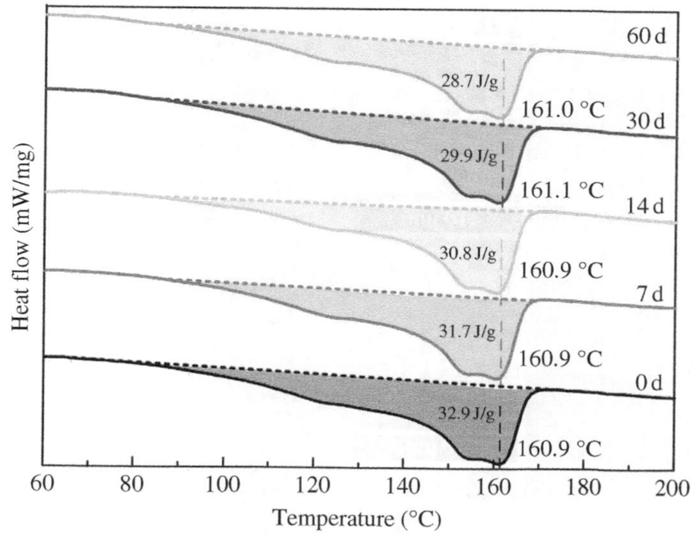

Figure 7.12 DSC curves of aged-PP samples. *Source:* Li et al. [34]/with permission from IEEE.

times can be obtained, as shown in Figure 7.12. It can be seen that the melting enthalpy was 32.9 J/g for the unaged sample. After aging for 60 days, the melting enthalpy decreased to 28.7 J/g, a reduction of 12.8%. This indicates that the spherical crystal structure of PP is destroyed.

7.3.1.4 SEM Test

The scanning electron microscope (SEM) of PP insulation is obtained, as shown in Figure 7.13. It can be observed that the micro-void appears after aging for 14d. The elastomer is etched to obtain the distribution of the elastomer in PP insulation. The SEM images are analyzed by ImageJ. The boundary of the elastomer is marked by yellow lines and the average size of elastomer is calculated. With the increase of the aging time, the average sizes of elastomer increase from 1.04 to 1.52 μm, a 46% increase. The decrease in the uniformity of the PP insulation indicates that the material has undergone severe oxidation.

7.3.2 Effect on Mechanical and Electrical Properties

7.3.2.1 Mechanical Test

Dumbbell-shaped specimen is stamped for tensile test with a length of 20 mm, a width of 4 mm, and a thickness of 1 mm. The elongation at break and the tensile strength of aged-PP samples are shown in Figure 7.14. It can be seen that with the increase of aging time, the elongation at break and the tensile strength of PP shows

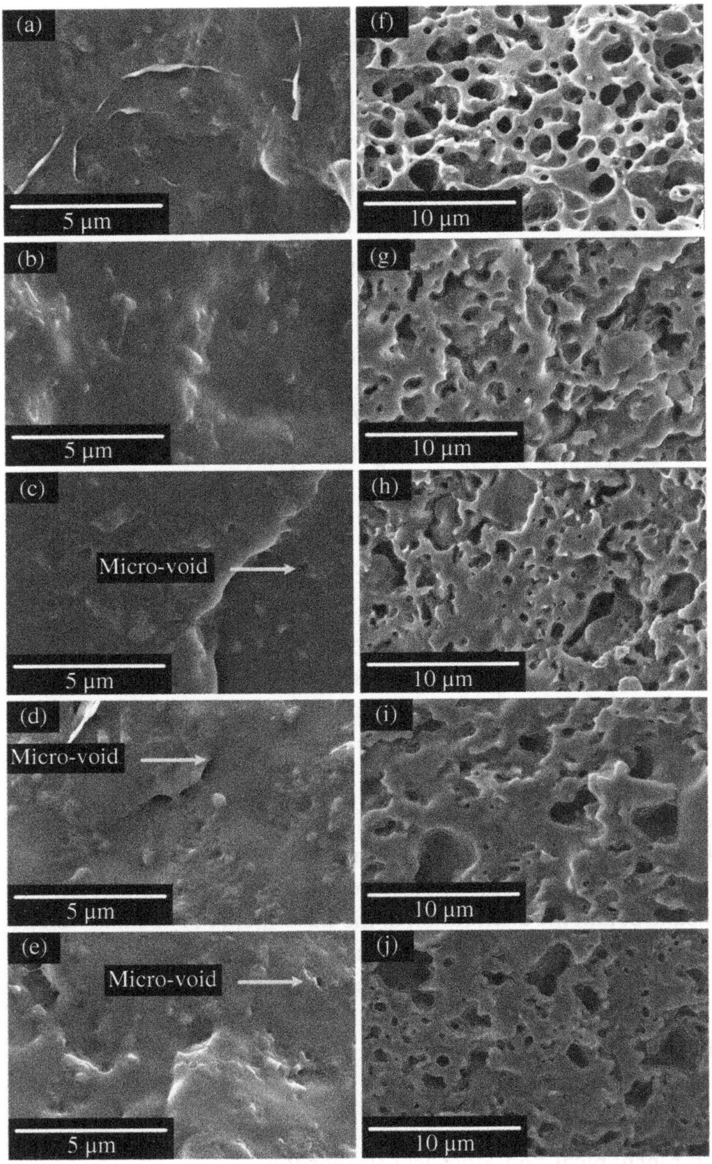

Figure 7.13 SEM images of PP insulation after aging for 0, 7, 14, 30, and 60 d: (a)–(e) unetched samples and (f)–(j) etched samples. *Source:* Li et al. [34]/with permission from IEEE.

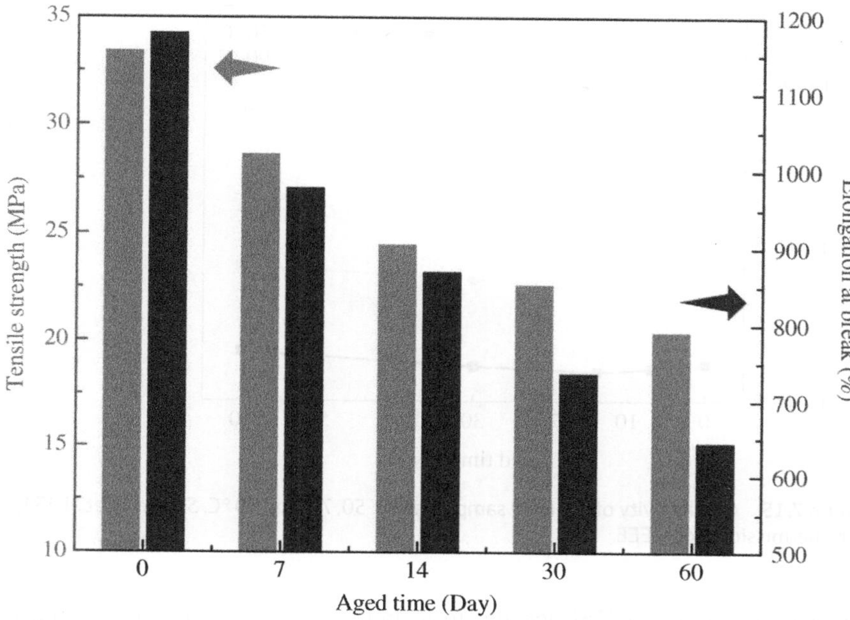

Figure 7.14 Tensile strength and elongation at break of aged-PP samples. *Source:* Li et al. [34]/with permission from IEEE.

a decreasing trend. When the aging time reaches 60 days, the elongation at the break of PP decreases by 35.1%. This is because, with the progress of aging, the spherical crystal structure of PP is destroyed and the amorphous region increases, resulting in a decrease in the mechanical properties of PP. In addition, the molecular chain breakage will increase the low-density region in PP, which will also lead to a decrease in the mechanical properties of PP.

7.3.2.2 Conductivity Test

Figure 7.15 shows the variation of the electrical conductivity of PP with different aging times and temperatures. The experiment is conducted at a field strength of 10 kV/mm for 1800 seconds. The last stable 100 seconds point is selected to calculate the average conductivity of PP [25]. According to Figure 7.15, with the temperature increases, the electrical conductivity of PP shows a rising trend. This is because an increase in temperature accelerates the carrier mobility of PP, resulting in an increase in conductivity. As the aging time increases, the electrical conductivity of PP shows a trend of first decreasing and then increasing. This is because in the early stage of aging, under the influence of thermal factors, the incomplete spherical crystal structure in PP will become more complete and the distribution

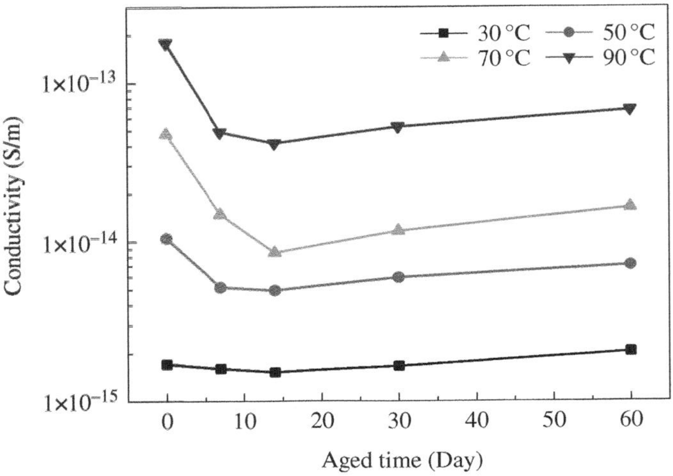

Figure 7.15 Conductivity of aged-PP samples at 30, 50, 70, and 90 °C. *Source:* Li et al. [34]/ with permission from IEEE.

will be more uniform [26], leading to a limitation on carrier mobility and a decrease in conductivity. As aging continues, the molecular chains of PP will be oxidized, the spherical crystal structure will be destroyed, and the amorphous region will increase which can be found in Figure 7.12 that the ΔH_m of PP is decreased, leading to an increase in conductivity.

7.3.2.3 FDS Test

The changes of the complex permittivity of PP at different aging times and temperatures are shown in Figure 5.20, which has been introduced in detail in Section 5.4. The changes in the real part of the complex permittivity of PP at different aging times and temperatures are shown in Figure 5.20a–e. It shows that the real part of the complex permittivity of PP increases with aging time. This indicates that the molecular chains of PP have undergone oxidation, resulting in the formation of polar groups or the breakage of molecular chains to produce polar small molecules [27], leading to an increase in the real part of the complex permittivity. As the frequency increases, the real part of the complex permittivity decreases, and when the frequency is very high, the real part of the complex permittivity tends to ε_∞. The real part of the complex permittivity of PP decreases with increasing temperature. This is a result of the density of PP decreasing due to thermal expansion and contraction, resulting in a decrease in the real part of the complex permittivity.

The variations of the imaginary part of the complex permittivity under different aging times and temperatures are shown in Figure 5.20f–g. The imaginary part of the complex permittivity in 10–10^4 Hz significantly increases with aging time, which can be attributed to the enhancement of the α-relaxation process associated with segmental chain motion [28]. In the frequency range of 10^{-2}–10^{-1} Hz, the imaginary part of the complex permittivity increases significantly with temperature, which can be attributed to the increase of DC conductivity.

7.3.2.4 AC Breakdown Test

The AC breakdown strength of aged-PP samples analyzed by Weibull plots is shown in Figure 7.16. The experiment employed a sphere-plate electrode configuration, with the high-voltage electrode connected to the sphere electrode and the ground electrode connected to the plate electrode. The thickness of the test sample is 0.2 mm. The samples are applied AC voltage increasing at a rate of 0.5 kV/s. The experimental data are analyzed with a Weibull plot.

The breakdown strength of unaged-PP is 99.02 kV/mm. After aging for 7, 14, 30, and 60 days, the breakdown strength of PP decreased to 98.31, 96.56, 93.49, and 87.89 kV/mm, respectively. The breakdown strength of PP aged for 60 days decreased by 11.2% compared to that of unaged PP. This is because during the aging process, the molecular chains of PP are broken, leading to the formation of local low-density regions and an increase in amorphous regions, resulting in a decrease in the breakdown strength of PP.

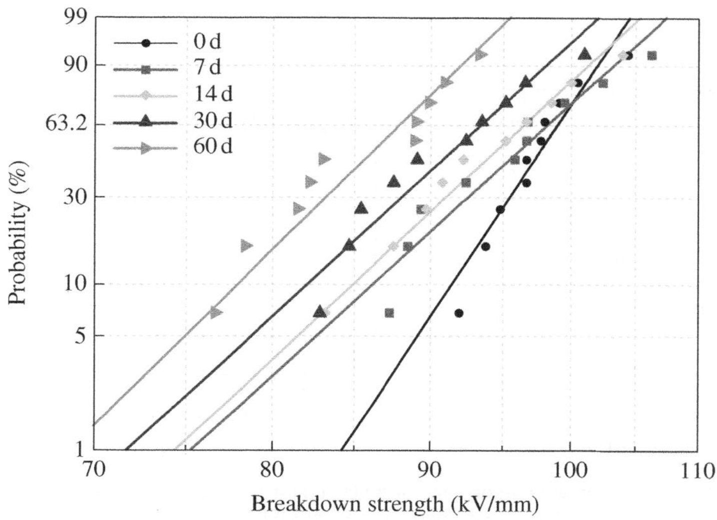

Figure 7.16 Effect of thermal aging on AC breakdown strength of PP blend insulation. *Source:* Li et al. [34]/with permission from IEEE.

7.3.3 Lifetime Prediction Under Thermal Stress

7.3.3.1 Lifetime Prediction Model

Traditional methods for detecting the aging status of cables require destructive sampling of the cables, followed by relevant experimental tests [29]. This method is costly and has an influence on the normal operation of the power system. FDS testing can obtain the dielectric performance parameters of cable insulation without the damage of the cable. Section 5.4 provides a detailed explanation of the changes in the complex permittivity of PP after aging. By conducting FDS tests on cable insulation and fitting them with the H–N model, parameters such as $\Delta\varepsilon_\alpha$ and τ_δ can be obtained. It is feasible to establish a connection between these parameters and other parameters for the evaluation of cable aging status without damaging the cable insulation.

The changes of the complex permittivity of PP at different aging times and temperatures are shown in Figure 5.20. The fitting results for α and δ relaxation have been introduced in Section 5.4 in detail. Figure 7.17 shows the relationship between the aging time and the FDS characteristics parameters of the PP insulation. During the aging process, the PP insulation undergoes oxidization, leading to an increase of the orientation polarization, thus increasing the $\Delta\varepsilon_\alpha$. In the early stages of aging, the spherulites in PP insulation undergo recrystallization, leading to a more complete structure which restricts the hopping of charge carriers, leading to the increase of the τ_δ. As aging time increases, the spherulite structure is damaged, making charge carriers easier to hop, leading to the decrease of the τ_δ. It is more significant to analyze the hopping polarization in higher

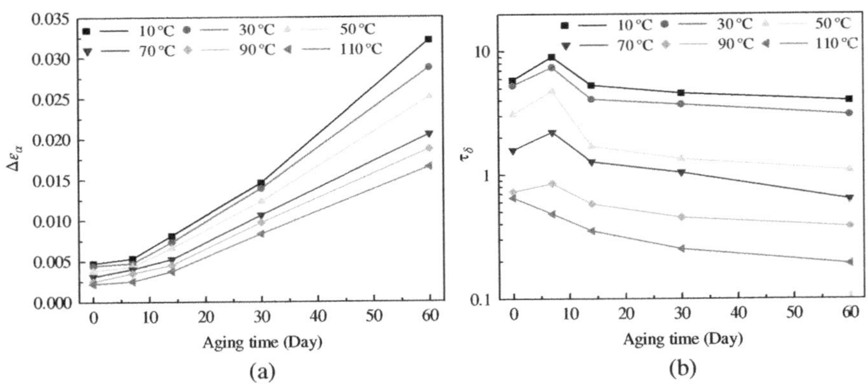

Figure 7.17 Relationship between aging time and FDS characteristics parameters: (a) $\Delta\varepsilon_\alpha$ and (b) τ_δ. *Source:* Li et al. [34]/with permission from IEEE.

temperatures since the carrier is more like to hop at higher temperatures. As a result, the τ_δ in 110 °C and $\Delta\varepsilon_\alpha$ are chosen as the characteristics parameters of the FDS to analyze the aging state of PP cable insulation.

The relationship between characteristics parameters and elongation at break (K), characteristics parameters, and breakdown strength (E_{bd}) are shown in Figure 7.18. It can be observed that the elongation at break shows a rapid decline, which can be attributed to the increase of the average size of the elastomer. The breakdown strength of PP elastomer insulation shows a roughly stable trend after aging for seven hours. However, it shows an obvious decline after aging for 14 hours, which can be attributed to the appearance of the micro-void. To further investigate the relationship between characteristics parameters and insulation parameters, the fitting curve is shown in Figure 7.18, and the fitting equation is shown in Table 7.2. The goodness-of-fit (GFI) index is chosen to assess the accuracy of the fitting curve.

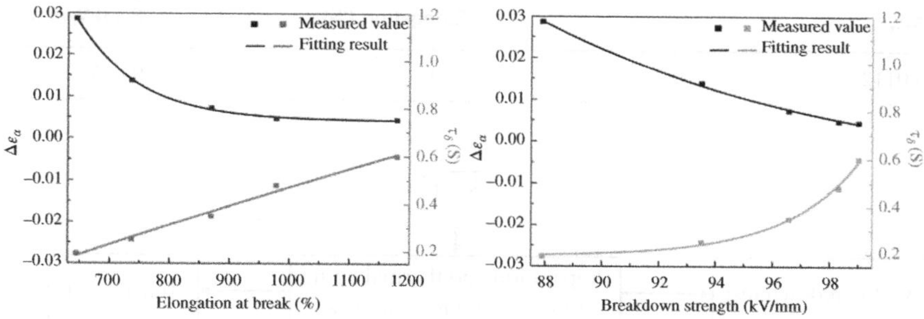

Figure 7.18 Fitting curves between FDS characteristics parameters ($\Delta\varepsilon_\delta$ and τ_δ) and insulation parameters (elongation at break and breakdown strength). *Source:* Li et al. [34]/with permission from IEEE.

Table 7.2 Fitting equation between characteristics parameters and insulation parameters. *Source:* Li et al. [34]/with permission from IEEE.

x	Y	Fitting equation	GFI
Elongation (K) (%)	$\Delta\varepsilon_\alpha$	$y = 13.71^*\exp(-x/101.8) + 0.004$	0.991
	τ_δ	$y = -3.52^*\exp(-x/3391.3) + 3.09$	0.981
E_{bd} (kV/mm)	$\Delta\varepsilon_\alpha$	$y = 113.2^*\exp(-x/11.02) - 0.01$	0.997
	τ_δ	$y = 1.44^*10^{-17*}\exp(x/2.61) + 0.19$	0.991

7.3.3.2 Validation of Prediction Model

To verify the accuracy of the fitting equations in Table 7.3, the PP insulation aged for 21 days is selected for verification and named S_1. The FDS of S_1 is tested and fitted with the H–N equation. The fitting results of the characteristic parameters are shown in Table 7.3. The values in Table 7.3 are substituted into the fitting equations in Table 7.2. The calculation results are $K_1 = 768.77\%$, $K_2 = 763.98\%$, $E_{bd1} = 94.58$ kV/mm, and $E_{bd2} = 94.92$ kV/mm, respectively. The average value of the above aging time is expressed as $K = (K_1 + K_2)/2 = 766.375\%$ and $E_{bd} = (E_{bd1} + E_{bd2})/2 = 94.75$ kV/mm, which has a relative error of 0.497% and 0.316% with the actual value. It indicates that the characteristics parameters related to orientational polarization and hopping polarization can be effectively applied for conditional assessment of the aging state of PP cable insulation. Figure 7.19 shows the flow chart of the condition assessment.

Table 7.3 The fitting result of S_1.

$\Delta\varepsilon_\alpha$	τ_δ	Elongation (%)	E_{bd} (kV/mm)
0.0112	0.28	770.2	94.45

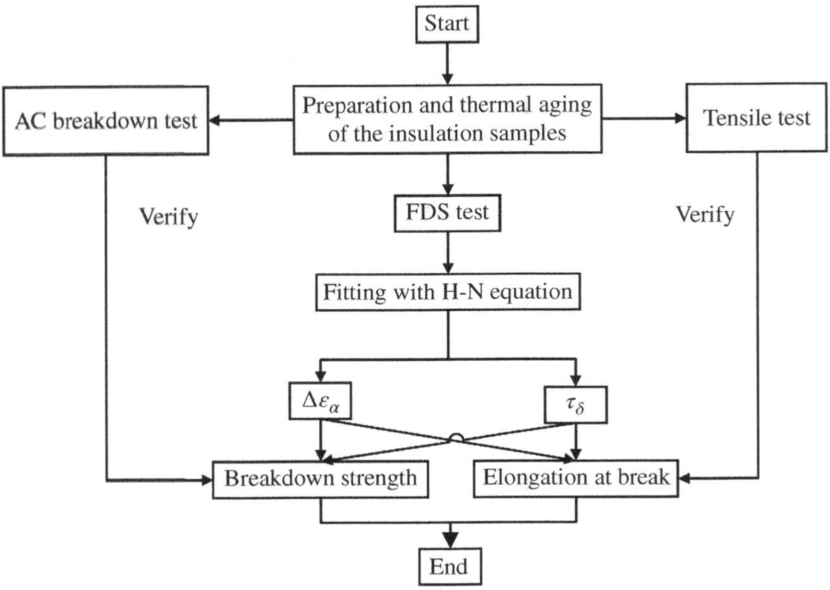

Figure 7.19 The flow chart of condition assessment of PP insulation based on FDS. *Source:* Li et al. [34]/with permission from IEEE.

7.4 Electrical–Thermal Aging

7.4.1 Breakdown Under Electrical–Thermal Stress

Based on the dual-stress electrothermal accelerated life experimental model, a test apparatus for the time–pressure characteristics of insulating materials has been constructed. The schematic diagram of the apparatus system is shown in Figure 7.20. It is primarily divided into four modules: the pressure application device, the circuit protection device, the ten-channel breakdown device, and the heating device.

The time–pressure test system can independently apply voltage to ten-test specimens at a certain temperature and record the time it takes for the specimens to fail from the application of voltage. This system mainly consists of four modules, and the composition and functions of each module are as follows:

1) The voltage application device consists of a long-time withstand voltage transformer and an AC power control console. The model of the long-time withstand voltage transformer is YDJW-12, with an output voltage range of 0–25 kV. The AC power control console integrates a voltage regulation device and a timer device, serving as the operational hub of the entire system. It allows for the adjustment of the applied voltage, controls the relay's circuit-breaking current, and manages the timer to record the total time from voltage application to circuit disconnection.

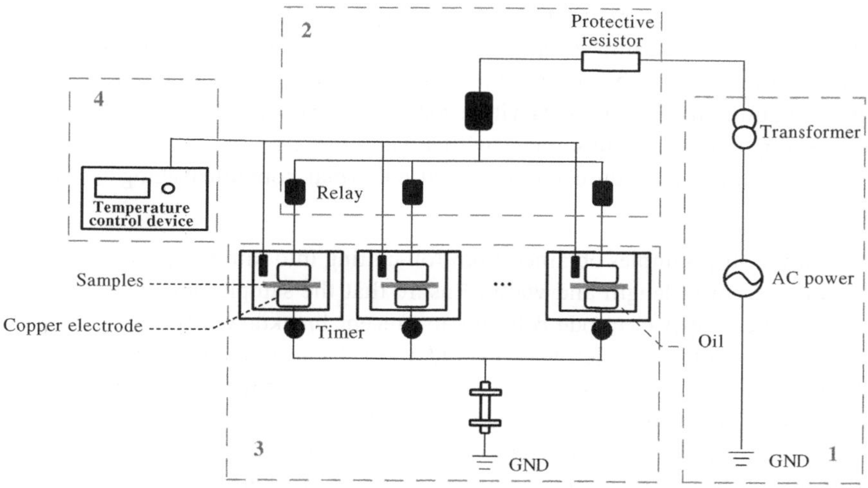

Figure 7.20 Material "time–pressure" testing device circuit.

2) The circuit protection device comprises a main relay, ten branch relays, and protective resistors. The main relay is connected to the transformer side and is responsible for disconnecting the circuit to protect against excessive total current in the loop. The ten branch relays operate based on the current value of each branch. When the insulation of a test sample fails, and the current exceeds a preset threshold, a signal is transmitted to cut the circuit. The corresponding relay quickly disconnects the circuit connection. The current limit for branch relays is set at 30 mA, while the current limit for the main relay is 20 A. The protective resistor is a water resistor, composed of a 1:1 mixture of deionized water and natural water, with a resistance value of 300 kΩ. It serves to limit the short-circuit current after the sample breakdown, preventing damage to the circuit and equipment due to excessive current.
3) The ten-channel breakdown device consists of column-to-column copper electrodes, double-layer beakers, and transformer oil. The column-to-column copper electrodes have a diameter of 25 mm, a thickness of 4 mm, and a rounded edge with a radius of 2.5 mm. The electrodes and test specimens are placed on the inner side of the double-layer beaker. Placing the ten test specimens separately helps prevent the breakdown of one specimen from affecting the others. The beaker is filled with silicone oil to prevent air breakdown and surface flashover of the test specimens during the breakdown process.
4) The heating device consists of a water bath circulation unit and double-layer beakers. The water bath circulation unit, model HH-501, has a temperature range of 0 to 99.9 °C. It is connected to the ten double-layer beakers through hoses to maintain the required temperature for the experiments.

To ensure a high correlation between the experimental results and actual operational cables, cable samples with a thickness of 0.3 mm were selected for the test. To avoid errors caused by uneven surface smoothness of samples made from cable granules, unused cable products with a voltage rating of 220 kV were processed by roll-cutting to obtain strip samples with a width of 5 cm.

The operating procedure for the electric-thermal combined aging test is as follows:

1) Cut the strip samples obtained from roll-cutting into thin samples, each measuring 5 cm in length and width. Ensure that the samples are slightly larger than the copper electrode columns to prevent breakdown from occurring at the edges of the samples, which could affect the accuracy of the data.
2) According to the wiring instructions, connect the power supply line, high-voltage line, ground wire, and signal wire, as well as the water bath circulation device, to ten sets of double-layered beakers with jackets.
3) Turn on the control panel, configure the timer's operating mode, and set the relay's operating current.

4) Smooth the surfaces of the upper and lower copper electrodes using sandpaper. Place the sample thin slices between the two electrodes and rotate the spacing nut on the electrode holder to clamp the sample. Then, insert the electrodes and the sample into the double-layered beaker with a jacket.
5) Turn on the water bath circulation device and set the temperature required for the experiment. Measure the temperature of the transformer oil in each beaker using a thermocouple temperature sensor. Continue heating until the transformer oil in all beakers reaches the desired temperature, with a maximum temperature difference not exceeding 2 °C. Maintain this temperature for 30 minutes before proceeding to the next step.
6) Power up the controller, rotate the dial to gradually increase the voltage on the transformer evenly to the desired level, and start the timer.
7) After ten samples have experienced breakdown, turn off the power and the water bath circulation device. Record the breakdown times for all ten samples. Several tests were conducted so that the insulation breakdown time of specimens was no less than ten per set of conditions. Based on the experimental results, box plots of the breakdown time of the specimen for different temperatures are shown.

The ALT mainly focuses on the VEC, which is an important parameter for the evaluation of the insulation electrical aging performance and is usually expressed as a value of n. In Section 7.2.2, the *E–t* characteristics of the insulation have already been presented. The curve is divided into three stages, each with a constant value of n. For this specimen, t_1 is usually less than 1000 seconds at high electric field strengths and has a small value of n_1. Specimens are considered to have a uniform life index from t_1 to t_2, with t_2 usually considered to be ten years or more. To ensure the accuracy and ease of residual life prediction, experimental data with failure times between t_1 and t_2 are used. It is possible to complete the insulation design with a certain safety margin. Figure 7.21 represents the fitting functions of the "*E–t*" characteristics of insulating materials at different temperatures. It can be observed that the experimental data can be reasonably well fitted with the theoretical model. At the same applied voltage, as the temperature increases, the insulation failure time decreases exponentially, indicating a faster insulation failure process.

Based on the average breakdown time of insulation material under various electric field strengths, and in conjunction with the *E–t* characteristics of the insulation material, a nonlinear fitting of the relationship between the average time to failure of the insulation material and voltage is performed using an inverse power function as shown in Figure 7.22. This is done for long-term breakdown experiments of specimens at different temperatures, and the fitting results are depicted in a box plot. Based on Eq. (7.19), the different VEC values for the material at various

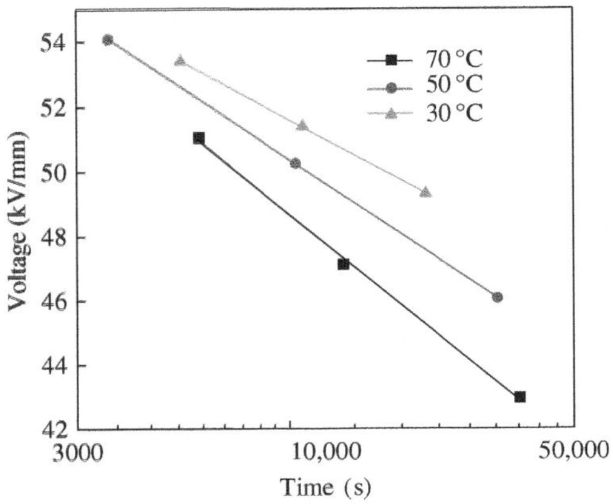

Figure 7.21 Fitting functions of "E–t" characteristics for samples at different temperatures.

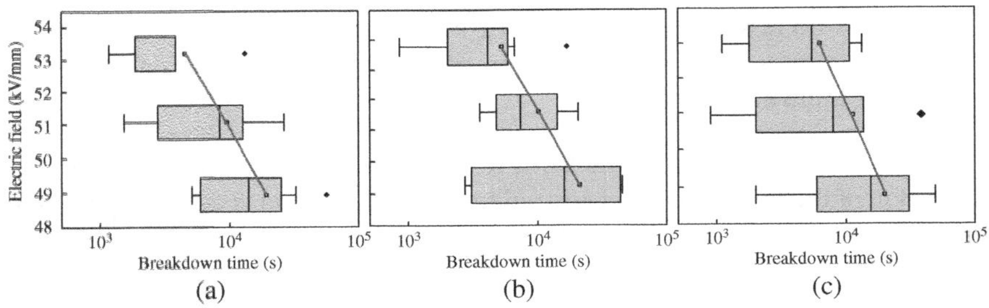

Figure 7.22 E–t characteristics at different temperatures for specimens. (a) 30 °C, (b) 50 °C, (c) 70 °C. *Source:* Li et al. [35]/MDPI / CC BY 4.0.

Table 7.4 VEC of specimens at different temperatures.

Temperature (°C)	VEC
30	16.90
50	15.72
70	13.35

temperatures can be calculated. Table 7.4 shows the VEC of specimens at different temperatures.

There is a correlation between the failure time and the applied voltage of the insulation. Both the time to failure and its dispersion increase significantly as the voltage decreases. Temperature also directly affects the E–t characteristics of insulation. VEC decreases with increasing temperature, as seen from the IPM fitting results and Table 7.4. The breakdown of insulating materials is mainly caused by the impingement of electrons on molecular chains under electric fields and breakage of molecular chains under thermal oxidative aging. The increase in temperature leads to an increase in the thermal movement of the molecules, which increases the rate at which the electrons hit the molecular chains. Furthermore, temperature accelerates the thermal oxidative aging process of insulating materials and accelerates the degradation of their electrical properties [30]. Therefore, an increase in temperature accelerates the process of molecular chain breakage, resulting in insulation failure, explaining the experimental phenomenon of a decrease in VEC with increasing temperature at the microscopic level.

Efforts have shown that the VEC of insulating materials can be obtained by various experimental methods, except for constant voltage tests. The relationship between the tree initiation time and the aging life curve of polyethylene has been investigated and shows a very similar variation pattern [31]. Other efforts found that the VEC was stable in the range of 10–14 at room temperature by studying the tree-starting time of insulation of 110 kV AC cables at different voltages [32]. A convenient method for calculating the VEC of specimen insulation has been proposed based on the equivalence of the damage curve trajectory and the step-up voltage test [33]. The above efforts on the *E*–*t* characteristics of insulating material all show similar experimental results, demonstrating that using VEC to characterize the voltage resistance of material under a single electrical stress is a correct method.

7.4.2 Lifetime Models and Prediction

In Section 7.2.2.3, several models for the analysis of electric-thermal combined aging were discussed, including the Simoni model, the RAMU model, the Fallou model, and the Crine model. These models require the determination of various constants within the model after aging and breakdown of the samples. They not only consider the separate effects of electrical and thermal factors but also account for their combined influence. The Simoni model, the RAMU model, and the Fallou model are more or less empirical in nature, while the Crine model tends to be more of a physical model. Therefore, it is not recommended to use the Crine model.

These three models were fitted with parameters based on the experimental results of the specimens *E*–*t* characteristics. The critical temperature T_0 is 20 °C, and the critical field strength E_0 is chosen as 10 kV/mm, conditions where the aging of the insulation is negligible. Figure 7.23a1–a3 shows the model fits for the remaining lifetime under the electrothermal stress of the specimen insulation. The vertical coordinate uses a logarithmic coordinate system to represent the remaining time. The pattern of variation in the remaining life calculated by the three models is similar, with all decreasing with increasing temperature and

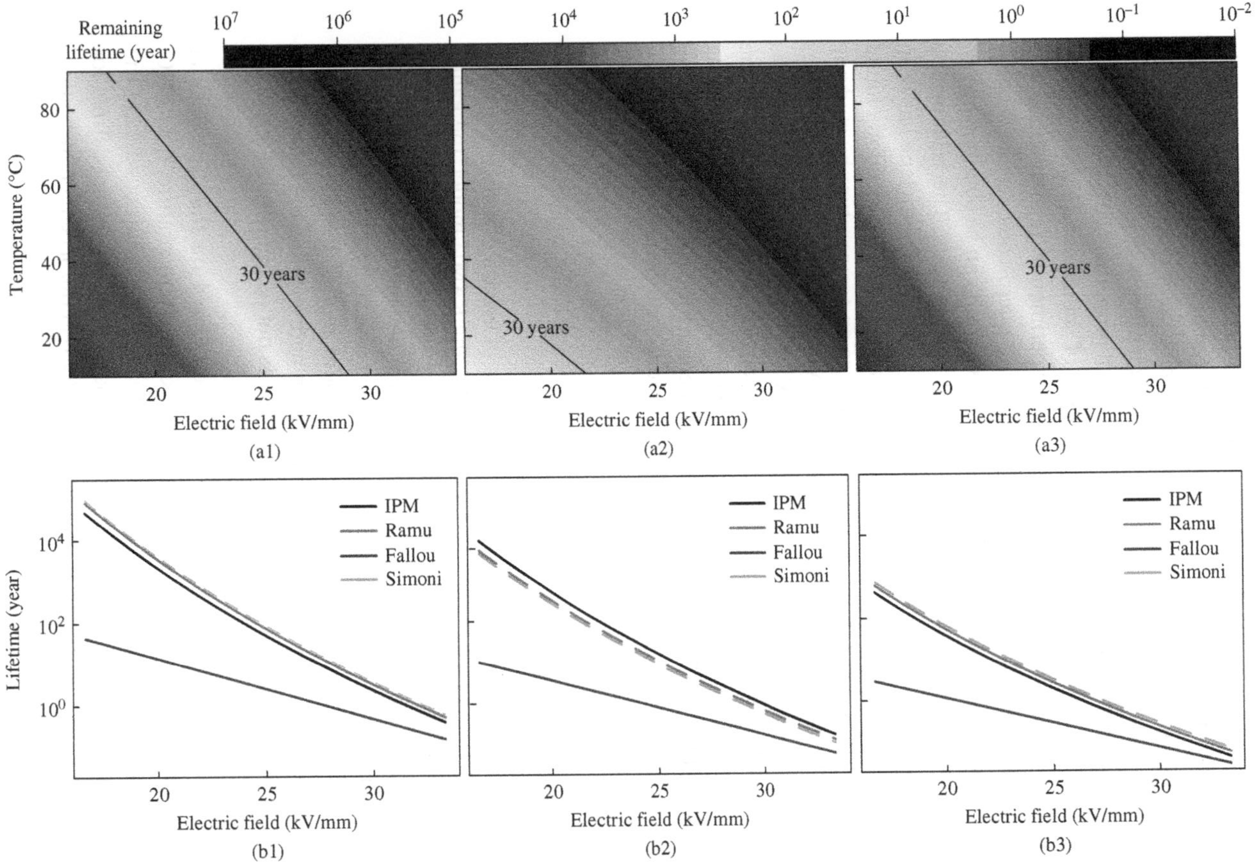

Figure 7.23 (a1)–(a3) Remaining lifetime fitting results for the RAMU, Fallou, and Simoni models. (b1)–(b3) Comparison of remaining lifetime results for different models at 30, 50, and 70 °C. *Source:* Li et al. [35]/MDPI / CC BY 4.0.

electric field strength. The remaining lifetime is low and varies more gently at lower temperatures and electric field strengths.

The fit results for the RAMU model are similar to those of the Simoni model, as shown in Figure 7.23a1 and a3. The remaining life calculated by the Fallou model is smaller than the results of the other models, and the difference is particularly significant at low electric fields. In order to analyze the accuracy of the prediction results of the electrothermal model, the IPM model at different temperatures is taken for comparative analysis, as it is the reference electrical lifetime model for cables in international standards [33]. Figure 7.23b1–b3 shows the lifetime prediction curves of the IPM at 30, 50, and 70 °C to obtain prediction models similar to the lifetime model prediction results under a single stress. The remaining lifetime predictions calculated by the Simoni and RAMU models are similar to the results obtained by the IPM. In contrast, the remaining lifetime results obtained by the Fallou model deviate significantly from the IPM, and the deviation further increases as the electric field decreases.

The RAMU and Simoni remaining lifetime models are compared and analyzed to investigate the factors leading to similar fitting results. The results show that after normalization of the model parameters, the two equations have a similar structure, as both are obtained by adding the temperature parameter to the inverse power function of the electrical stress. On the other hand, the Fallou model is a two-factor remaining lifetime model derived from an exponential function of electrical aging. Therein, the electric field is located at the exponential position, and the coefficient is negative. In contrast, in the IPM model, the residual lifetime calculated by the exponential function model is less affected by the electric field at low electric fields. Therefore, in the life assessment of cable insulation, when there are large fluctuations in the cable insulation temperature, the use of IPM will make the prediction results have a large deviation, while the use of Simoni and RAMU models for lifetime prediction has a better accuracy.

Figure 7.24 displays multiple electric-thermal stress models fitted to it. The plotted surface represents the correlation between insulation failure time and field strength (electric stress) and temperature. It is evident that the insulation failure time of cable materials exhibits a nonlinear relationship with both the insulation field strength and the temperature. This demonstrates that using the RAMU and Simoni lifetime models can be effective for predicting the remaining lifespan of samples under electric-thermal dual stresses.

From the 3D plot of failure time–temperature–field strength, it is apparent that, whether it is the Simoni, RAMU, or Fallou model, the insulation failure time increases as the temperature decreases and the field strength weakens. Moreover, it is notable that at lower temperatures and weaker field strengths, the failure time exhibits more significant variations with changes in field strength and temperature.

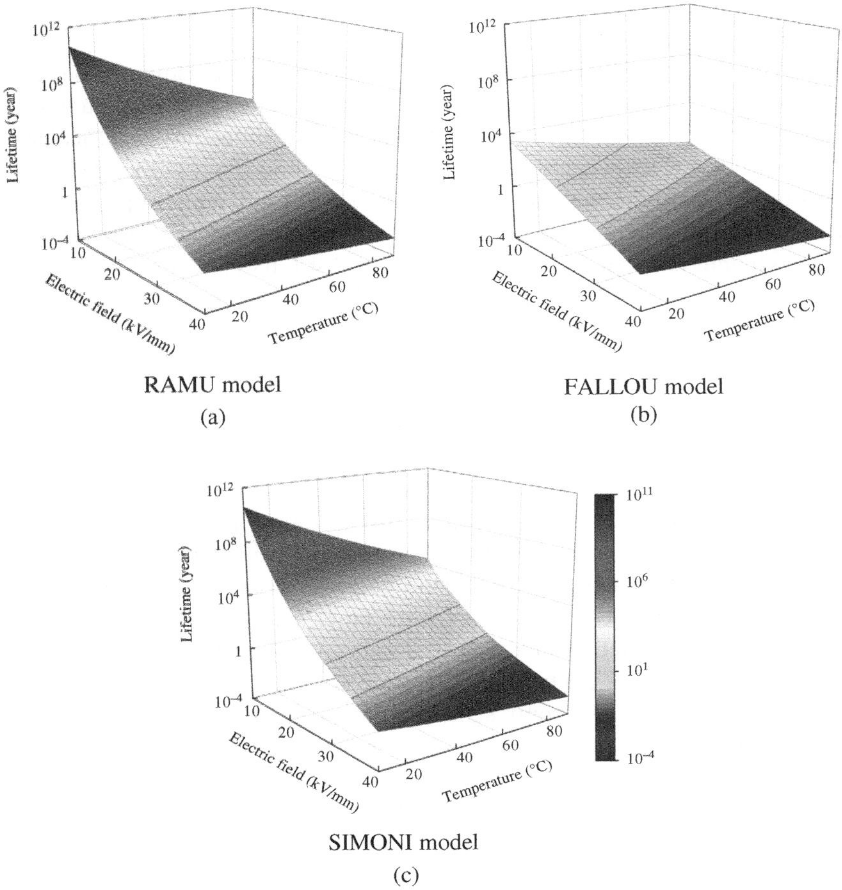

Figure 7.24 Fitting results of the three-dimensional function graph of sample failure time, temperature, and electric field strength under electric-thermal (a) RAMU model, (b) FALLOU model, (c) SIMONI moddel. *Source:* Li et al. [35]/MDPI / CC BY 4.0.

7.5 Conclusions

The future of polypropylene insulated cable aging and lifespan assessment is full of opportunities. Although the causes and manifestations of polypropylene cable aging are complex, most of its mechanisms are related to elevated internal material temperatures and chemical bond breakage, resulting in material oxidation and degradation failure. The hydrogen on the tertiary carbon atom of polypropylene is easily to break away from the chain of polypropylene at high temperatures, leading to the increase of free radicals. Free radical chain reactions can cause the break of molecular chains of polypropylene and the generation of polar groups, resulting in a decrease in electrical properties of the polypropylene. In addition, the break of molecular chains leads to the destruction of polypropylene spherulites, making it easier for elastomers to migrate and aggregate, resulting in an increase in the average size of elastomers and a decrease in the mechanical properties of polypropylene. A method for the condition assessment of thermally aged polypropylene cable insulation based on FDS is proposed by extracting the characteristic parameters of FDS. In the future, through innovative technologies, advancements in materials science, and stricter regulatory standards, we hope to enhance the reliability, safety, and durability of cables to meet the ever-growing demands in the fields of electricity and communication.

References

1 Mazzanti, G. (2015). Life estimation of HVDC cables under the time-varying electrothermal stress associated with load cycles. *IEEE Transactions on Power Delivery* 30 (2): 931–939.

2 Montanari, G.C. and Mazzanti, G. (1994). From thermodynamic to phenomenological multistress models for insulating materials without or with evidence of threshold. *Journal of Physics D: Applied Physics* 27: 1691–1702.

3 Mazzanti, G., Montanari, G.C., and Dissado, L.A. (2005). Electrical aging and life models: the role of space charge. *IEEE Transactions on Dielectrics and Electrical Insulation* 12 (5): 876–890.

4 Dissado, L.A. (2002). Understanding electrical trees in solids: from experiment to theory. *IEEE Transactions on Dielectrics and Electrical Insulation* 9 (4): 483–497.

5 Mazzanti, G. and Marzinotto, M. *Extruded Cables for High Voltage Direct Current Transmission: Advance in Research and Development*. Hoboken, NJ, USA: Wiley.

6 Dissado, L.A., Mazzanti, G., and Montanari, G.C. (1997). The role of trapped space charges in the electrical aging of insulating materials. *IEEE Transactions on Dielectrics and Electrical Insulation* 4 (5): 496–506.

7 Glasstone, S., Laidler, K.J., and Eyring, H.E. (1941). *The Theory of Rate Processes*. New York, NY, USA: McGraw Hill.

8 Lewis, T.J., Llewellyn, J.P., and van der Sluijs, M.J. (1994). Electrically induced mechanical strain in insulating dielectrics. *Proceedings of IEEE Conference on Electrical Insulation and Dielectric Phenomena (CEIDP)* (23–26 October 1994). Arlington, TX, USA, 328–333.

9 Dakin, T.W. (1948). Electrical insulation deterioration treated as a chemical rate phenomenon. *Transactions of American Institute of Electrical Engineers* 67: 113–122.

10 Dissado, L.A. and Fothergill, J.C. (1992). *Electrical Degradation and Breakdown in Polymers*. London, UK: Peter Peregrinus.

11 Crine, J.P. (1997). A molecular model to evaluate the impact of aging on space charges in polymer dielectrics. *IEEE Transactions on Dielectrics and Electrical Insulation* 4 (5): 487–495.

12 Dissado, L.A., Mazzanti, G., and Montanari, G.C. (2001). Elemental strain and trapped space charge in thermoelectrical aging of insulating materials. Part 1: elemental strain under thermo-electrical-mechanical stress. *IEEE Transactions on Dielectrics and Electrical Insulation* 8 (6): 959–965.

13 Dissado, L.A., Mazzanti, G., and Montanari, G.C. (2001). Elemental strain and trapped space charge in thermoelectrical aging of insulating materials. Life modeling. *IEEE Transactions on Dielectrics and Electrical Insulation* 8 (6): 966–971.

14 Mazzanti, G., Stomeo, G., and Mancini, S. (2016). State of the art in insulation of Gas Insulated Substations: main issues, achievements, and trends. *IEEE Electrical Insulation Magazine* 32 (5): 18–31.

15 Alghamdi, H.A., Chen, G. and Vaughan, A.S. (2016). Simulate the effect of trapped charges and trap cross-section on aging process. *Proceedings of IEEE Conference on Electrical Insulation and Dielectric Phenomena* (16–19 October 2016), Toronto, ON, Canada, 27–30.

16 Alghamdi, H., Chen, G., and Vaughan, A.S. (2015). The electro-mechanical effect from charge dynamics on polymeric insulation lifetime. *AIP Advances* 5 (12): 127126.

17 Crine, J.P. (2005). On the interpretation of some electrical aging and relaxation phenomena in solid dielectrics. *IEEE Transactions on Dielectrics and Electrical Insulation* 12 (6): 1089–1107.

18 Montanari, G.C., Mazzanti, G., and Simoni, L. (2002). Progress in electrothermal life modeling of electrical insulation during the last decades. *IEEE Transactions on Dielectrics and Electrical Insulation* 9 (5): 730–745.

19 Hirose, H. (1987). A method to estimate lifetime of solid electrical insulation. *IEEE Transactions on Electrical Insulation* 22 (6): 745–753.

20 Montanari, G.C. and Mazzanti, G. (1999). Insulation aging models. In: *Encyclopedia of Electrical and Electronics Engineering*, vol. 10 (ed. J. Webster), 308–319. New York, NY, USA: Wiley (updated in 2007).

21 Cygan, P. and Laghari, J.R. (1990). Models for insulation aging under electrical and thermal multistress. *IEEE Transactions on Electrical Insulation* 25 (5): 923–934.

22 Mandlik, M. and Ramu, T.S. (2014). Moisture aided degradation of oil impregnated paper insulation in power transformers. *IEEE Transactions on Dielectrics and Electrical Insulation* 21 (1): 186–193.

23 Bhutta, M.S., Yang, L., Ma, Z. et al. (2020). Simulation of thickness controlled DC breakdown of XLPE regulated by space charge & molecular chain movement. *IEEE Transactions on Dielectrics and Electrical Insulation* 27 (4): 1143–1151.

24 Liu, Y. and Cao, X. (2017). Insulation performance evaluation of HV AC/DC XLPE cables by 0.1 Hz tan δ test on circumferentially peeled samples. *IEEE Transactions on Dielectrics and Electrical Insulation* 24 (6): 3941–3950.

25 Chen, M., Zhang, H., Wang, Y. et al. (2020). Space charge dynamics of acetophenone and cumyl alcohol and their synergistic effect in LDPE. *IEEE Transactions on Dielectrics and Electrical Insulation* 27 (1): 67–75.

26 Xu, Z., Yang, M., Peng, H. et al. (2023). Influence of Combined electrothermal aging on dielectric and thermal properties of HVAC XLPE cable. *Energies* 16 (8): 3353.

27 Singh, B. and Sharma, N. (2008). Mechanistic implications of plastic degradation. *Polymer Degradation and Stability* 93 (3): 561–584.

28 Kao, K.C. (2004). *Dielectric Phenomena in Solids*. Elsevier.

29 Werelius, P., Tharning, P., Eriksson, R. et al. (2001). Dielectric spectroscopy for diagnosis of water tree deterioration in XLPE cables. *IEEE Transactions on Dielectrics and Electrical Insulation* 8 (1): 27–42.

30 Alghamdi, A.S. and Desuqi, R.K. (2020). A study of expected lifetime of XLPE insulation cables working at elevated temperatures by applying accelerated thermal ageing. *Heliyon* 6 (1), Art. no. 03120.

31 Liu, X., Yu, Q., Liu, M. et al. (2017). DC electrical breakdown dependence on the radial position of specimens within HVDC XLPE cable insulation. *IEEE Transactions on Dielectrics and Electrical Insulation* 24 (3): 1476–1484.

32 Su, Y., Liu, Y., and Zhong, L. (2019). Evaluation of voltage endurance characteristics for new and aged XLPE cable insulation by electrical treeing test. *IEEE Transactions on Dielectrics and Electrical Insulation* 26 (1): 72–80.

33 Bian, H., Yang, L., Ma, Z. et al. (2020). Improved physical model of electrical lifetime estimation for crosslinked polyethylene AC cable. *IEEE Transactions on Dielectrics and Electrical Insulation* 27 (1): 132–139.

34 Li, Z., Zhang, M., Zhao, S., & Du, B. (2024). Condition Assessment of Thermally-Aged Polypropylene Cable Insulation Based on Frequency Domain Spectroscopy. *IEEE Transactions on Dielectrics and Electrical Insulation*, 1–1. https://doi.org/10.1109/tdei.2024.3389637.

35 Li, Z., Dong, Y., Wu, Y., Meng, Z., Song, P., Zhu, M., Li, X., & Du, B. (2024). Breakdown Performance Evaluation and Lifetime Prediction of XLPE Insulation in HVAC Cables. *Energies*, 17 (6), 1337. https://doi.org/10.3390/en17061337.

Index

Polypropylene Cable Insulation, First Edition. Boxue Du and Zhonglei Li.

Published 2025 by John Wiley & Sons, Inc.

d

e

Printed and bound by CPI Group (UK) Ltd, Croydon, CR0 4YY
19/09/2026
14965091-0001